D1559494

Transmission Lines and Wave Propagation

3rd edition

Philip C. Magnusson
Gerald C. Alexander
Vijai K. Tripathi
Department of Electrical and Computer Engineering
Oregon State University
Corvallis, Oregon

CRC Press
Boca Raton New York

Library of Congress Cataloging-in-Publication Data

Magnusson, Philip Cooper.
Transmission lines and wave propagation/Philip Cooper Magnusson, Gerald C. Alexander, Vijai Kumar Tripathi. —3rd ed.
p. cm.
Includes bibliographical references and index.
ISBN 0-8493-4279-1
1. Electric lines. 2. Electromagnetic waves. I. Alexander, Gerald C. II. Tripathi, Vijai Kumar. III. Title.
TK3221.M24 1992
621.319'2—dc20 91-28063
CIP

Direct all inquiries to CRC Press LLC, 2000 N.W. Corporate Blvd., Boca Raton, Florida 33431.

International Standard Book Number 0-8493-4279-1
Library of Congress Card Number 91-28063
Printed in the United States of America 6 7 8 9 0
Printed on acid-free paper

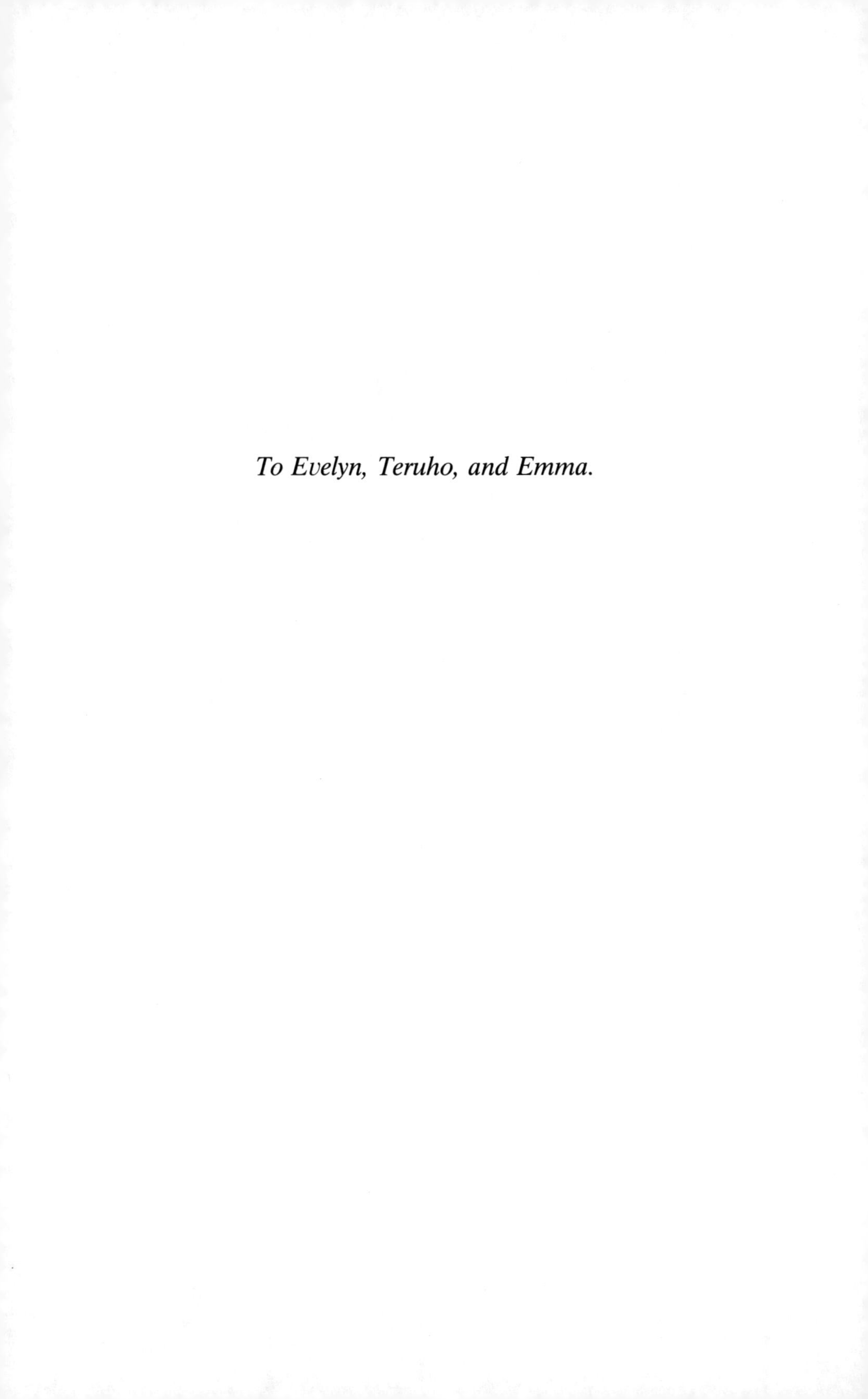

To Evelyn, Teruho, and Emma.

Contents

Preface

A transmission line is more than a set of long, parallel wires. To the electrical engineer it is a distributed-parameter physical system, one in which voltage and current, whether conveying communication signals or moving bulk power, must be regarded as continuous functions of location along the given line. This is in contrast to lumped-parameter networks of resistors, inductors, and capacitors, in which a discrete number of currents and a discrete number of voltage drops describe fully the system behavior.

Electrical analysis of a transmission line involves two independent variables, time and distance, and, thanks to linearity, mathematics may be applied advantageously. This presentation will be a mathematical one, but physical interpretation will be emphasized.

The first seven chapters examine transmission-line response in terms of voltage and current, using distributed-circuit parameters. Chapter 8 is devoted to the ladder-network delay line or artificial line, a lumped-parameter network which, at sufficiently low frequencies, has properties corresponding to those of a distributed parameter line. Chapters 9 through 14 deal with dynamic field theory, which involves three spatial dimensions rather than one, and hence four independent variables, applying Maxwell's equations to transmission lines to supplement the treatment of the earlier chapters, and then to waveguides.

It should be noted that transmission line and waveguide behavior may be readily studied experimentally with such conventional instruments (of appropriate frequency range) as voltmeters, oscilloscopes, swept-frequency generators, bridges, and slotted lines. Observed waveforms and standing-wave patterns, and measured impedances, may be compared with those predicted from theory.

Changes which were incorporated in this revision include addition of material on (a) the Bergeron graphical analysis of reflection behavior, (b) the *ABCD* transmission matrix, (c) earth currents and formulas for the impedance of ground-return circuits, and (d) a descriptive treatment of the elliptical cylindrical waveguide. The first chapter has been expanded to emphasize the dependence of the line parameters on frequency and temperature.

The mathematical level of most of the presentation is well within the scope of preparation provided in most undergraduate programs in electrical engineering. In the treatment of the step-function response of certain systems, namely (1) a lossy transmission line (Chap. 5), (2) a ladder-network artificial line (Sec. 8-3), and (3) a hollow waveguide (Sec. 12-5c), Laplace transforms involving irrational and transcendental functions arise; but, in those instances, the interpretation of results rather than mathematical theory has been emphasized. Those analyses could be omitted at the option of the instructor without interfering with the structure of the remainder of the book.

The contents of this book are based largely on a course on transmission systems required of upper division students in electrical engineering at Oregon State University. It was originally taught by the late Professor Arthur L. Albert. Each of the co-authors has taught the course at various times, and has contributed his experience to this revision.

P.C.M.

Transmission Lines and Wave Propagation

3rd edition

CHAPTER 1
Introduction

"A system of conductors suitable for conduction of electric power or signals between two or more termini."* This is one definition, and a concise one, among many which have been written for the noun "transmission line."[7, 10] From the standpoint of applications, one will find that the term covers a tremendous range in power level, frequencies, line lengths, and modes of construction.

1-1 TRANSMISSION SYSTEMS: GENERAL COMMENTS

This text deals with transmission lines and wave propagation primarily from an engineering-mathematical point of view, and emphasizes aspects of importance in communication systems. Such study is enhanced if one considers the growth of the surrounding technological pattern, and that will be the theme of this section.

The earliest device to use electrical transmission lines was one that required little power to make it function, could be built of simple and rather crudely fashioned mechanical parts, and was modest in terms of the frequency spectrum needed, less than 100 Hz. It was the telegraph; Morse sent the first public message in 1844. Wire telegraphy over land was followed by submarine-cable telegraphy. The English Channel was crossed in 1851; a successful cable was laid across the Atlantic Ocean in 1866.[2, 4]

Telephony, more demanding in the complexity and minimum quality of sending-end and receiving-end instruments, and requiring a wider frequency band (2 kHz or more, depending on the quality sought), came several years later. Bell's patent was dated 1876.

Radio, or "wireless," followed. Marconi sent signals across the Atlantic Ocean in 1901. At first glance, it would seem that radio dispensed

*S. P. Parker, ed., *McGraw-Hill Encyclopaedia of Engineering*, McGraw-Hill, New York, 1983, 1121. Reprinted with permission of McGraw-Hill, Inc.

with transmission lines. But because of the relatively short wavelengths commonly used (from some hundreds of meters down to centimeters), physically short sections of radio-frequency line, such as those joining transmitters to their antennas, may be electrically long. By this is meant that their lengths are on the order of an eighth of a wavelength or more. Thus radio systems include, inadvertently, transmission lines.

A tremendous increase in the rate of handling messages by a given transmission line was made possible by *carrier-frequency* transmission, which was introduced commercially in 1918. Carrier telephony employs techniques similar to those of radiotelephony. The signal is combined with a sinusoidal wave of much higher frequency, the carrier, by the process of modulation; a band of frequencies close to the carrier is transmitted, and at the receiving end of the line, the audio-frequency form of the signal is recovered by demodulation. Many signals, modulated with differing carrier frequencies, may be transmitted simultaneously. Appropriate filters, or frequency-selective networks, are needed to separate the individual modulated signals from the composite received wave.* The transmission of video signals follows basically the same plan, although the signal bandwidth required is of the order of 6 MHz, a value many times that for audio signals.

Up to this stage of development the basic form of communication line was that of two parallel conductors.† Three principal types had evolved: (1) the open-wire line, consisting of two identical wires with a spacing of about 100 times the conductor diameter, with air as the surrounding medium; (2) the cabled pair, consisting of two identical wires with solid insulation, twisted together with a center-to-center spacing of about twice the conductor diameter, and encased, commonly with many other similar pairs, by a metallic sheath; and (3) the coaxial cable, consisting of a solid or stranded center conductor surrounded by a cylindrical return conductor.

The microwave spectrum (approximately 3 GHz and above) was entered during the development of radar in the latter 1930s. In this frequency range a single, hollow, metallic tube, or *waveguide*, proved to be a practical, low-loss means to transmit energy. Since that time other varia-

---

*An interesting sidelight of the last several decades has been the widespread use of power transmission lines as carrier-frequency communication channels for telemetering, dispatching, and control.[9] In recent years microwave radio or fiber optics links have tended to displace such installations.

†Two important variations which have been extensively used are (1) the use of the physical ground as one of the conductors, particularly in telegraphy systems (App. D), and (2) the phantom system, in which four wires accommodate three telephone channels by means of bridge circuits at the terminals.[1,4,6] These techniques yield some economy for audio-frequency circuits, but are not suitable for carrier-frequency application.

tions have been tried and adopted, particularly the use of dielectric media (fiber optics) for guide structures.

1-2 THE CIRCUIT-THEORY APPROACH TO TRANSMISSION-LINE ANALYSIS

Electromagnetic-energy propagation along transmission lines will be considered initially in terms of voltage and current waves traveling on two uniform conductors. This is the distributed-parameter-circuit point of view, in contrast to the more general one of electric and magnetic fields.

Voltage and current on a transmission line may be regarded at this stage simply as quantities measurable with conventional instruments such as voltmeters, ammeters, and oscilloscopes. They are functions of two independent variables, time t and distance along the line z.* Whatever the length of the line and the type of termination, and whatever the waveform or time-varying character of the energy source or sources, those functions, $v(z,t)$ and $i(z,t)$, are constrained to follow differential equations based on Kirchhoff's laws.

As with lumped circuits, two general plans of analysis with rather different scopes may be developed: (1) A complete solution (transient and steady-state components) for voltages and currents as functions of time may be sought for a selected driving function and particular set of initial conditions; or (2) evaluation of the steady-state component of the response to a sinusoidal source may be the objective.

Complete solutions are, in all but idealized transmission-line cases, mathematically complicated, but they are of practical importance, for example, in the study of systems using abruptly rising voltages for pulse-coded signals, or for such purposes as timing or triggering. The step function is commonly used as the mode of time variation of the source for such analyses. This function is mathematically convenient and it is reasonably representative of the outputs of pulse-forming circuits or switched sources.

The transmission of intelligence is characterized by a signal that varies in a nonrepetitive fashion. Theoretically, in accordance with the Fourier integral, the complete frequency spectrum (except possibly some discrete values) is present in such a signal. Actual transmission processes will

---

*The lowercase letter z was chosen for the distance coordinate because the distributed-circuit equations about to be derived will be supplemented with field equations. Some of the latter will be written in terms of cylindrical coordinates, and z, the axial coordinate, is the direction of movement of the fields. In no instance will lowercase z represent an impedance.

distort a composite wave or change the proportions and relative phases among its frequency components, but in practice the intelligence put into the signal is recoverable if distortion is kept small within some limited band of frequencies, even though distortion elsewhere in the spectrum is extreme. For a "smoothly varying, nonabrupt" signal, such as the usual analog form of an audio signal, steady-state alternating-current analysis is a means of obtaining useful and concise results.

a. Line Parameters

The transmission properties of a two-conductor line are traditionally described in terms of four distributed-circuit parameters: l, the distributed inductance in henrys per unit length of line; c, the distributed capacitance in farads per unit length of line; r, the distributed resistance in ohms per unit length of line (the sum of the resistances of the separate conductors per unit length of line); and g, the distributed shunt conductance in siemens (mhos) per unit length of line (insulation-loss-equivalent leakage from one conductor to the other).* The distributed inductance[5]—the ratio of flux linkages (weber turns) per unit length to current (amperes)—is necessarily nonzero because the current paths are physically separated. The distributed capacitance is necessarily nonzero because the conductors have nonzero cross-sectional areas, and hence nonzero perimeters, or surface areas per unit length.

(1) *Functional Dependence on Frequency.* Under sinusoidal excitation these parameters have distinct values, which may be measured experimentally. If r is nonzero, both r and l will be found to vary with frequency when the range that is measured is extended sufficiently; the same is true of g and c when g is nonzero. A frequency-domain (Laplace transform) approach is in order when the voltage and current response on a lossy line is to be found for excitation conditions other than sinusoidal steady state.

For purposes of analysis of a two-wire line (conductors a and b) l may be regarded as the sum of three component inductances: l' is the value which the inductance would have if the conductors had infinite conductivity (no magnetic flux within the conductors), and l_a and l_b are the increments which result because of finite conductivity of the respective conductors. In general l' is frequency-independent, whereas l_a and l_b are not.

---

*Lower-case letters will be used to designate distributed parameters; capital letters will be used for lumped parameters. Italic lower-case r will be reserved for *radius*.

At zero frequency, the resistance per unit length of each conductor is inversely proportional to the conductivity* and the cross-sectional area. But as shown in Chap. 14 specifically for the circular and annular conductors, resistance increases with frequency, and the incremental inductances l_a and l_b decrease with frequency, owing to *skin effect*, which is a nonuniformity of current density caused by time-varying magnetic flux within the conductors. Conductors of other cross sections have been found experimentally to have similar properties.[8]

Solid dielectric materials and line-support structures have properties which cause the voltage-related losses, and hence the conductance parameter, to increase with frequency. For usual insulation materials and within the intended frequency range, the variation in capacitance is slight, commonly not detectable in routine measurements.

(2) *Linearity.* In practical lines, under normal operating conditions, the parameters are essentially constant with respect to the magnitudes of voltage and current; this makes the system, for mathematical purposes, *linear*. A consequence of linearity is that there are no interactions among voltages and currents of different frequencies.†

---

*As noted in Appendix F conductivity is a function of temperature (the resistivity of copper increases approximately 0.4% per degree Celsius); ambient temperature varies with time (seasonally, daily, randomly, for example) unless externally controlled, and current through a conductor generates heat. Such changes with time are slow compared with the durations and repetition periods of voltage and current pulses and waves, and hence *rates of change* of temperature (and hence conductivity) are not ordinarily incorporated into traveling-wave analyses.

For numerical application of results in system design and prediction of performance, *anticipated ranges of temperature* are important. The following quotation is indicative of practice: "The temperature range for buried cables is normally assumed to vary from 32 to 78°F and 0 to 110°F for open-wire and aerial cables with an annual mean temperature of 55°F. The maximum temperature of open-wire and aerial cable may be 25°F greater than the ambient-air temperature." (From Wolff, J. H., Wire and cable transmission characteristics, in *Communication System Engineering Handbook*, Hamsher, D. H., Ed., McGraw-Hill, New York, 1967, chap. 11, 11-5. Reprinted with permission of McGraw-Hill, Inc.) Corresponding Celsius values are 0 to 26°C, −18° to 43°C, 13°C, and 14°C.

†Some common origins of nonlinear behavior in electric circuits and transmission lines may be noted: (a) saturation in ferromagnetic materials (steel-cored conductors and transformer cores, for example), (b) materials which have dielectric saturation and hysteresis properties which are analogous to the properties of ferromagnetic materials (accordingly such dielectric materials, $BaTiO_3$, for example, are known colloquially as *ferroelectrics*,[3] even though they contain no iron), (c) corona, which is a form of partial breakdown of insulation, and (d) arcing, usually associated with accidental short circuits.

Nonlinear resistances will be discussed briefly in Secs. 3-1a and 3-2d.

(3) *Evaluation of Parameters.* Precise, closed-form prediction of the parameters from field theory is practical only when the cross section has a high degree of symmetry; formulas are derived in Secs. 11-1 through 11-3 for l' and c of (a) the coaxial cable with a circular inner conductor and an annular outer one, (b) two parallel circular conductors, and (c) a tubular-shielded pair of wires. (Equations 11-24, 11-25, 11-77, 11-79, and 11-96.)

Other configurations may be examined numerically, and manufacturers' catalogs and handbooks list average values for standard types of cables and open-wire conductors.

The parameters of a given specimen of line may be found experimentally from measurements of impedance with a signal generator and a bridge; one procedure is outlined in Sec. 7-3.

b. Specific Mathematical Approaches

In order that the situation may be mathematically tractable, some simplifying assumptions are necessary. Three avenues of interest may be explored advantageously using different approximations.

(1) *Lossless Line: Transmitted and Reflected Waves.* The elementary aspects of wave propagation and reflection may be easily surveyed if one assumes (a) that the line itself is lossless (r and g both equated to zero, so l and c are then frequency-independent) and (b) that all lumped impedances connected to the line (the internal impedance of the generator and a load impedance, for example) are purely resistive (Chaps. 2 and 3). Voltage and current on the line then obey the *wave equation* of classical physics, a second-order partial differential equation with constant coefficients, one which has the same basic form of traveling-wave solution for any driving function. The terminal or boundary relationships, in turn, are simple in that they change only the magnitudes of the waves, not their form.

(2) *Lossy Line: Sinusoidal Steady State.* Much may be learned about the effect of a line, lossy or lossless, or a system of lines and lumped networks, on audio- and radio-frequency signals other than pulses, by examination (as a function of frequency) of the steady-state component of the response to a sinusoidal source. Under this restriction the lossy-line problem may be formulated in the time domain as a partial differential equation (known in classical physics as the *equation of telegraphy*) in which the line parameters are constant coefficients. (For a numerical application, they would be evaluated for the particular frequency considered.)

The traveling-wave form of the steady-state solution for a lossy line is developed in Chap. 4; standing waves on sinusoidally excited lossless and lossy lines are studied in Chaps. 6 and 7.

(3) *Lossy Line: Step-Function Response.* The manner of propagation on a lossy line of a pulse-type signal, or of the front of the wave produced by switching, or by accidental short-circuiting or open-circuiting, is oftentimes of interest. Adequate portrayal of such functions inherently involves frequency ranges over which skin effect is pronounced. This may be taken into account by formulating the problem on the Laplace transform basis, in which the traditional four parameters are replaced by two functions: an *operational distributed impedance* and an *operational distributed admittance*. By suitable approximation of those functions, inverse transforms for step-function response may be obtained in closed form; this is developed in Chap. 5.

1-3 TRAVELING-WAVE FIELDS — LINES AND WAVEGUIDES

The transfer of electromagnetic energy is accompanied by observable effects (mechanical forces and induced voltages, for example), which may be spatially removed from directly energized conducting bodies such as the wires forming a transmission line. Spatially continuous mathematical functions known as *electric fields* and *magnetic fields* are postulated as mental aids in the systematizing of macroscopic phenomena of that type.

a. Electromagnetic Field Parameters and Constraints

A set of time-dependent differential equations, based on experimental observations and known as Maxwell's equations, states relationships among these fields, from which the manner of field propagation in a given geometrical situation may be derived. Differences in any one or more of the following properties may distinguish one medium electromagnetically from another: the electric-field *permittivity* (ϵ) and *conductivity* (σ), and the magnetic-field *permeability* (μ). Analysis in this text will be limited to media in which those parameters are *linear* (independent of field magnitude), *isotropic* (independent of field direction), *time-invariant*, and *frequency-independent*. (Chap. 9.)

At the boundary between two media of different properties, Maxwell's equations reduce to requirements of continuity of the normal components of two fields ($\boldsymbol{D}$, electric flux density, and $\boldsymbol{B}$, magnetic flux density), and equality across the bounding surface of the tangential components of two others (*E*, electric field intensity, and *H*, magnetic field intensity).

b. Geometrical Scope of Analysis

Attention will be given primarily to steady-state sinusoidally varying fields in lossless media, with emphasis on the application of boundary conditions. The following will be considered in some detail.

(1) *Plane Waves of Infinite Extent* (Chap. 10). Maxwell's equations for postulated electromagnetic fields which vary spatially in only one direction reduce to the one-dimensional wave equation when the medium is lossless, or to an equation analogous to the equation of telegraphy when the conductivity of the medium is nonzero. This geometrically simple case is useful for examining such phenomena as polarization and reflection.

(2) *Transmission-Line Fields* (Chap. 11). Lines are composed of two or more parallel conductors, and field theory serves to relate the circuit parameters of distributed inductance and capacitance to the cross-sectional geometry. It also provides an approximate method for calculating the high-frequency resistive losses (skin effect and proximity effect) in a low-loss line.

(3) *Fields within Hollow, Uniconductor Waveguides* (Chaps. 12 and 13). Energy propagation is possible within a single hollow conductor if the frequency is sufficiently high. For some geometrically simple interior cross sections a mathematical description of the fields in closed form is possible. Two such cases will be studied in detail, the rectangular and the circular cross sections; a summary of results will be given for the elliptical cross section.

1-4 CLOSURE

The purpose of Sec. 1-1 was to introduce, partly in historical perspective, some aspects of transmission lines and their applications. The body of this book is concerned with the mathematical analysis, as outlined in Secs. 1-2 and 1-3, of the electromagnetic circuit and field phenomena which make those applications possible.

REFERENCES

1. Albert, A. L., *Electrical Communication*, 3rd ed., Wiley, New York, 1950.
2. Dibner, B., *The Atlantic Cable*, 2nd ed., Burndy Library, Inc., Norwalk, CT, 1959; Blaisdell Publishing Company, New York, 1964.
3. Beaty, H. W., Ed., *Electrical Engineering Materials Reference Guide*, McGraw-Hill, New York, 1990, 13-15 to 13-19.
4. *Encyclopædia Britannica,* Encyclopædia Britannica, Inc., Chicago, 1972. See the entries on "Telegraphy" and "Telephony."
5. Higgins, T. J., The origins and developments of the concepts of inductance, skin effect, and proximity effect, *Amer. J. Phys.*, 9, 337, 1941.
6. O'Neill, E. F., Ed., *A History of Engineering and Science in the Bell System. Transmission Technology (1925–1975),* AT&T Bell Laboratories, 1985.
7. Jay, F., Ed., *IEEE Standard Dictionary of Electrical and Electronics Terms*, 4th ed., The Institute of Electrical and Electronics Engineers, Inc., New York, 1988, 1037.
8. Kennelly, A. E., Laws, F. A., and Pierce, P. H., Experimental researches on skin effect in conductors, *AIEE Trans.*, 34, part II, 1953, 1915.
9. Gohari, J., Power line carrier, in *Standard Handbook for Electrical Engineers*, 12th ed., Fink, D. G. and Beaty, H. W., Eds., McGraw-Hill, New York, 1987, 16-71.
10. *Webster's Third New International Dictionary*, G. and C. Merriam Company, Springfield, MA, 1961, 2429.

CHAPTER 2
Wave Propagation on an Infinite Lossless Line

A lossless line, extending toward infinity in both directions, will be considered first. (Among the four line parameters listed in Sec. 1-2a, the two which cause loss, r and g, will be assumed to be zero.) This choice will, with minimal algebraic distraction along the way, give results that emphasize the propagation properties.

2-1 PARTIAL DIFFERENTIAL EQUATIONS OF LOSSLESS LINE

Let the positive directions of current, voltage, and distance along the line be assumed as indicated in Fig. 2-1. It is essential that the positive directions of all three quantities be specified if ambiguity is to be avoided.

a. Voltage-Current Interdependence

Whenever the voltage across the incremental capacitance $c\,\Delta z$ located between $z_1 - \frac{1}{2}\Delta z$ and $z_1 + \frac{1}{2}\Delta z$ is changing, the currents at the two locations differ by the resulting displacement current ($c\,\partial v(z,t)/\partial t$ per unit length) from one conductor to the other. In like manner, the voltage between conductors at one of those locations differs from that at the other by the rate at which the magnetic flux linkage $(l\,\Delta z)i(z,t)$ of that Δz longitudinal span is changing with respect to time.

To reduce the foregoing to equations, one may note first that the voltage in the vicinity of z_1 may be stated in a Taylor series as

$$v(z,t) = v(z_1,t) + (z - z_1)\left[\frac{\partial v(z,t)}{\partial z}\right]_{z=z_1} + \frac{(z - z_1)^2}{2}\left[\frac{\partial^2 v(z,t)}{\partial z^2}\right]_{z=z_1} + \cdots \qquad (2\text{-}1)$$

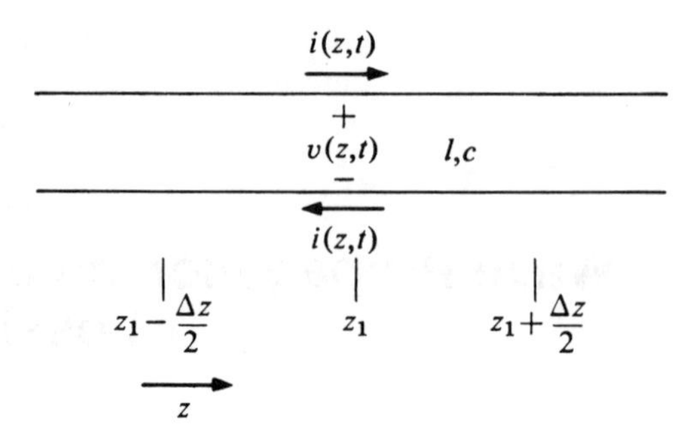

Figure 2-1. Coordinate system for two-conductor transmission line.

If this function is averaged with respect to $z - z_1$ over the interval from $-\frac{1}{2}\Delta z$ to $\frac{1}{2}\Delta z$, the following is obtained:

$$v_{\Delta z}(z_1, t) = \frac{1}{\Delta z}\int_{(-1/2)\Delta z}^{(1/2)\Delta z} v(z,t)\,d(z - z_1)$$

$$= \frac{1}{\Delta z}\left\{(z - z_1)v(z_1,t) + \frac{(z - z_1)^2}{2}\cdot\left[\frac{\partial v(z,t)}{\partial z}\right]_{z=z_1}\right.$$

$$\left. + \frac{(z - z_1)^3}{6}\cdot\left[\frac{\partial^2 v(z,t)}{\partial z^2}\right]_{z=z_1} + \cdots\right\}_{(-1/2)\Delta z}^{(1/2)\Delta z}$$

$$= \frac{1}{\Delta z}\left\{\Delta z\cdot v(z_1,t) + 0\cdot\left[\frac{\partial v(z,t)}{\partial z}\right]_{z=z_1}\right.$$

$$\left. + \frac{2(\frac{1}{2}\Delta z)^3}{6}\cdot\left[\frac{\partial v^2(z,t)}{\partial z^2}\right]_{z=z_1} + \cdots\right\} \tag{2-2}$$

In classical analysis the derivatives are assumed to be continuous, so that as Δz approaches zero, the average voltage over the Δz interval approaches the value at the center. Then

$$i(z_1 + \tfrac{1}{2}\Delta z, t) = i(z_1 - \tfrac{1}{2}\Delta z, t) - c\,\Delta z\frac{\partial v(z_1,t)}{\partial t} \qquad \Delta z \to 0 \tag{2-3}$$

If $i(z,t)$ is expanded about z_1 in a Taylor's series as v in Eq. 2-1, and the derivatives are assumed to be continuous, the following results:

$$i\left(z_1+\tfrac{1}{2}\Delta z,t\right)-i\left(z_1-\tfrac{1}{2}\Delta z,t\right)=\Delta z\left[\frac{\partial i(z,t)}{\partial z}\right]_{z=z_1} \qquad \Delta z\to 0 \tag{2-4}$$

Substitution of Eq. 2-4 in Eq. 2-3 yields

$$(\Delta z)\left[\frac{\partial i(z,t)}{\partial z}\right]_{z=z_1}=-c\,\Delta z\frac{\partial v(z_1,t)}{\partial t} \qquad \Delta z\to 0 \tag{2-5}$$

The subscripts 1 may be omitted, since this relationship applies at all locations on the line, and Δz may be divided from both sides, yielding

$$\frac{\partial i(z,t)}{\partial z}=-c\frac{\partial v(z,t)}{\partial t} \tag{2-6}$$

Correspondingly

$$v\left(z_1+\tfrac{1}{2}\Delta z,t\right)=v\left(z_1-\tfrac{1}{2}\Delta z,t\right)-l\,\Delta z\frac{\partial i(z_1,t)}{\partial t} \qquad \Delta z\to 0 \tag{2-7}$$

If one follows the same mathematical steps as before, this reduces to

$$\frac{\partial v(z,t)}{\partial z}=-l\frac{\partial i(z,t)}{\partial t} \tag{2-8}$$

b. Reduction to Wave Equations

Equations 2-6 and 2-8 are the basic equations which interrelate v and i with t and z. The function $i(z,t)$ may be eliminated from the set by differentiating Eq. 2-6 with respect to t and Eq. 2-8 with respect to z:

$$\frac{\partial}{\partial t}\left[\frac{\partial i(z,t)}{\partial z}\right]=-c\frac{\partial^2 v(z,t)}{\partial t^2} \tag{2-9}$$

$$\frac{\partial^2 v(z,t)}{\partial z^2}=-l\frac{\partial}{\partial z}\left[\frac{\partial i(z,t)}{\partial t}\right] \tag{2-10}$$

According to classical function theory, the order of differentiation of $i(z, t)$ with respect to z and t in Eqs. 2-9 and 2-10 is immaterial if $\partial i/\partial t, \partial i/\partial z$ and the second-order derivatives of i in those equations are continuous.[2] The *theory of distributions*[1, 3] extends this property to functions with discontinuities, such as the step function and the impulse function. Hence

$$\frac{\partial}{\partial z}\left[\frac{\partial i(z,t)}{\partial t}\right] = \frac{\partial}{\partial t}\left[\frac{\partial i(z,t)}{\partial z}\right] \tag{2-11}$$

By substituting Eqs. 2-10 and 2-11 in Eq. 2-9, the following is obtained:

$$\frac{\partial^2 v(z,t)}{\partial z^2} = lc\frac{\partial^2 v(z,t)}{\partial t^2} \tag{2-12}$$

In like manner, an equation in terms of current may be derived:

$$\frac{\partial^2 i(z,t)}{\partial z^2} = lc\frac{\partial^2 i(z,t)}{\partial t^2} \tag{2-13}$$

Equations 2-12 and 2-13 are known in classical physics as *wave equations*; they are identical except for the replacement of v by i. When one of those functions has been found, the other may be found from it by means of either Eq. 2-6 or 2-8.

2-2 TRAVELING-WAVE SOLUTIONS TO THE WAVE EQUATION

An intuitive approach will be used in the construction of solutions to the wave equation, and emphasis will be placed on interpretation of results in the transmission-line setting rather than on the theoretical aspects of differential equations.

a. Derivation of General Traveling-Wave Functions

The constraint that the wave equation itself imposes on a possible solution is a mild one, namely, that the second derivatives of the function with respect to t and z be directly proportional to each other. The proportionality may be simplified to an equality by the following change of variable. Let

$$w = z\sqrt{lc} \tag{2-14}$$

$$\frac{dw}{dz} = \sqrt{lc}$$

Hence
$$\frac{\partial v(z,t)}{\partial z} = \frac{\partial v}{\partial w} \cdot \frac{dw}{dz} = \sqrt{lc}\,\frac{\partial v}{\partial w}$$

$$\frac{\partial^2 v(z,t)}{\partial z^2} = lc\frac{\partial^2 v}{\partial w^2} \tag{2-15}$$

Substitution of Eq. 2-15 will reduce Eq. 2-12 to

$$\frac{\partial^2 v}{\partial w^2} = \frac{\partial^2 v}{\partial t^2} \tag{2-16}$$

Any function in which t and w appear only in the particular combination $t - w$ would have first derivatives, with respect to those variables, equal in magnitude but of opposite signs, but the second derivatives would be equal. Such a function would therefore satisfy Eq. 2-16. In algebraic form, let

$$k_A = t - w \tag{2-17}$$

and
$$v_A = f_A(k_A) \tag{2-18}$$

The derivatives called for in Eq. 2-16 may be derived as follows:

$$\frac{\partial v_A}{\partial t} = \frac{df_A(k_A)}{dk_A} \cdot \frac{\partial k_A}{\partial t} = \frac{df_A(k_A)}{dk_A}$$

$$\frac{\partial^2 v_A}{\partial t^2} = \frac{d^2 f_A(k_A)}{dk_A^2} \tag{2-19}$$

$$\frac{\partial v_A}{\partial w} = \frac{df_A(k_A)}{dk_A} \cdot \frac{\partial k_A}{\partial w} = -\frac{df_A(k_A)}{dk_A}$$

$$\frac{\partial^2 v_A}{\partial w^2} = \frac{d^2 f_A(k_A)}{dk_A^2} \tag{2-20}$$

Thus Eqs. 2-19 and 2-20 satisfy Eq. 2-16. After substitution of Eqs. 2-17 and 2-14, Eq. 2-18 may be rewritten

$$v_A(z,t) = f_A(t - z\sqrt{lc}) \tag{2-21}$$

Likewise any function in which t and w appear only in the particular combination $t + w$ will have equal second derivatives with respect to t and w, and will also satisfy Eq. 2-16. Let

$$k_B = t + w \tag{2-22}$$

and

$$v_B = f_B(k_B) \tag{2-23}$$

Differentiations and substitutions corresponding to Eqs. 2-19 and 2-20 may be made with v_B and k_B, thereby demonstrating an identity in Eq. 2-16. Substitution of Eqs. 2-22 and 2-14 in Eq. 2-23 yields

$$v_B(z,t) = f_B(t + z\sqrt{lc}) \tag{2-24}$$

b. Traveling-Wave Property: Velocity of Propagation

The behavior of the functions described in Eqs. 2-21 and 2-24 may be studied by selecting some particular point or feature on the wave (zero-crossing, maximum, etc.) and noting what change δz in location on the line should be made for an incremental change δt in time, in order to identify the same point on the wave. This result is achieved by keeping the argument of f_A or f_B constant. Thus

$$f_A\left[(t + \delta t) - (z + \delta z_A)\sqrt{lc}\right] = f_A(t - z\sqrt{lc}) \tag{2-25}$$

provided

$$\delta z_A = \frac{\delta t}{\sqrt{lc}} \tag{2-26}$$

Likewise

$$f_B\left[(t + \delta t) + (z + \delta z_B)\sqrt{lc}\right] = f_B(t + z\sqrt{lc}) \tag{2-27}$$

provided

$$\delta z_B = -\frac{\delta t}{\sqrt{lc}} \tag{2-28}$$

The foregoing is true regardless of what initial combination of t and z is used as a test point on the wave. Thus the function f_A, if plotted as a function of z for consecutive values of time as in Fig. 2-2(a), appears to

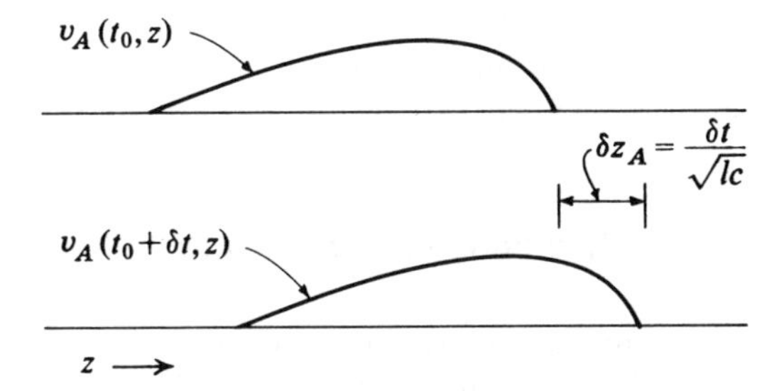

(a) Arbitrary function of $t - z\sqrt{lc}$

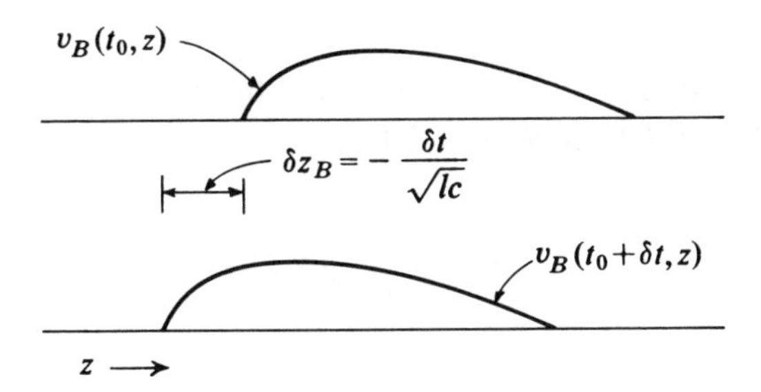

(b) Arbitrary function of $t + z\sqrt{lc}$

Figure 2-2. Traveling-wave voltage (Eqs. 2-21 and 2-24) as functions of z at two discrete times.

move in the $+z$ direction, but undergoes no change in shape. Hence f_A is said to be a *traveling-wave function*. Similarly the function f_B, shown in Fig. 2-2(b), appears to move in the $-z$ direction. The corresponding velocity will be designated as the *velocity of propagation*, ν. Let

$$\nu = \left| \frac{\delta z}{\delta t} \right| \tag{2-29}$$

Thus, from Eqs. 2-26 and 2-28,

$$\nu = \frac{1}{\sqrt{lc}} \quad \text{(lossless line)} \tag{2-30}$$

If SI units are used, that is, l is in henrys per meter and c in farads per meter, ν will have the unit meters per second.

In a linear system, such as has been assumed here, any linear combination of solutions to a differential equation is itself a solution. In other words, $v_A(z,t)$ and $v_B(z,t)$ may exist simultaneously on the line.

c. Voltage-to-Current Relationship: Characteristic Impedance

The current functions i_A and i_B that accompany v_A and v_B may be found with the aid of Eq. 2-8. Equations 2-14, 2-17, 2-18, and 2-21 will be used to find i_A:

$$\frac{\partial v_A}{\partial z} = -\sqrt{lc}\,\frac{df_A}{dk_A} \tag{2-31}$$

Substitution of Eq. 2-31 in Eq. 2-8, followed by integration with respect to t and a return to the original variables, yields

$$i_A(z,t) = \sqrt{\frac{c}{l}}\,f_A(t - z\sqrt{lc})$$

$$= \sqrt{\frac{c}{l}}\,v_A(z,t) \tag{2-32}$$

Equations 2-14, 2-22, 2-23, and 2-24 will be used for i_B:

$$\frac{\partial v_B}{\partial z} = \sqrt{lc}\,\frac{df_B}{dk_B} \tag{2-33}$$

$$i_B(z,t) = -\sqrt{\frac{c}{l}}\,f_B(t + z\sqrt{lc})$$

$$= -\sqrt{\frac{c}{l}}\,v_B(z,t) \tag{2-34}$$

Thus i_A and i_B are directly proportional to v_A and v_B, respectively, for all combinations of t and z. The quantity $\sqrt{l/c}$, the reciprocal of which appears in Eqs. 2-32 and 2-34, has the unit ohms in the SI system; it relates the magnitudes of voltage and current of a traveling-wave pair and has been customarily assigned the name *characteristic impedance* and the symbol Z_0:

$$Z_0 = \sqrt{\frac{l}{c}} \quad \text{(lossless line)} \tag{2-35}$$

For the lossless line the characteristic impedance is purely resistive.*

The negative sign on the right-hand side of Eq. 2-34 is important. Regardless of how the coordinate system may be chosen (see Prob. 2-6), the voltage and current of one traveling-wave pair will be related by $+Z_0$ and that of the other pair will be related by $-Z_0$. This will be discussed in Sec. 2-2d in terms of power and the direction of energy movement.

d. Analysis of Sinusoidal Traveling Waves

One particular mode of variation for the traveling-wave functions, the sinusoidal function, will be commented on in some detail because it is also applicable (with minor modification) to the line with losses.

The general function f_A in Eq. 2-21 may be readily replaced by the cosine function. Let

$$v_1(z,t) = V_{1M} \cos\left[\omega\left(t - t_1 - \frac{z}{\nu}\right)\right] \tag{2-36}$$

From Eq. 2-32

$$i_1(z,t) = \frac{V_{1M}}{Z_0} \cos\left[\omega\left(t - t_1 - \frac{z}{\nu}\right)\right] \tag{2-37}$$

Here V_{1M} will be understood to be simply a magnitude in volts, and ω is the angular frequency in radians per second. The quantity t_1 is an arbitrary offset or delay time, which permits the choosing of the origins in t and z without constraining the solution. Figure 2-3 shows v_1 and i_1 as functions of *time*, as viewed at two locations on a lossless line.

---

*In the case of a lossy line (see Chaps. 4 and 5), the simple relationships indicated in Eqs. 2-32 and 2-34 do not apply, nor, except for a wave varying sinusoidally with time, is a distinct velocity of propagation for the complete wave discernible. Nonsinusoidal voltage or current waves will change shape (i.e., suffer *distortion*) between different locations on a lossy line. Characteristic impedance for a lossy line may be stated as (1) the complex ratio between the phasors representing a voltage-and-current traveling-wave pair, or (2) an operational function relating voltage and current transforms.

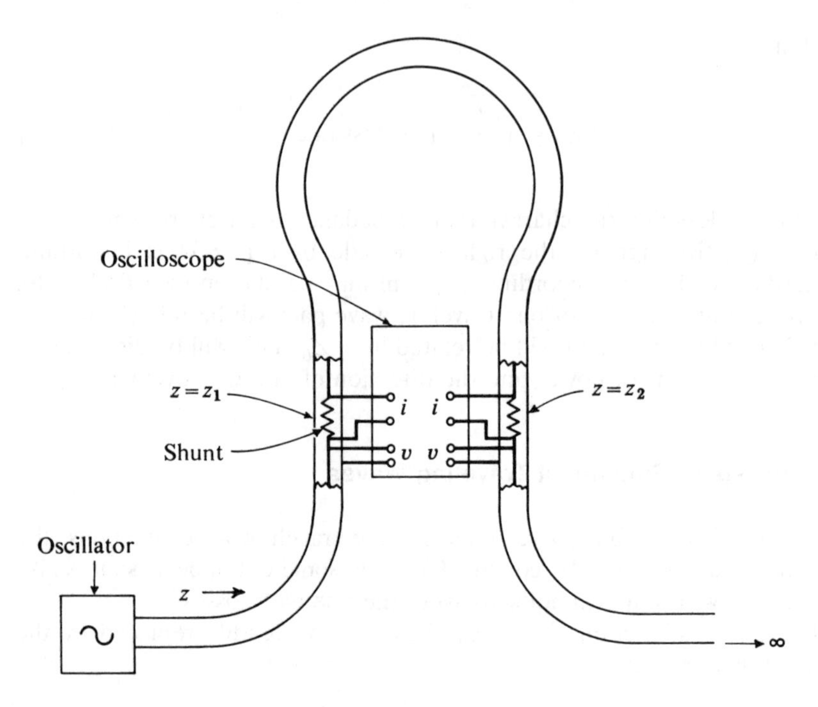

(a) Measurement circuit

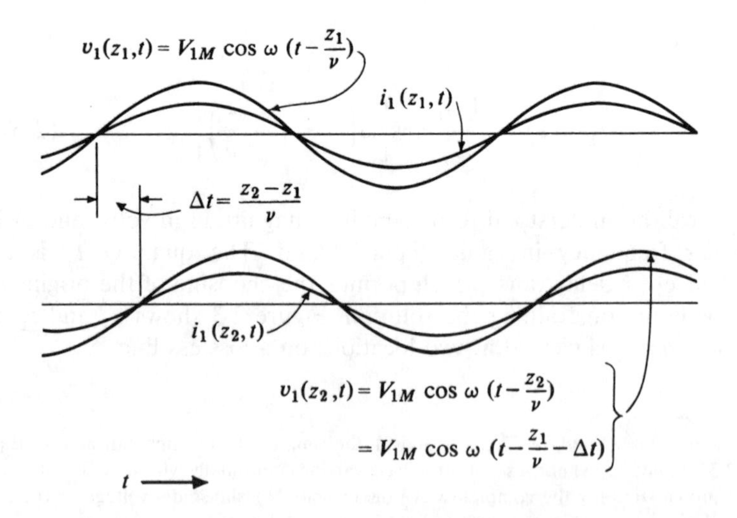

(b) Oscillographic waves

Figure 2-3. Steady-state sinusoidal traveling waves observed as functions of time at two locations on lossless cable.

Equations 2-24 and 2-34 may be similarly specified as cosine functions. Let

$$\begin{aligned} v_2(z,t) &= V_{2M}\cos\left[\omega\left(t - t_2 + \frac{z}{\nu}\right)\right] \\ i_2(z,t) &= -\frac{V_{2M}}{Z_0}\cos\left[\omega\left(t - t_2 + \frac{z}{\nu}\right)\right] \end{aligned} \tag{2-38}$$

Since these are continuous functions of t and z, one could readily find the derivatives needed for direct substitution in Eqs. 2-6, 2-8, 2-12, or 2-13. These functions (Eqs. 2-36 through 2-38) may be studied from the standpoint of direction of power flow. Instantaneous power is the product of instantaneous voltage and instantaneous current, and when only $v_1(z,t)$ and $i_1(z,t)$ are present this becomes

$$p_1(z,t) = v_1(z,t)i_1(z,t) \tag{2-39}$$

Substitution of Eqs. 2-36 and 2-37 yields

$$\begin{aligned} p_1(z,t) &= \frac{V_{1M}^2}{Z_0}\cos^2\left[\omega\left(t - \frac{z}{\nu} - t_1\right)\right] \\ &= \frac{V_{1M}^2}{2Z_0}\left\{1 + \cos\left[2\omega\left(t - \frac{z}{\nu} - t_1\right)\right]\right\} \end{aligned} \tag{2-40}$$

Similarly, if only $v_2(z,t)$ and $i_2(z,t)$ are present, the instantaneous power is, with substitution of Eqs. 2-38,

$$p_2(z,t) = v_2(z,t)i_2(z,t) \tag{2-41}$$

$$p_2(z,t) = -\frac{V_{2M}^2}{2Z_0}\left\{1 + \cos\left[2\omega\left(t + \frac{z}{\nu} - t_2\right)\right]\right\} \tag{2-42}$$

Corresponding time-averaged values $P_1(z)$ and $P_2(z)$ are easily obtained:

$$P_1(z) = \frac{V_{1M}^2}{2Z_0} \tag{2-43}$$

$$P_2(z) = -\frac{V_{2M}^2}{2Z_0} \tag{2-44}$$

In terms of the polarity convention adopted in Fig. 2-1, a positive value of power, which arises when the instantaneous current and voltage at a given location are both positive or both negative, corresponds to a movement of energy in the increasing direction of z. Thus the fact that $P_1(z)$ is positive indicates that energy is moving in the positive direction of z, the same as the direction of travel of the $v_1(z,t)$ and $i_1(z,t)$ waves. On the other hand, $P_2(z)$ is negative for the reason that the instantaneous voltage $v_2(z,t)$ at a given location is positive only when the instantaneous current $i_2(z,t)$ is negative, and vice versa. This indicates a movement of energy in the negative direction of z, which is the direction of travel of the $v_2(z,t)$ and $i_2(z,t)$ waves. The same principle applies to the general functions v_A, i_A and v_B, i_B.

e. Significance of Coordinate-Direction Designation

It should be noted that the speed of propagation and the magnitude of the characteristic impedance are each the same for both directions of travel. This is reasonable because physical phenomena may be expected to proceed independently of any observer's choice of coordinate system, and the properties of the line appear to be the same when viewed from either direction.

On the other hand, the algebraic signs associated with the *mathematical descriptions* of such quantities as voltage, current, and power may be expected to differ if changes are made in the coordinate scheme. Specifically, if the coordinate plan of Fig. 2-4 is used (see Prob. 2-6), the following results, paralleling Eqs. 2-21, 2-24, 2-32, and 2-34, apply:

$$v_C(z',t) = f_C\left(t - \frac{z'}{\nu}\right) \tag{2-45}$$

$$i_C(z',t) = -\frac{1}{Z_0} f_C\left(t - \frac{z'}{\nu}\right) \tag{2-46}$$

$$v_D(z',t) = f_D\left(t + \frac{z'}{\nu}\right) \tag{2-47}$$

$$i_D(z',t) = \frac{1}{Z_0} f_D\left(t + \frac{z'}{\nu}\right) \tag{2-48}$$

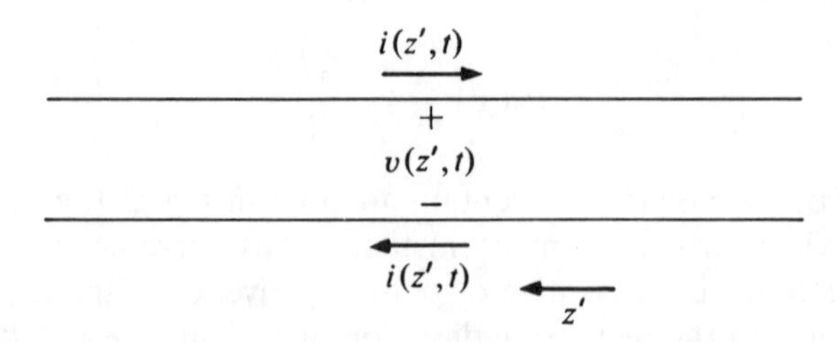

Figure 2-4. Alternative coordinate system for two-conductor transmission line.

In any event, a sketch in which the assumed positive polarities and directions have been marked is necessary for physical interpretation of mathematical results.

2-3 CONCLUSIONS

A uniform lossless electrical transmission line may be described by the following two parameters, each stated on a per-unit-of-length basis: inductance l and capacitance c.

On a lossless line, the voltage and current functions must satisfy the following differential equations (the coordinate system is shown in Fig. 2-1):

$$\frac{\partial i(z,t)}{\partial z} = -c\frac{\partial v(z,t)}{\partial t} \tag{2-6}$$

$$\frac{\partial v(z,t)}{\partial z} = -l\frac{\partial i(z,t)}{\partial t} \tag{2-8}$$

Either current or voltage may be eliminated from this set to yield a *wave equation* in terms of the other function:

$$\frac{\partial^2 v(z,t)}{\partial z^2} = lc\frac{\partial^2 v(z,t)}{\partial t^2} \tag{2-12}$$

$$\frac{\partial^2 i(z,t)}{\partial z^2} = lc\frac{\partial^2 i(z,t)}{\partial t^2} \tag{2-13}$$

Traveling-wave functions satisfy these equations. In such functions the argument has the form $t - z\sqrt{lc}$ or $t + z\sqrt{lc}$; any identified point on the function for which the argument value is fixed will appear to move in the $+z$ or $-z$ directions, respectively, at the velocity of propagation

$$\nu = \frac{1}{\sqrt{lc}} \quad \text{(lossless line)} \tag{2-30}$$

The magnitude of a voltage wave traveling in one direction is at all times and locations equal to the product of the magnitude of the accompanying current wave and the characteristic impedance, Z_0:

$$Z_0 = \sqrt{\frac{l}{c}} \quad \text{(lossless line)} \tag{2-35}$$

However the coordinate plan is chosen (Fig. 2-1, Fig. 2-4, or any other variation) the voltage and current of one traveling-wave pair will be related by $+Z_0$ and the other pair will be related by $-Z_0$. Energy is transferred in the direction in which the given voltage- and current-wave pair travels.

PROBLEMS

2-1. An open-wire line with negligible losses has the following parameters: $l = 1.75 \times 10^{-6}$ H/m and $c = 6.60 \times 10^{-12}$ F/m. What is the velocity of propagation and what is the characteristic impedance?

2-2. A given cable with negligible losses has a characteristic impedance of 50 Ω and a velocity of propagation of 2×10^8 m/s. Find l and c.

2-3. Sketch a copy of Fig. 2-3(b) and, on the same abscissa scale, redraw the v and i functions if the frequency is doubled, but the distance $z_2 - z_1$ remains the same. At what frequency, in terms of $z_2 - z_1$ and the other parameters, would $v_1(z_2, t) = -v_1(z_1, t)$ in the steady state?

2-4. Sketch a copy of Fig. 2-3(b) and, on the same abscissa scale, add curves of instantaneous power at locations z_1 and z_2.

2-5. Write expressions for instantaneous power and time-averaged power [$p(z,t)$ and $P(z)$] if traveling-wave pairs $v_1(z,t), i_1(z,t)$ and $v_2(z,t), i_2(z,t)$ are both present on a lossless line.

2-6. Let the polarity-and-direction convention shown in Fig. 2-4 be adopted. Rewrite Eqs. 2-6 and 2-8 so as to be consistent with this convention. Note which of the succeeding equations are altered thereby, and compare the various mathematical results with those obtained with Fig. 2-1. Comment on the agreement or disagreement when each set of mathematical results is interpreted in terms of the polarity-and-direction diagram from which it was derived.

REFERENCES

1. Courant, R., *Methods of Mathematical Physics*, vol. II, *Partial Differential Equations*, Interscience, New York, 1962.
2. Woods, F. S., *Advanced Calculus*, Ginn and Company, Boston, 1934, 68.
3. Zemanian, A. H., *Distribution Theory and Transform Analysis*. New York: McGraw-Hill, New York, 1965.

CHAPTER 3
Step-Function Waves on a Terminated Lossless Line

The infinitely long uniform line is a fiction of mathematics; practical lines are uniform only over finite distances. Such lines are necessarily terminated; in addition they may branch, or lumped-impedance networks may be inserted in tandem between sections of uniform line.

Ohm's law and Kirchhoff's laws must be obeyed at each such transition point. In order to meet all impedance requirements simultaneously, a set of voltage and current waves departing on each of the lines joined to the discontinuity is usually necessary.

3-1 BOUNDARY CONDITIONS AND REFLECTION PROCESS

For an initial illustration, consider the lossless line shown in Fig. 3-1 with a voltage source and a resistive termination, but with neither initial charge nor initial current on the line.

a. Terminal Characteristics of Generators and Loads

Traveling-wave problems arise in diverse physical situations, and corresponding mathematical representations of the source and load assemblies are needed. The following are among many that arise: (a) routine transmission of signals over distances of many kilometers, (b) routine movement of pulses within a printed circuit board a few centimeters square, or within an integrated circuit chip, and (c) response of protective devices to lightning or other high-voltage surges.

(1) *Equivalent Sources.* Many physical source assemblies may be satisfactorily represented by a linear resistive Thévénin equivalent source, that is, by an ideal voltage source in series with a linear resistor. For example,

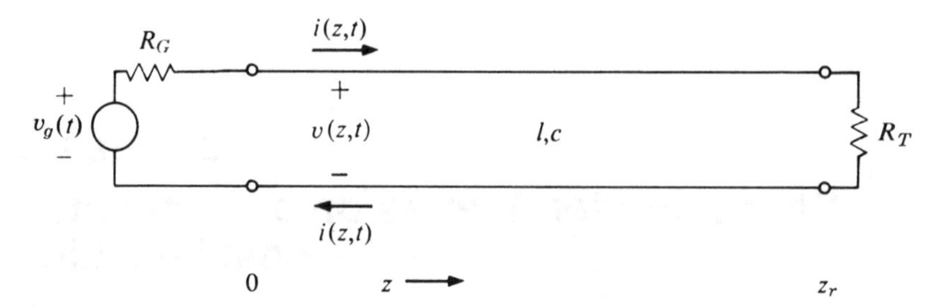

Figure 3-1. Lossless transmission line with resistive termination.

electronic oscillators and pulse-generator test sets ordinarily include a buffering resistive network, or *attenuator*, between the generating circuit and the output terminals.* This makes the function of voltage generation essentially independent of the impedance connected to the terminals (eliminates "pulling"), and thereby validates the Thévénin equivalent network.

Some other sources have nonlinear output characteristics. One example would be a solid-state amplifier, such as a MOSFET (metal-oxide-semi-conductor-field-effect transistor), connected directly to the line. Such behavior might be described in the form of a plot of voltage versus current, or by means of an empirical equation. (Examples in Fig. 3-7 and Prob. 3-6.)

(2) *Equivalent Terminations.* In many situations the termination is effectively a linear lumped resistor. This is usual for telephone lines and pulse transmission lines.

On the other hand, a solid-state diode rectifier is an elementary example of a nonlinear termination. It presents very low resistance to current in one direction, but very high resistance to the opposite polarity; the transition is nonabrupt, but most of it takes place within a range of perhaps one volt. Again a plot of voltage versus current, or an empirical equation, would present the information needed for traveling-wave analysis.

---

*The variation of resistance due to skin effect (Chap. 14) is a function of the product $\omega\mu\sigma$; high resistivity (very low conductivity) materials, such as carbon film, are used in the resistors in attenuators and line terminations, and variation of resistance with frequency is negligible within the frequency range of interest. (This is in contrast to the behavior of open-wire line or cable conductors; these are made of low resistivity metals, usually copper or aluminum. The resulting effects on step-function response, when account is taken of loss, are examined in Chap. 5.)

Surge arresters, connected from power lines to the earth, include nonlinear resistors which pass continually increasing amounts of current for successive increments of voltage. Silicon carbide has been used for this application for many years; more recently zinc oxide has come into general usage.[4]

The remainder of this chapter will be devoted primarily to transmission lines with linear resistive terminations; Sec. 3-2d on the Bergeron diagram will include some material on nonlinear terminations.

b. Traveling Waves and Reflection Coefficients

Voltage-and-current wave sets traveling in each direction may be expected; in this situation those traveling to the right in Fig. 3-1, functions of $t - z/v$, may be designated as *outgoing* (subscript O), and those traveling to the left, functions of $t + z/v$, as *returning* (subscript R). Each of these may be regarded in the general case as the summation of an infinite number of component traveling-wave functions, launched at successively delayed starting times.

Let
$$v(z,t) = v_O(z,t) + v_R(z,t) \tag{3-1}$$

$$i(z,t) = i_O(z,t) + i_R(z,t) \tag{3-2}$$

In accordance with Eqs. 2-32 or 2-34, the current components may be expressed in terms of the traveling-wave voltages and the characteristic impedance:

$$i_O(z,t) = \frac{v_O(z,t)}{Z_0} \tag{3-3a}$$

$$i_R(z,t) = -\frac{v_R(z,t)}{Z_0} \tag{3-3b}$$

Equations 3-3 are substituted in Eq. 3-2:

$$i(z,t) = \frac{v_O(z,t)}{Z_0} - \frac{v_R(z,t)}{Z_0} \tag{3-4}$$

At the right-hand terminal, Ohm's law must be obeyed by the resultant voltage and current, $v(z_r, t)$ and $i(z_r, t)$, at all values of time regardless of the specific waveform of the *incident* or arriving voltage, $v_O(z, t)$:

$$v(z_r, t) = R_T i(z_r, t) \tag{3-5}$$

When Eqs. 3-1 and 3-4 are substituted in Eq. 3-5, the result is

$$v_O(z_r, t) + v_R(z_r, t) = \frac{R_T}{Z_0}[v_O(z_r, t) - v_R(z_r, t)]$$

This may be multiplied by Z_0 and solved for $v_R(z_r, t)$ in terms of $v_O(z_r, t)$:

$$v_R(z_r, t) = v_O(z_r, t)\frac{R_T - Z_0}{R_T + Z_0}$$

Thus the *reflected*, or departing voltage $v_R(z, t)$ is specifically related to $v_O(z, t)$. Let ρ_T, the *reflection coefficient* at $z = z_r$, be defined as

$$\rho_T = \frac{R_T - Z_0}{R_T + Z_0} \tag{3-6}$$

Hence

$$v_R(z_r, t) = \rho_T v_O(z_r, t) \tag{3-7}$$

A similar result is obtained by applying Ohm's and Kirchhoff's laws at the left-hand terminal, where $v_R(z, t)$ is the incident voltage and $v_O(z, t)$ is the sum of (1) the outgoing wave produced directly by the source and (2) the reflected wave caused by the arrival of v_R:

$$v(0, t) = v_g(t) - R_G i(0, t) \tag{3-8}$$

When Eqs. 3-1 and 3-4 are substituted in Eq. 3-8, the result is

$$v_O(0, t) + v_R(0, t) = v_g(t) - \frac{R_G}{Z_0}[v_O(0, t) - v_R(0, t)]$$

This may be multiplied by Z_0 and solved for $v_O(0, t)$ in terms of $v_g(t)$ and $v_R(0, t)$.

$$v_O(0, t) = \frac{Z_0 v_g(t) + (R_G - Z_0) v_R(0, t)}{R_G + Z_0}$$

Let ρ_G, the reflection coefficient at the generator, be defined in a manner corresponding to ρ_T:

$$\rho_G = \frac{R_G - Z_0}{R_G + Z_0} \tag{3-9}$$

The preceding equation for $v_O(0, t)$ may be restated as

$$v_O(0, t) = \frac{Z_0}{R_G + Z_0} v_g(t) + \rho_G v_R(0, t) \tag{3-10}$$

3-2 RESPONSES OF SINGLE-SECTION LINES

To apply the foregoing results to some specific problems, Eqs. 3-10 and 3-7 may be used alternately to assemble consecutive component traveling-wave voltage functions. Let the first of these, the outgoing wave produced directly by the source, be designated by $v_{O1}(z, t)$. A step-function voltage source will be assumed in the examples which follow, primarily for brevity in mathematical expressions and for ease in plotting:

$$v_g(t) = V_0 U(t) \tag{3-11}$$

The moment that t exceeds zero, the step function v_{O1}, traveling to the right, will become nonzero at the left-hand end of the line. No returning function will become nonzero until the front of v_{O1} has reached the right-hand end of the line. Thus the first traveling-wave component, a function of $t - z/v$ which will satisfy Eqs. 3-10 and 3-11 at $z = 0$, is

$$v_{O1}(z, t) = \frac{V_0 Z_0}{R_G + Z_0} U\left(t - \frac{z}{v}\right) \tag{3-12}$$

Complete solutions will be derived for several particular combinations of R_G and R_T.

a. Line Terminated with Characteristic Impedance

If the terminating resistance R_T is equal to the characteristic impedance Z_0, the reflection coefficient ρ_T vanishes, and the first returning wave $v_{R1}(z, t)$ and all subsequent traveling-wave components are zero. Hence, from Eqs. 3-12 and 3-3,

$$v(z,t) = \frac{V_0 Z_0}{R_G + Z_0} U\left(t - \frac{z}{\nu}\right) \tag{3-13}$$

$$i(z,t) = \frac{V_0}{R_G + Z_0} U\left(t - \frac{z}{\nu}\right) \tag{3-14}$$

The unit function with the argument $t - z/\nu$ is particularly significant, since it indicates that at each location z on the line, the voltage and current are both zero up to the instant that t equals the quotient of the particular value of z and the velocity ν, but that after that instant the value of voltage or current indicated by the multiplying constant is present at that point on the line and remains there indefinitely afterward.

Thus a wave front of voltage and current progresses along the line from the sending end ($z = 0$) to the terminated end ($z = z_r$) at the velocity of propagation ν. The proportion between voltage and current in the traveling wave is exactly the same as that required by the terminating resistor; hence no reflection occurs and the transient portion of the response is completed in z_r/ν seconds. Plots of voltage and current at two points along the line are shown in Fig. 3-2. The time required for the wave front to travel the length of the line, z_r, may be designated by t_r; thus

$$t_r = z_r/\nu \tag{3-15}$$

b. Response Involving One Reflection

A slightly more complicated situation is that in which R_G is equal to the characteristic impedance but R_T is not. The function $v_{R1}(z,t)$, the reflected wave which results when $v_{O1}(z,t)$ (Eq. 3-12) arrives at z_r, becomes nonzero when $t = t_r$.

$$v_{R1}(z_r,t) = \rho_T v_{O1}(z_r,t) \tag{3-7}$$

In order to meet the requirements of Eq. 3-7, let

$$v_{R1}(z,t) = \frac{\rho_T V_0 Z_0}{R_G + Z_0} U\left(t + \frac{z}{\nu} - t_{R1}\right) \tag{3-16}$$

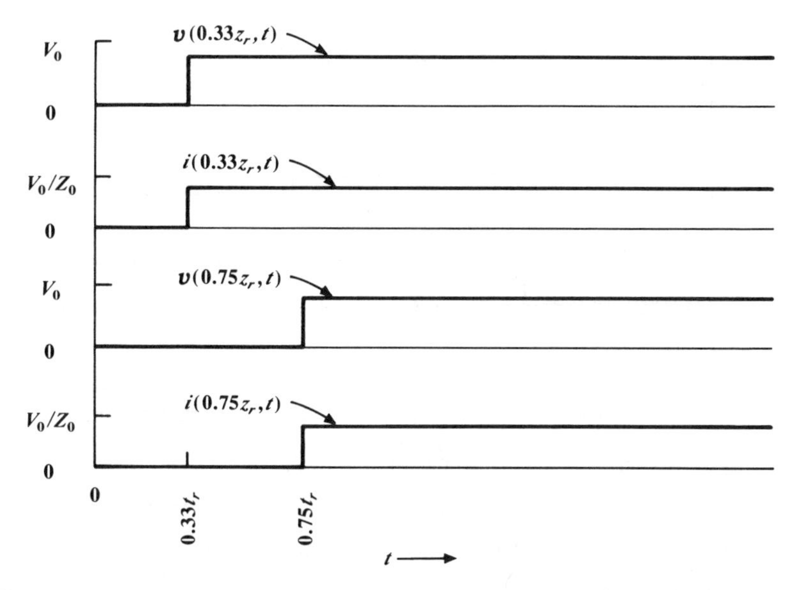

Figure 3-2. Voltage and current as functions of time at two locations on lossless transmission line. Reflectionless case: $R_G = 0.3Z_0$, $R_T = Z_0$. $[Z_0/(R_G + Z_0) = 0.769$, $\rho_T = 0]$.

Here t_{R1} is whatever delay time is necessary to satisfy the time-synchronization aspect of Eq. 3-7:

$$U(t + z_r/v - t_{R1}) = U(t - z_r/v)$$

Hence

$$t_{R1} = 2z_r/v$$

$$= 2t_r \tag{3-17}$$

Equation 3-17 may be substituted into Eq. 3-16:

$$v_{R1}(z, t) = \frac{\rho_T V_0 Z_0}{R_G + Z_0} U\left(t + \frac{z}{v} - 2t_r\right) \tag{3-18}$$

Because R_G was assumed equal to Z_0, the sending-end reflection coefficient ρ_G is zero and the second outgoing wave, $v_{O2}(z, t)$, vanishes. Equations 3-12 and 3-18 may be added together and simplified to

$$v(z, t) = \frac{V_0}{2}\left[U\left(t - \frac{z}{v}\right) + \rho_T U\left(t + \frac{z}{v} - 2t_r\right)\right] \tag{3-19}$$

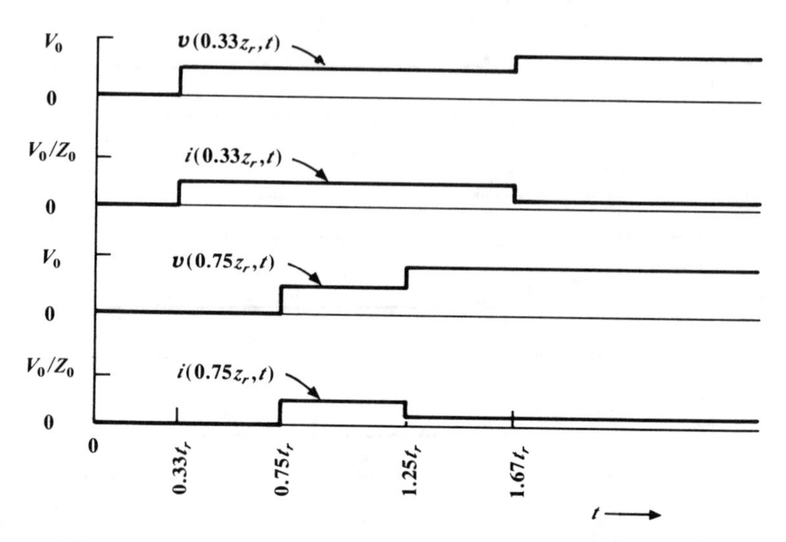

Figure 3-3. Voltage and current as functions of time on lossless transmission line. Single reflection case: $R_G = Z_0$, $R_T = 5Z_0$. ($\rho_G = 0$, $\rho_T = 0.667$).

In accordance with Eq. 3-4,

$$i(z,t) = \frac{V_0}{2Z_0}\left[U\left(t - \frac{z}{\nu}\right) - \rho_T U\left(t + \frac{z}{\nu} - 2t_r\right)\right] \tag{3-20}$$

Plots of voltage and current as functions of time for two distinct locations on the line are given in Fig. 3-3.

The transient aspect of the response ends with the arrival of the reflected waves at the generator end ($t = 2t_r$), by which time the current at all points on the line has become $V_0/(R_G + R_T)$, and the voltage at all points is $V_0 R_T/(R_G + R_T)$.

c. Response with Infinite Succession of Reflections

A more general situation is that in which neither R_T nor R_G equals the characteristic impedance; then an infinite succession of reflected waves results. In accordance with Eq. 3-10, the second outgoing wave, $v_{O2}(z,t)$, is related to $v_{R1}(z,t)$ in the manner

$$v_{O2}(0,t) = \rho_G v_{R1}(0,t) \tag{3-21}$$

The returning wave v_{R1} will reach the sending end, $z = 0$, at $t = 2t_r$;

hence that value should be the delay time for v_{O2}. From Eqs. 3-18 and 3-21,

$$v_{O2}(z,t) = \frac{\rho_G \rho_T V_0 Z_0}{R_G + Z_0} U\left(t - \frac{z}{\nu} - 2t_r\right) \tag{3-22}$$

The second returning wave, v_{R2}, must be synchronized with v_{O2} at $z = z_r$ at the instant $t = 3t_r$; hence its delay time (referred to $z = 0$) is $4t_r$. The sums of the first four voltage and current functions are

$$v(z,t) = \frac{V_0 Z_0}{R_G + Z_0}\left[U\left(t - \frac{z}{\nu}\right) + \rho_T U\left(t - 2t_r + \frac{z}{\nu}\right)\right.$$
$$\left. + \rho_G \rho_T U\left(t - 2t_r - \frac{z}{\nu}\right) + \rho_G \rho_T^2 U\left(t - 4t_r + \frac{z}{\nu}\right) + \cdots\right] \tag{3-23}$$

$$i(z,t) = \frac{V_0}{R_G + Z_0}\left[U\left(t - \frac{z}{\nu}\right) - \rho_T U\left(t - 2t_r + \frac{z}{\nu}\right)\right.$$
$$\left. + \rho_G \rho_T U\left(t - 2t_r - \frac{z}{\nu}\right) - \rho_G \rho_T^2 U\left(t - 4t_r + \frac{z}{\nu}\right) + \cdots\right] \tag{3-24}$$

Thus the response consists of an infinite series of voltage and current "layers" which are added successively as the wave front sweeps from generator to terminated end and back. Unless both reflection coefficients have unity magnitudes, the magnitudes of the added layers will progressively diminish. As $t \to \infty$ and the residual traveling-wave set becomes infinitesimal, the accumulated values of voltage and current (with a step-function source) may be viewed simply as DC quantities, limited by the resistors R_G and R_T. Thus

$$i(z,\infty) \to \frac{V_0}{R_G + R_T} \tag{3-25}$$

$$v(z,\infty) \to \frac{V_0 R_T}{R_G + R_T} \tag{3-26}$$

The *lattice diagram* devised by Bewley[2,3] provides a convenient graphical means of keeping track of multiple reflections. Its application to this relatively simple problem is shown in Fig. 3-4. The time–distance locus of a wave traveling in the increasing z direction is a line sloping downward to the right, whereas the locus of one traveling in the decreasing z direction

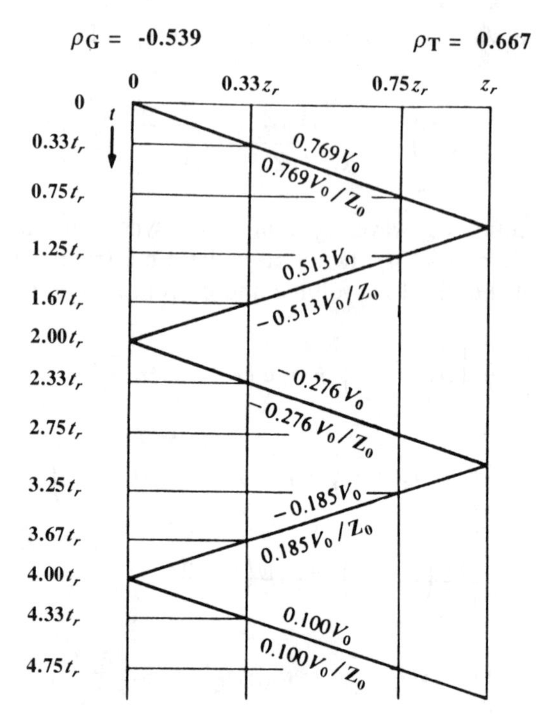

Figure 3-4. Lattice diagram for lossless transmission line with unmatched terminations: $R_G = 0.3Z_0$, $R_T = 5Z_0$.

is a line sloping downward to the left. The ordinate and abscissa scales and the slopes of the traveling-wave loci are related such that $\Delta t = |\Delta z|/v$. The value of each voltage wave has been written above the corresponding locus and the value of the accompanying current wave below it. Each reflection introduces the multiplier ρ_T or ρ_G in the voltage function and $-\rho_T$ or $-\rho_G$ in the current function.

Voltage and current responses at any particular location may be determined by drawing a vertical line through the z value desired and noting the successive values of time corresponding to intersections of the sloping loci with the given vertical line. At the arrival time of the front of each reflected wave, that wave is added to what has been accumulated from preceding waves. This is illustrated in Fig. 3-5.

A step-function source was used here for simplicity, but it may be noted that had any other driving function been used (audio or video signal, for example), each traveling-wave component response at a given location on the line would have had the same shape as the driving function. This is

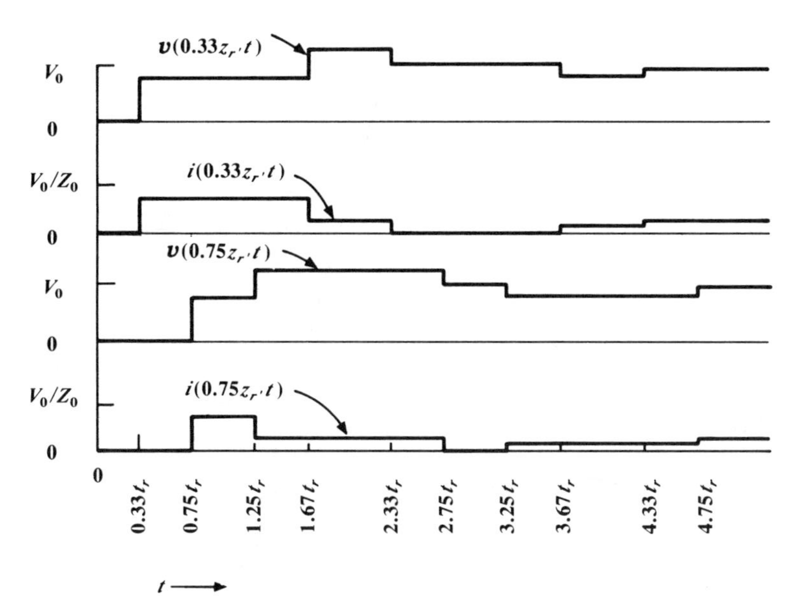

Figure 3-5. Voltage and current as functions of time on lossless transmission line. Case with infinite sequence of reflections: $R_G = 0.3Z_0$, $R_T = 5Z_0$. ($\rho_G = -0.539$, $\rho_T = 0.667$).

because (1) the sole effect of the lossless line is to introduce a time delay inversely proportional to the velocity of propagation and (2) the effect of resistive non-Z_0 terminations is to produce reflected voltages and currents which, at the respective terminations, are instantaneously in direct proportion to the incident voltages and currents. The second and subsequent outgoing waves, if nonzero, constitute, at the receiving end, *echos* of the original signal. In the case of video operation, *ghost images* would be produced.

Reflected waves are undesirable in communication systems generally; hence terminations there are designed so as to minimize (within practical tolerances for manufacturing) the corresponding reflection coefficients. Such lines and networks are said to be *matched*.

d. Bergeron Diagram: Linear and Nonlinear Resistive Terminations

The reflected wave sets which arise alternately at the two ends of the line may be found, and cumulatively summed, by a graphical construction attributed to Bergeron.[1] (An example is displayed in Fig. 3-6; it is drawn to

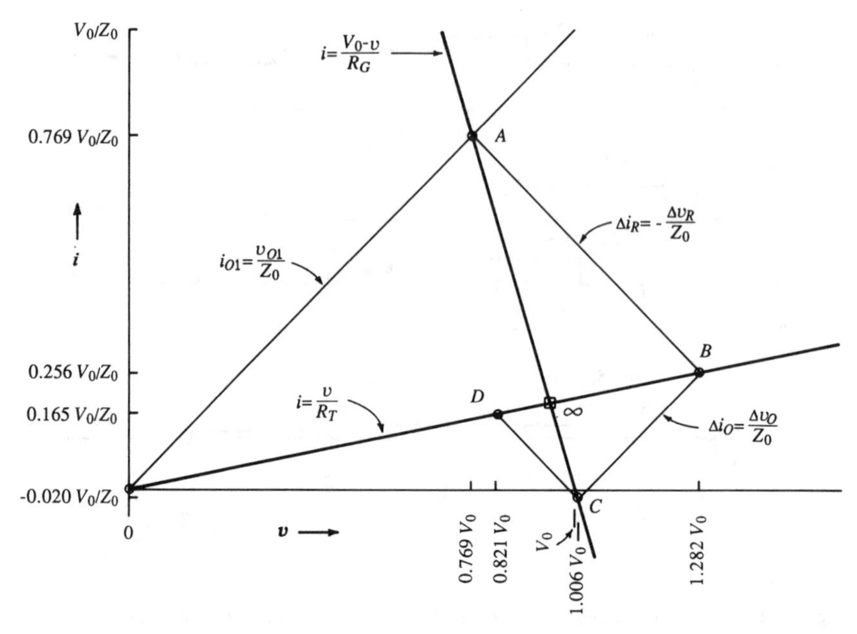

Figure 3-6. Bergeron diagram for lossless transmission line with unmatched terminations: $R_G = 0.3Z_0$, $R_T = 5Z_0$. $v_g(t) = V_0U(t)$.

- A Locus point for $z = 0$, $0 < t < 2t_r$. $v = v_{O1}$; $i = i_{O1}$.
- B Locus for $z = z_r$, $t_r < t < 3t_r$. $v = v_{O1} + v_{R1}$; $i = i_{O1} + i_{R1}$.
- C Locus for $z = 0$, $2t_r < t < 4t_r$. $v = v_{O1} + v_{R1} + v_{O2}$; $i = i_{O1} + i_{R1} + i_{O2}$.
- D Locus for $z = z_r$, $3t_r < t < 5t_r$. $v = v_{O1} + v_{R1} + v_{O2} + v_{R2}$; $i = i_{O1} + i_{R1} + i_{O2} + i_{R2}$.
- ∞ Locus for $t \to \infty$; all values of z.

scale for the same situation as was examined in Figs. 3-4 and 3-5, and will be explained in the following paragraphs.) This technique will also accommodate nonlinear resistive terminations; then the magnitude of the reflected voltage or current depends functionally on the magnitude of the resultant voltage impressed on the termination, rather than varying in direct proportion to the arriving traveling wave.

(1) *Linear Resistive Termination.* Equations 3-5 and 3-8 are linear equations which relate resultant voltage to resultant current at the two ends of the transmission circuit. They have been plotted as loci in Fig. 3-6. Equations 3-3*a* and 3-3*b* relate voltage to current for traveling-wave sets which move in the two opposite directions.

In the present example the circuit has been assumed to be initially quiescent (voltage and current both zero immediately before $t = 0$); hence the locus for i_{O1} vs. v_{O1} will emanate from the origin. The intersection of the loci of Eqs. 3-3a and 3-8, point A, corresponds to the solution of that pair of simultaneous equations for v_{O1} and i_{O1} in terms of V_0, R_G, and Z_0, and is equivalent to the first term on the right-hand side of Eq. 3-10.

The locus of Eq. 3-3b, for the first pair of reflected waves leaving the far end ($z = z_r$), passes through point A; its intersection with the locus of Eq. 3-5, point B, yields the resultant voltage and current at $z = z_r$ immediately after the arrival of v_{O1} and i_{O1} ($t = t_r +$). In terms of the plotted values,

$$v_{R1} = v(B) - v(A)$$

$$= (1.282 - 0.769)V_0 = 0.513V_0 \tag{3-27}$$

$$i_{R1} = i(B) - i(A)$$

$$= (0.256 - 0.769)V_0/Z_0 = -0.513V_0/Z_0 \tag{3-28}$$

The locus of Eq. 3-3a for the second outgoing voltage-and-current pair is parallel to that for v_{O1} and i_{O1}, but it passes through point B. Its intersection with Eq. 3-8, point C, yields the resultant voltage and current at $z = 0$ immediately after the arrival of v_{R1} and i_{R1} ($t = 2t_r +$):

$$v_{O2} = v(C) - v(B)$$

$$= (1.006 - 1.282)V_0 = -0.276V_0 \tag{3-29}$$

$$i_{02} = i(C) - i(B)$$

$$= (-0.020 - 0.256)V_0/Z_0 = -0.276V_0/Z_0 \tag{3-30}$$

This process may be continued for as many intervals of t_r as one wishes. The intersection of the loci of Eqs. 3-5 and 3-8 yields the limiting values which will be approached by v and i as time increases without limit, Eqs. 3-25 and 3-26.

One should note that values of v and i at points along any of the loci other than the intersections of two loci have no interpretation in the response functions (see Fig. 3-5). The ranges of time within which a given value of voltage or current is applicable are noted in the caption for Fig. 3-6.

The voltage function at some location $z = z_1$, where $z_1 < z_r$ (examples in Fig. 3-5), may be noted. Let

$$t_1 = z_1/\nu \tag{3-31}$$

$$\begin{aligned} v(z_1, t) &= 0 && t < t_1 \\ &= v(A) && t_1 < t < 2t_r - t_1 \\ &= v(B) && 2t_r - t_1 < t < 2t_r + t_1 \\ &= v(C) && 2t_r + t_1 < t < 4t_r - t_1 \\ &= v(D) && 4t_r - t_1 < t < 4t_r + t_1 \\ &\;\vdots \\ &\to v(\infty) && t \to \infty \end{aligned} \tag{3-32}$$

(2) ***Nonlinear Resistive Terminations.*** Figure 3-7 is the Bergeron diagram for a lossless transmission line with particular nonlinear resistances at each end. Loci of resultant current vs. resultant voltage have been

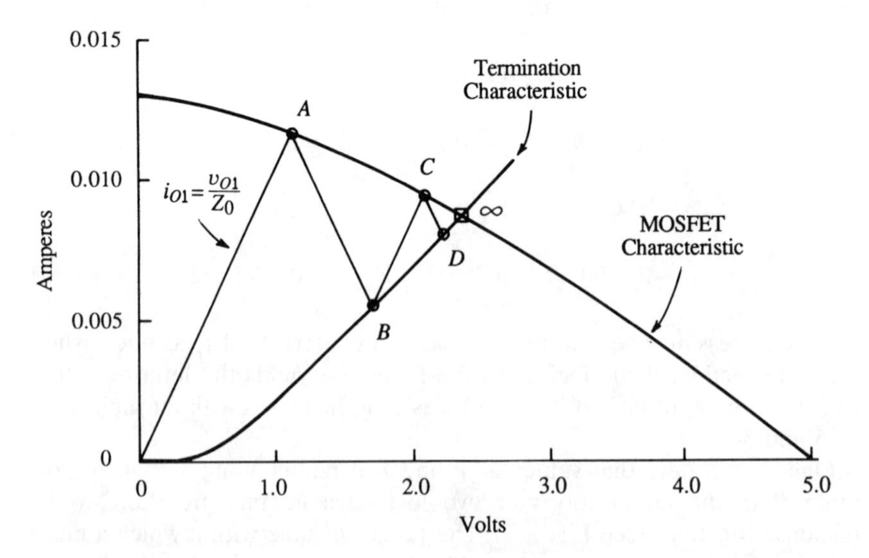

Figure 3-7. Bergeron diagram for MOSFET-driven lossless line ($Z_0 = 100\ \Omega$) terminated with a diode rectifier in series with a 200-Ω resistor.

plotted for both terminations. Because the relationships are nonlinear, the coordinate axes must be scaled in absolute rather than normalized values.

If the line is initially quiescent, the characteristic admittance (Eq. 3-3*a*) locus for the first outgoing traveling-wave set emanates from the origin, as in Fig. 3-6. The magnitude of that voltage-and-current pair is fixed by the intersection of the locus of Eq. 3-3*a* with the i-vs.-v locus of the source, which is point A. This process, and the subsequent procedure of finding points B, C, etc., and finally point ∞, parallel that for the linear case.

Plots of voltage and current, at one or more locations, as functions of time, may be readily assembled.

3-3 REFLECTIONS ON LINES WITH DISCONTINUITIES OR BRANCHES

The preceding examples have all involved a single section of transmission line, with lumped resistors of various magnitudes at its ends. Greater complexity arises when two sections of line with differing characteristic impedances are joined in tandem, perhaps through a lumped-impedance network, or when several lines converge at a single junction. The arrival of a traveling-wave set on one line at such a junction may be expected to yield a reflected, departing wave set on the same line, and a transmitted, departing wave set on each of the other lines. This may be expressed in matrix form by the following:

$$[v_D] = [\rho][v_A] \tag{3-33}$$

The subscript A denotes "arriving", and D denotes "departing."

a. Reflection-and-Transmission-Coefficients Matrix

For a junction which involves two lines, for example, the ρ-matrix in Eq. 3-33 has the expanded form

$$[\rho] = \begin{bmatrix} \rho_{11} & \rho_{12} \\ \rho_{21} & \rho_{22} \end{bmatrix} \tag{3-34}$$

The terms on the main diagonal, ρ_{11} and ρ_{22}, are the *reflection coefficients* for waves arriving on the respective lines, whereas the other terms will be designated as *transmission coefficients*. In keeping with the subscript convention for matrix elements, each of these is the value which, when multiplied by the voltage arriving at the junction on one line

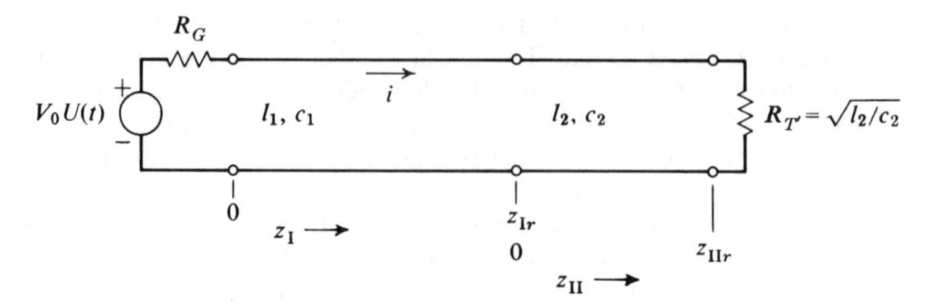

Figure 3-8. Tandem connection of two lossless lines.

(designated by the second subscript on ρ), will yield the voltage which departs on the other line (first subscript on ρ).

b. Tandem Connection of Two Lines

A situation closely related to those in Secs. 3-2b and c is that shown in Fig. 3-8, in which two lines of different characteristic impedance are connected in tandem.

(1) *Reflection and Transmission Coefficients.* Let

$$Z_{01} = \sqrt{\frac{l_1}{c_1}} \tag{3-35}$$

$$Z_{02} = \sqrt{\frac{l_2}{c_2}} \tag{3-36}$$

So far as the immediate wave reflection process for the left-hand line is concerned, the behavior is exactly the same as if the right-hand line were replaced with a lumped resistance $R_T = Z_{02}$. In accordance with Eq. 3-6,

$$\rho_{11} = \frac{Z_{02} - Z_{01}}{Z_{02} + Z_{01}} \tag{3-37}$$

Similarly,

$$\rho_{22} = \frac{Z_{01} - Z_{02}}{Z_{01} + Z_{02}} \tag{3-38}$$

The voltage on the left face of the junction must equal that on its right face; hence in this instance (in which the lines are joined together with no intermediate network) each transmitted voltage wave is equal to the sum of the incident and reflected voltages on the other line:

$$\rho_{21} = 1 + \rho_{11} = \frac{2Z_{02}}{Z_{02} + Z_{01}} \tag{3-39}$$

$$\rho_{12} = 1 + \rho_{22} = \frac{2Z_{01}}{Z_{01} + Z_{02}} \tag{3-40}$$

(In more general situations, such as those discussed in Secs. 3-3c and e, network calculations must be made in order to determine (1) the resultant impedance terminating each incoming line and (2) the proportionate part of the sum of the incident and reflected voltages that is transmitted onto the other line.)

(2) Velocities and Travel Times; Lattice Diagram. Let

$$\nu_1 = \frac{1}{\sqrt{l_1 c_1}} \tag{3-41}$$

$$\nu_2 = \frac{1}{\sqrt{l_2 c_2}} \tag{3-42}$$

In order to accommodate inequality between these velocities, distance along the two sections will be designated by the variables z_{I} and z_{II}, as shown in Fig. 3-8. The respective travel times are

$$t_{\mathrm{I}r} = \frac{z_{\mathrm{I}r}}{\nu_1} \tag{3-43}$$

$$t_{\mathrm{II}r} = \frac{z_{\mathrm{II}r}}{\nu_2} \tag{3-44}$$

In the specific case shown in Fig. 3-8, the right-hand line is terminated with a resistance equal to its characteristic impedance, thereby suppressing reflections from that point.

Equations 3-23 and 3-24 would be applicable to the left-hand section of line if Z_0, ν, z, t_r, and ρ_T were replaced by Z_{01}, ν_1, z_{I}, $t_{\text{I}r}$, and ρ_{11}. The following equations would describe the voltage and current in the right-hand section of line:

$$v(z_{\text{II}}, t) = \frac{V_0 Z_{01}}{R_G + Z_{01}} \Bigg[\rho_{21} U\left(t - t_{\text{I}r} - \frac{z_{\text{II}}}{\nu_2}\right) + \rho_{21}\rho_{11}\rho_G U\left(t - 3t_{\text{I}r} - \frac{z_{\text{II}}}{\nu_2}\right) + \rho_{21}\rho_{11}^2\rho_G^2 U\left(t - 5t_{\text{I}r} - \frac{z_{\text{II}}}{\nu_2}\right) + \cdots \Bigg] \quad \text{(3-45)}$$

The following is valid because, in this instance, all the waves on line 2 are traveling to the right.

$$i(z_{\text{II}}, t) = \frac{v(z_{\text{II}}, t)}{Z_{02}} \qquad \rho_{T'} = 0 \qquad \text{(3-46)}$$

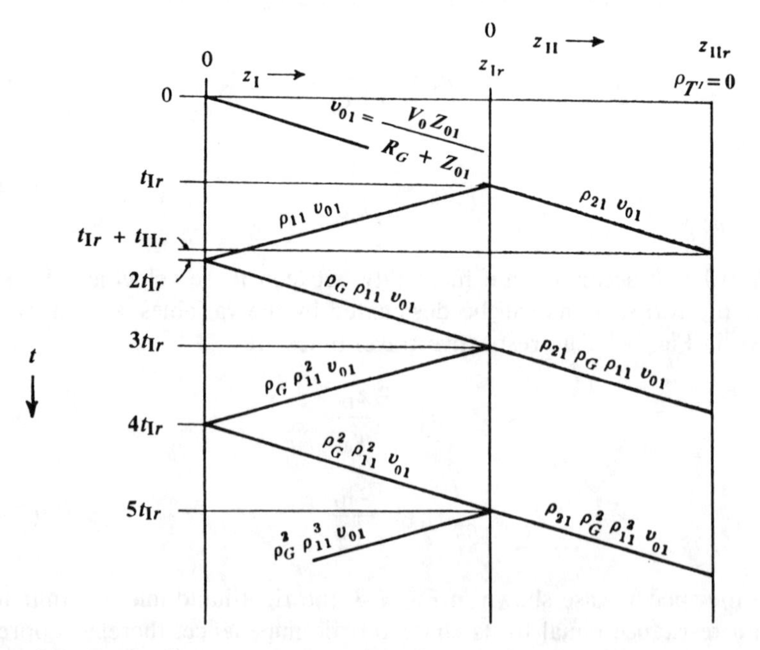

Figure 3-9. Lattice diagram for tandem-connected lines.

If, in the construction of the lattice diagram, the distances in z_{I} and z_{II} represented by a given abscissa increment (meters per centimeter, or other proportion) are chosen proportional to the respective velocities, one may circumvent the need for different slopes of the segments of the traveling-wave loci in the two lines. Figure 3-9 was drawn on this basis.

Example—Assume that $\nu_1 = 2.5 \times 10^8$ m/s, and $\nu_2 = 3.0 \times 10^8$ m/s. If the abscissa scale for z_{I} were chosen as 50 m/cm (or 5.0×10^3:1), the abscissa scale for z_{II} should be (50)(3.0/2.5) = 60 m/cm. (Depending on the total line length to be displayed and the width of the graph paper, a smaller or larger value of abscissa scale for z_{I}, with the proportionate abscissa scale for z_{II}, may be appropriate.)

c. Lines Connected through a Resistive Network

The assembly shown in Fig. 3-10 will illustrate the calculation of reflection coefficients and transmission coefficients when two lines are joined through a simple network. The resultant terminating impedances R_{1T} and R_{2T}, which are viewed by waves approaching the junction on the respective lines, may be found from the equivalent circuits shown in Fig. 3-11:

$$R_{1T} = R_A + \frac{R_C Z_{02}}{R_C + Z_{02}} \tag{3-47}$$

In accordance with Eq. 3-6,

$$\rho_{11} = \frac{R_{1T} - Z_{01}}{R_{1T} + Z_{01}} \tag{3-48}$$

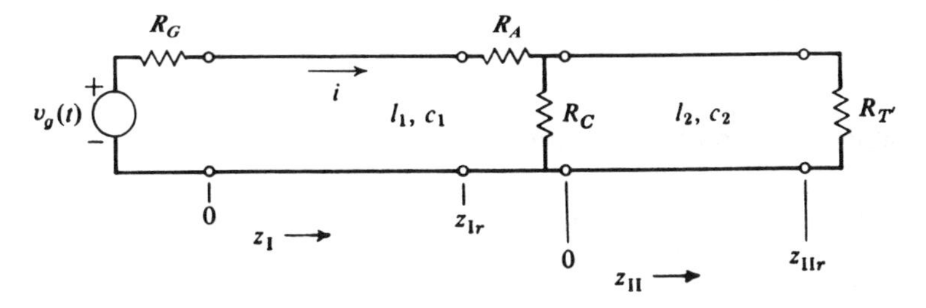

Figure 3-10. Tandem connection of two lossless lines through resistive network.

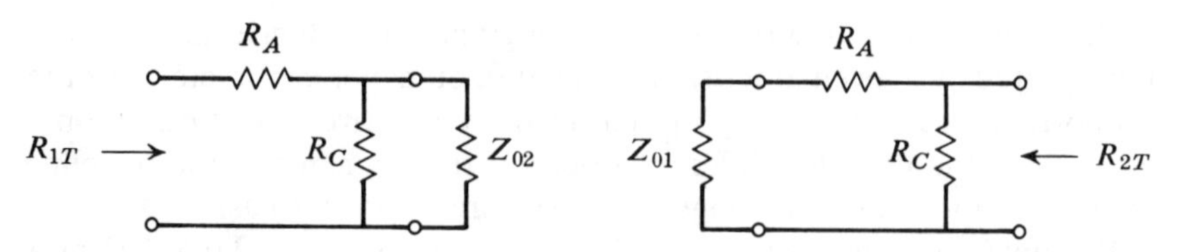

Figure 3-11. Effective terminating resistances for calculation of reflection coefficients.

Likewise

$$R_{2T} = \frac{R_C(R_A + Z_{01})}{R_C + R_A + Z_{01}} \tag{3-49}$$

$$\rho_{22} = \frac{R_{2T} - Z_{02}}{R_{2T} + Z_{02}} \tag{3-50}$$

For a wave arriving on line 2, the total voltage across R_C, $1 + \rho_{22}$ times the incident voltage, is divided between R_A and Z_{01} to yield the voltage departing on line 1:

$$\rho_{12} = \frac{Z_{01}}{Z_{01} + R_A} \cdot (1 + \rho_{22}) \tag{3-51}$$

Likewise for a wave arriving on line 1, the total voltage of $1 + \rho_{11}$ times the incident voltage is divided between R_A and the parallel combination of R_C and Z_{02}. This reduces to

$$\rho_{21} = \frac{R_C Z_{02}}{R_C + Z_{02}} \cdot \frac{1}{R_{1T}} \cdot (1 + \rho_{11}) \tag{3-52}$$

The results of the example in Sec. 3-3b are applicable here if the ρ coefficients as given in Eqs. 3-47 through 3-52 are used instead of Eqs. 3-37 through 3-40.

d. Minimum-Loss Pad (Impedance-Matching Network)

By proper choice of R_A and R_C in terms of Z_{01} and Z_{02} (and moving R_C to the left-hand side of the network if $Z_{02} > Z_{01}$), the reflection coefficients ρ_{11} and ρ_{22} may be made to vanish. The network is then known as a *minimum-loss pad*. Let R_{1T} in Eq. 3-47 equal Z_{01}, and R_{2T} in

Eq. 3-49 equal Z_{02} (this is the condition for an *image match*):

$$Z_{01} = R_A + \frac{R_C Z_{02}}{R_C + Z_{02}} \tag{3-53}$$

$$Z_{02} = \frac{R_C(R_A + Z_{01})}{R_C + R_A + Z_{01}} \tag{3-54}$$

If Eq. 3-53 is solved for R_A and this result substituted in Eq. 3-54, R_C may be found:

$$R_C = Z_{02}\sqrt{\frac{Z_{01}}{Z_{01} - Z_{02}}} \tag{3-55}$$

Substitution of this result in Eq. 3-53 yields

$$R_A = \sqrt{Z_{01}(Z_{01} - Z_{02})} \tag{3-56}$$

As noted above, R_A and R_C were chosen so that the reflection coefficients would vanish. The transmission coefficients, Eqs. 3-51 and 3-52, reduce to

$$\rho_{12} = \frac{Z_{01}}{Z_{01} + R_A} \tag{3-57}$$

$$\rho_{21} = \frac{R_C Z_{02}}{(R_C + Z_{02})Z_{01}} \tag{3-58}$$

e. Multiple-Line Junction

Branching of a line, as illustrated in Fig. 3-12, increases the complexity since two transmitted waves plus a reflected wave may be expected whenever a wave on any of the three line sections reaches the junction. If the characteristic impedances differ, the reflection and transmission coefficients applicable to the three lines will differ. Some of them are computed here. Let

$$R_{1T} = \frac{Z_{02}Z_{03}}{Z_{02} + Z_{03}} \tag{3-59}$$

$$R_{2T} = \frac{Z_{01}Z_{03}}{Z_{01} + Z_{03}} \tag{3-60}$$

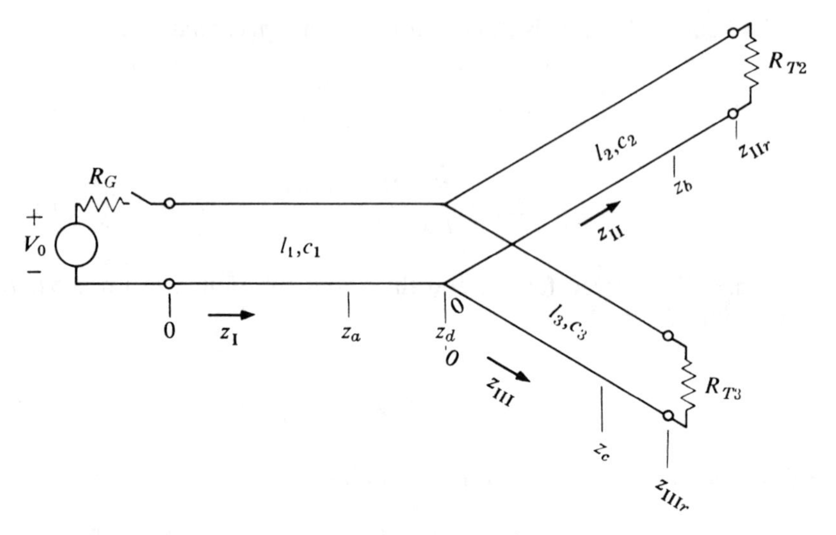

Figure 3-12. Branched transmission line (see Prob. 3-14).

and so forth:

$$\rho_{11} = \frac{R_{1T} - Z_{01}}{R_{1T} + Z_{01}} \tag{3-61}$$

$$\rho_{31} = \rho_{21} = 1 + \rho_{11}$$

$$= \frac{2R_{1T}}{R_{1T} + Z_{01}} \tag{3-62}$$

$$\rho_{12} = \rho_{32} = \frac{2R_{2T}}{R_{2T} + Z_{02}} \tag{3-63}$$

A relatively simple example is given in Prob. 3-14.

3-4 CONCLUSIONS

Reflected waves will result when a traveling-wave pair reaches the end of a uniform section of lossless line unless the termination presents an impedance R_T that is purely resistive and equal to the characteristic impedance of the line $Z_0 = \sqrt{l/c}$. The ratio of the reflected voltage to the incident voltage at the termination is known as the *reflection coefficient* ρ_T. For a resistive termination

$$\rho_T = \frac{R_T - Z_0}{R_T + Z_0} \tag{3-6}$$

The same principle is applicable to returning waves reaching the source, if the generator impedance is other than a pure resistance equal to Z_0. If neither the terminating impedance nor the generator impedance is equal to Z_0, an infinite number of reflections will follow any change in the voltage or current function impressed by the source. The *lattice diagram*, illustrated in Figs. 3-4 and 3-9, presents graphically the time sequence of the arrivals of successive voltage and current traveling waves at any specific location.

The *Bergeron diagram* provides a graphical solution for the resultant voltage and current following each reflection, and it will accommodate nonlinear resistive terminations.

If two or more lines are joined at a junction, whether directly or through a lumped-resistance network, a wave arriving on any one line may be expected to give rise to departing waves on all lines. The ratio of a departing voltage wave to the corresponding incident wave, if they are on different lines, is known as the *transmission coefficient*. For the purpose of computing reflection coefficients and transmission coefficients, each such line may be replaced by a lumped resistance equal to its characteristic impedance.

One such network, known as a *minimum-loss pad* (Fig. 3-10 and Eqs. 3-55 and 3-56), will yield an impedance match, that is, the reflection coefficients for waves arriving on either line will be zero.

PROBLEMS

3-1. Sketch waveshapes corresponding to those in Fig. 3-3 for the case in which $R_G = Z_0$ and $R_T = 0.2Z_0$.

3-2. See Figs. 3-4 and 3-5. Draw a sketch similar to Fig. 3-5 for $v(t)$ and $i(t)$ at $z = z_r$, and give values of the ordinates in terms of V_0 and Z_0.

3-3. Circuit as shown in Fig. 3-1. $R_G = 4Z_0$, $R_T \to \infty$. Let $v_g(t) = V_0 U(t)$. Draw a lattice diagram showing numerical values for the first three traveling waves of v and i (in terms of V_0 and Z_0). Sketch $v(0.1z_r, t)$ and $i(0.1z_r, t)$, and find the limits of v and i for $t \to \infty$.

3-4. Draw to scale the Bergeron diagram for the situation considered in Prob. 3-3.

3-5. Assume that the resistor R_T in Fig. 3-1 is replaced by a diode rectifier, the conductance characteristic of which may be approximated by the following empirical equation:

$$i_r = 1.0 \times 10^{-8} \exp(23.0 v_r) \text{ A}$$

where v_r is in volts. Let $R_G = 25.0\ \Omega$, $Z_0 = 50.0\ \Omega$, and $v_g(t) = 0.6\,U(t)$ V. Draw to scale the Bergeron diagram for $0 \le t \le 5t_r$, and plot $v(0, t)$ and $v(z_r, t)$ for that interval. What value does $v(0, t)$ approach as $t \to \infty$?

3-6. Assume that the ideal voltage source and the resistor R_G in Fig. 3-1 have been replaced by a MOSFET which is switched "On" at $t = 0$, and which energizes the line in a manner which may be approximated by the following empirical equation:

$$i(0, t) = 1.3 \times 10^{-2} - 5.2 \times 10^{-4}[v(0, t)]^2 \text{ A}$$

($0 \leq v \leq 5.0$ V), where v is in volts. Let $Z_0 = 100\ \Omega$ and $R_T \to \infty$. Draw to scale the Bergeron diagram for $0 \leq t \leq 5t_r$, and plot $v(z_r, t)$ for that interval.

3-7. Given a lossless transmission line with source and resistive termination as shown in Fig. 3-1. A finite portion of the voltage observed at $z_1 = 240$ m is shown in Fig. 3-13, and a shorter portion of the corresponding current response is also shown.

(a) Find numerical values for the following: the velocity of propagation, the characteristic impedance, the length of the line, the reflection coefficient at $z = z_r$, and the sending-end reflection coefficient.

(b) Draw a lattice diagram which includes the first set of waves reflected from the sending end; give numerical values for each traveling-wave component of voltage and current and for the time of each reflection.

(c) Complete the current function over the same period of time as the voltage function is shown, and give numerical values for the resultant values. Find the time at which the last shown change in voltage occurs.

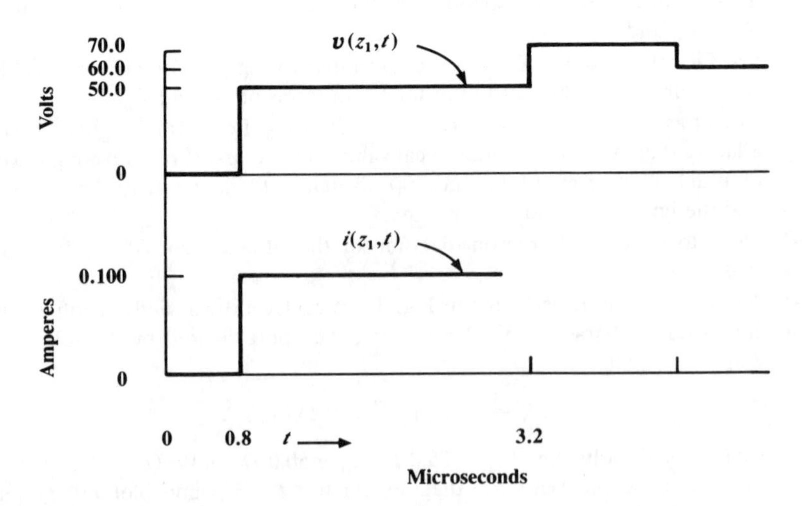

Figure 3-13. Step-function response on unmatched lossless line (see Prob. 3-7).

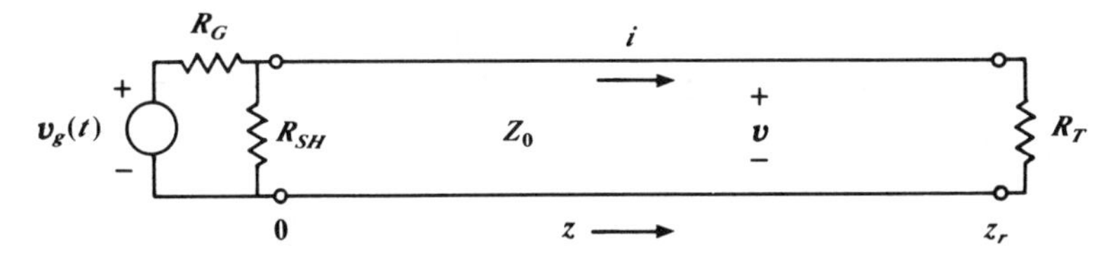

Figure 3-14. Lossless line with adjustable source resistance (see Prob. 3-8).

3-8. Given a lossless line as shown in Fig. 3-14. $R_G = 75\ \Omega$, $Z_0 = 50\ \Omega$, $R_T = 25\ \Omega$, $z_r = 600$ m, $v = 2.0 \times 10^8$ m/s, $v_g(t) = 20.0U(t)$ V.
   (a) Find the value of R_{SH} which will suppress reflections from the sending end.
   (b) Draw a lattice diagram for $0 \le t \le 10.0\ \mu$s, and sketch $v(z_1, t)$ and $i(z_1, t)$ for the same interval. $z_1 = 0.25z_r$.

3-9. Consider the composite line shown in Fig. 3-8. Let $Z_{01} = 72\ \Omega$, $Z_{02} = 50\ \Omega$, and $R_G = 50\ \Omega$. Compute the relative magnitudes of the components shown on the lattice diagram in Fig. 3-9, and sketch voltage and current as observed at the midpoint of each line as functions of time.

3-10. Consider the composite line shown in Fig. 3-10. Let $Z_{01} = 72\ \Omega$, $Z_{02} = 50\ \Omega$, and $R_G = 72\ \Omega$. Find the values of R_A and R_C for the network to be an impedance-matching pad. Assume that $R_{T'}$ is an open circuit. Draw a lattice diagram, evaluate the magnitudes of the various components of voltage and current, and sketch the latter as observed at the midpoint of each line as functions of time. Let $v_g(t) = V_0U(t)$.

3-11. Given the system shown in Fig. 3-10. Both lines are assumed to be lossless. Let $R_C \to \infty$, $R_A = 43\ \Omega$, $R_G = 50\ \Omega$, $R_{T'} = 0$, $Z_{01} = 93\ \Omega$, $Z_{02} = 50\ \Omega$, $\nu_1 = 2.5 \times 10^8$ m/s, and $\nu_2 = 2.0 \times 10^8$ m/s. The first outgoing wave set reaches the junction at $t = 0.40\ \mu$s, and the far end of the line at $0.70\ \mu$s. $v_g(t) = 10.0U(t)$ V.
   (a) Find the reflection coefficients and the transmission coefficients at the junction of the lines (numerical values).
   (b) Draw a lattice diagram for $0 \le t \le 1.2\ \mu$s, with numerical values for the traveling-wave components.
   (c) Find the length of each section of line.

3-12. Given the circuit shown in Fig. 3-10. $Z_{01} = 93.0\ \Omega$, $Z_{02} = 50.0\ \Omega$, $R_G = 93.0\ \Omega$, $R_A = 0$, $R_C = 50.0\ \Omega$, $R_{T'} = R_1 + R_2$, where $R_1 = 63.2\ \Omega$ and $R_2 = 73.5\ \Omega$, $z_{Ir} = 150$ m, $z_{IIr} = 140$ m, $\nu_1 = 2.50 \times 10^8$ m/s, $\nu_2 = 2.00 \times 10^8$ m/s, $v_g(t) = 6.00U(t)$ V.
   (a) Find the reflection coefficients and the transmission coefficients at the junction of the lines.
   (b) Draw a lattice diagram for $0 \le t \le 1.5 \times 10^{-6}$ s, with numerical values for the traveling-wave components.

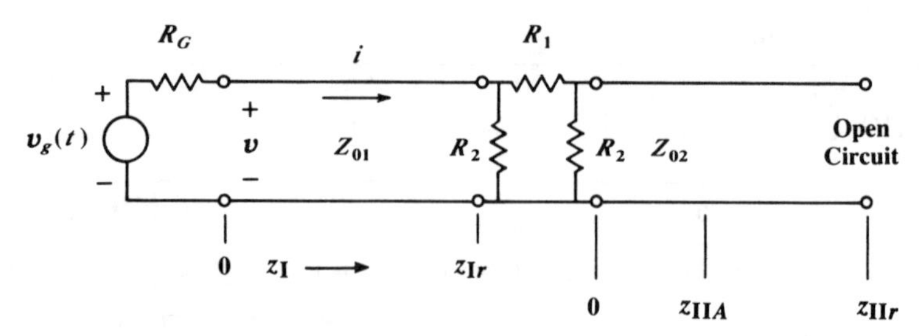

Figure 3-15. Tandem-connected lines with "Pi" resistive network (see Prob. 3-13).

(c) Plot the voltage across R_2, $v_{R2}(t)$, for the same time interval.

(d) Find the limit which v_{R2} approaches as $t \to \infty$.

3-13. Assume that both sections of line shown in Fig. 3-15 are lossless, and that the velocity of propagation is 2.0×10^8 m/s in both sections. $R_G = 50\ \Omega$, $Z_{01} = 60\ \Omega$, $Z_{02} = 60\ \Omega$, $R_1 = 80\ \Omega$, $R_2 = 120\ \Omega$, $z_{Ir} = 250$ m, $z_{IIr} =$ 200 m, $z_{IIA} = 80$ m, $v_g(t) = 15.0U(t)$ V.

(a) Draw circuit diagrams (two) which show the effective terminations (for traveling waves) of the two lines at the resistive network, and compute the values of the reflection coefficients and transmission coefficients.

(b) Draw a lattice diagram for $0 \leq t \leq 5.0 \times 10^{-6}$ s. Locate point z_{IIA} approximately on the diagram and show graphically how the arrival times of the first two wave sets may be determined.

3-14. Assume the following relationships among the parameters in Fig. 3-12. $l_1 = l_2 = l_3$, $c_1 = c_2 = c_3$, $R_G = \sqrt{l_1/c_1}$, $R_{T2} = \sqrt{l_2/c_2}$, and $R_{T3} = 10\sqrt{l_3/c_3}$. Assume that the lines are initially uncharged, and that the source V_0 is a battery. If the switch closes at t equals zero, find $v(t)$ and $i(t)$ at locations z_a, z_b, and z_c for values of time prior to the third reflection from R_{T3}. Sketch these functions.

REFERENCES

1. Bergeron, L., *Du Coup de Bélier en Hydraulique au Coup de Foudre en Electricité*. Dunod, Paris, 1949. English translation: *Water Hammer in Hydraulics and Wave Surges in Electricity*, The American Society of Mechanical Engineers, 1961.
2. Bewley, L. V., *Traveling Waves on Transmission Systems*, 2nd ed., Wiley, New York, 1951.
3. *Electrical Transmission and Distribution Reference Book*, Westinghouse Electric Corporation, East Pittsburgh, PA, 1950.
4. Fink, D. G. and Beaty, H. W., Eds., *Standard Handbook for Electrical Engineers*, 12th ed., McGraw-Hill, New York: 1987, 18–42 and 18–43.
5. Thrift, P. R., Pulse propagation in an unterminated environment, *Radio and Electron. Eng.*, 48, 181, 1978.

CHAPTER 4
Sinusoidal Traveling Waves on a Lossy Line

Practical line conductors have nonzero resistance; this affects propagation phenomena. Shunt conductance too should be included in a general study. These parameters alter the results just found, in that many current and voltage functions do not travel, with wave shapes intact, at a single, clearly defined velocity; instead, they change in appearance progressively as they move along the line. Fortunately the sinusoidal functions of $\omega(t - z/v)$ and $\omega(t + z/v)$ (where v may be a function of ω), if multiplied by exponential functions of distance, are exceptions to this discouraging generalization. Steady-state analysis of the lossy transmission line is built around those two functions.

Propagation behavior of the sinusoidal traveling waves may be described concisely with the aid of two complex quantities, the propagation function and the characteristic impedance. Each is derivable in terms of the four line parameters and the frequency, a total of five variables.

Certain limiting cases such as the lossless line yield much simpler results than the general expressions for the propagation function and the characteristic impedance. The asymptotes approached as frequency approaches zero or infinity are other limiting-case solutions. These results are helpful reference points from which to survey the multiparameter transmission line. Series expansions prove useful in showing quantitatively the effects caused by small deviations from the respective limiting-case conditions.

Explicit functions of time and distance will be used for the initial presentation, after which the phasor concept will be developed.

4-1 MATHEMATICAL SOLUTION FOR TRAVELING-WAVE PROPERTIES

Under the restriction that the excitation be sinusoidal, the parameters r, l, g, and c will have constant values corresponding to the chosen radian frequency ω. The partial differential equations with constant coefficients in Chap. 2 may be enlarged by the addition of terms which account for the effects of r and g. Then the same general plan of analysis will be followed, that of testing a candidate solution for the voltage function and determining the accompanying current function. This will be followed by interpretation of wave propagation in terms of direction and speed.

a. Partial Differential Equations of General Line

The derivation of the differential equations for the lossy line closely parallels that for the lossless line, so only the differences will be emphasized here. The polarity-and-direction convention given in Fig. 2-1 will be followed.

Equation 2-8, for the change of voltage along the line because of current flowing through the series inductance, is altered if one allows for the incremental conductor resistance $r\,\Delta z$:

$$\Delta z\left[\frac{\partial v(z,t)}{\partial z}\right]_{z=z_1} = -r\,\Delta z\, i(z_1,t) - l\,\Delta z\frac{\partial i(z_1,t)}{\partial t} \qquad \Delta z \to 0 \quad (4\text{-}1)$$

In like manner, shunt conductance will alter Eq. 2-6, because of current through the incremental shunt $g\,\Delta z$:

$$\Delta z\left[\frac{\partial i(z,t)}{\partial z}\right]_{z=z_1} = -g\,\Delta z\, v(z_1,t) - c\,\Delta z\frac{\partial v(z_1,t)}{\partial t} \qquad \Delta z \to 0 \quad (4\text{-}2)$$

Again the subscripts 1 may be omitted, and Δz may be divided out:

$$\frac{\partial v(z,t)}{\partial z} = -ri(z,t) - l\frac{\partial i(z,t)}{\partial t} \quad (4\text{-}3)$$

$$\frac{\partial i(z,t)}{\partial z} = -gv(z,t) - c\frac{\partial v(z,t)}{\partial t} \quad (4\text{-}4)$$

Except for the added terms in r and g, these parallel Eqs. 2-6 and 2-8. By differentiating Eq. 4-3 with respect to z, and Eq. 4-4 with respect to t, and substituting, $i(z,t)$ may be eliminated from the set:

$$\frac{\partial^2 v(z,t)}{\partial z^2} = rgv(z,t) + (rc + lg)\frac{\partial v(z,t)}{\partial t} + lc\frac{\partial^2 v(z,t)}{\partial t^2} \quad (4\text{-}5)$$

Similarly, $v(z, t)$ may be eliminated from Eqs. 4-3 and 4-4 to yield

$$\frac{\partial^2 i(z,t)}{\partial z^2} = rgi(z,t) + (rc + lg)\frac{\partial i(z,t)}{\partial t} + lc\frac{\partial^2 i(z,t)}{\partial t^2} \quad (4\text{-}6)$$

These equations are known in classical physics as the *equations of telegraphy*. They will reduce to wave equations (Eqs. 2-12 and 2-13) if r and g are set equal to zero.

b. Traveling-Wave Solutions for Voltage

From experience with lumped-parameter networks and the differential equations derived from them, one would suspect that a function which varied sinusoidally with time could be adjusted to fit the right-hand side of Eq. 4-5. A sinusoidal traveling-wave function would seem to be a possible solution, although from physical reasoning one would expect the amplitude to be reduced with distance because of losses in the resistance and shunt conductance. Let the following solution be assumed, in which α and β are unknown coefficients:

$$v_1(z,t) = V_{1M}\epsilon^{-\alpha z}\cos(\omega t - \beta z) \quad (4\text{-}7)$$

Before differentiating and substituting $v_1(z, t)$ in Eq. 4-5, it is convenient to replace the cosine function, in accordance with Euler's identity, by exponentials with complex arguments:

$$v_1(z,t) = \frac{V_{1M}}{2}\epsilon^{-\alpha z}\left[\epsilon^{j(\omega t - \beta z)} + \epsilon^{-j(\omega t - \beta z)}\right] \quad (4\text{-}8)$$

If Eq. 4-8 is differentiated to give the various partial derivatives called for in Eq. 4-5, the coefficients of $\epsilon^{j\omega t}$ and those of $\epsilon^{-j\omega t}$ prove to be conjugates of each other at every step. This is true of the voltage-and-current relationships in linear circuits generally; hence it suffices to perform the calculations on the coefficient of $\epsilon^{j\omega t}$; the corresponding coefficient of $\epsilon^{-j\omega t}$ in the final result can be written immediately as the conjugate. In practice the latter step need not be performed. Thus

$$\epsilon^{j(\omega t - \beta z)} = \cos(\omega t - \beta z) + j\sin(\omega t - \beta z) \quad (4\text{-}9)$$

$$\epsilon^{-j(\omega t - \beta z)} = \cos(\omega t - \beta z) - j\sin(\omega t - \beta z) \quad (4\text{-}10)$$

The *real part* of $\epsilon^{j(\omega t - \beta z)}$ is $\cos(\omega t - \beta z)$, the same as one-half the sum of the last two equations. This concept may be stated in mathematical form by introducing the operator Re; it means "real part of."* Thus

$$\mathrm{Re}[\epsilon^{j(\omega t-\beta z)}] = \cos(\omega t - \beta z) \tag{4-11}^{\dagger}$$

Equation 4-8, and hence Eq. 4-7, may be restated as follows:

$$\begin{aligned} v_1(z,t) &= \mathrm{Re}\left[V_{1M}\epsilon^{-\alpha z}\epsilon^{j(\omega t-\beta z)}\right] \\ &= \mathrm{Re}\left[V_{1M}\epsilon^{j\omega t}\epsilon^{-(\alpha+j\beta)z}\right] \end{aligned} \tag{4-14}$$

If the terms on the right-hand side of Eq. 4-5 are moved to the left-hand side, and Eq. 4-14 and its derivatives substituted, the following results:

$$\mathrm{Re}\left\{V_{1M}\epsilon^{j\omega t}\epsilon^{-(\alpha+j\beta)z}\left[(\alpha+j\beta)^2 - rg - (rc+lg)j\omega - lc(j\omega)^2\right]\right\} = 0 \tag{4-15}$$

It may be noted that $|\exp(j\omega t)| = 1.0$ for all values of t, and that $|\exp[-(\alpha+j\beta)z]|$ is nonzero for all values of z. Equation 4-15 must be satisfied for all values of t and z chosen independently; hence the quantity within the brackets must equal zero. The reduced equation may be solved for the quantity $\alpha + j\beta$:

$$\begin{aligned} \alpha + j\beta &= \sqrt{rg + (rc+lg)j\omega + lc(j\omega)^2} \\ &= \sqrt{(r+j\omega l)(g+j\omega c)} \end{aligned} \tag{4-16}$$

It is appropriate to designate $r + j\omega l$ as the *distributed complex impedance*, and $g + j\omega c$ as the *distributed complex admittance*, of the line. For a lossless line, these functions would be simply $j\omega l$ and $j\omega c$, respec-

---

*Correspondingly, Im indicates "imaginary part of."

†By examination of Eq. 4-9 it is apparent that, if a is a real constant or function,

$$\mathrm{Re}[a\epsilon^{j(\omega t-\beta z)}] = a\{\mathrm{Re}[\epsilon^{j(\omega t-\beta z)}]\} \tag{4-12}$$

However, if one multiplies Eq. 4-9 by j,

$$\mathrm{Re}[j\epsilon^{j(\omega t-\beta z)}] = -\sin(\omega t - \beta z) \tag{4-13}$$

Hence Eq. 4-12 would not be true if a were a complex quantity.

tively. Let

$$\gamma = \alpha + j\beta \tag{4-17}$$

The complex quantity γ is given the name *propagation function*.

Thus the postulated solution $v_1(z, t)$, Eq. 4-7, satisfies the equation of telegraphy, Eq. 4-5, provided α and β are chosen in accordance with Eq. 4-16.

In like manner, the following function $v_2(z, t)$ may be shown to satisfy Eq. 4-5:

$$v_2(z, t) = V_{2M}\epsilon^{\alpha z}\cos(\omega t + \beta z) \tag{4-18}$$

The values for α and β prove to be the same as those in Eq. 4-16.

c. Properties of the Propagation Function

The real and imaginary parts of the propagation function are called the *attenuation function* and the *phase function*, respectively.*

(1) *Dimensions and Units.* The term $r + j\omega l$ has the units ohms per unit of length, whereas $g + j\omega c$ has the units siemens per unit of length. Accordingly, the dimensions of γ are the inverse of length, and the same is true of its real and imaginary components.

The products αz and βz, as used in Eqs. 4-8 and 4-14, are exponents of the natural logarithmic base ϵ. Both products are dimensionless, as every exponent must be; αz is measured in units of *nepers*, while βz is measured in units of *radians*. Thus α and β, as found directly from Eq. 4-16, have the units of nepers per unit of length and radians per unit of length, respectively. Unfortunately, no unit name for γ, the complex sum of those two quantities, has been generally accepted.

For numerical work the phase constant β is often stated in degrees per unit of length, and the attenuation constant α in *decibels* per unit of length. To introduce the latter unit and to relate it to nepers, the maximum values of $v_1(z, t)$ at two locations, $V_M(z_1)$ and $V_M(z_2)$, where

---

*Numerical values for particular problems may be found by solving Eq. 4-16 explicitly for γ in polar form; thus

$$\gamma = \left[(r^2 + \omega^2 l^2)(g^2 + \omega^2 c^2)\right]^{1/4} \exp\left\{(j/2)\left[\tan^{-1}(\omega l/r) + \tan^{-1}(\omega c/g)\right]\right\} \tag{4-19}$$

The numerical γ may then be resolved into rectangular components.

$z_2 > z_1$, will be compared. From Eq. 4-7,

$$V_M(z_1) = V_{1M}\epsilon^{-\alpha z_1}$$
$$V_M(z_2) = V_{1M}\epsilon^{-\alpha z_2}$$

Dividing the first of these equations by the second, and taking the natural logarithm of both sides, yields

$$\alpha(z_2 - z_1) = \ln\left[\frac{V_M(z_1)}{V_M(z_2)}\right] \quad \text{(nepers, N)} \tag{4-20}$$

The same attenuation may be stated in decibels by means of the following equation:

$$\alpha_{\text{dB}}(z_2 - z_1) = 20\log_{10}\left[\frac{V_M(z_1)}{V_M(z_2)}\right] \quad \text{(decibels, dB)} \tag{4-21}^*$$

---

*The original concept of the decibel unit was as a measure of the *power loss* in a line or network,

$$\text{Loss in dB} = 10\log_{10}\left(\frac{P_{\text{in}}}{P_{\text{out}}}\right) \tag{4-22}$$

In many experimental situations it is more convenient to measure voltage than to measure power, and Eq. 4-22 may be restated in terms of the corresponding voltage ratio. If the input and output voltages are measured *at the same impedance level*, which is true for a section of uniform transmission line with a reflectionless termination, the following applies:

$$\frac{P(z_1)}{P(z_2)} = \frac{V_M^2(z_1)}{V_M^2(z_2)} \tag{4-23}$$

Hence Eqs. 4-21 and 4-22 are equivalent.

Should the input and output voltages be measured at different impedance levels, such as at the terminals of the pad shown in Fig. 3-10 and described in Eqs. 3-55 and 3-56, the impedance ratio must be taken into account. In terms of that problem, in which Z_{01} and Z_{02} are purely resistive and a step-function signal is considered, let p_1 and v_1 be measured at the left-hand terminals of the network and p_2 and v_2 at the right-hand terminals:

$$\frac{p_1}{p_2} = \frac{v_1^2}{v_2^2} \cdot \frac{Z_{02}}{Z_{01}} \tag{4-24}$$

For a wave going through the network from left to right:

$$\text{Loss in dB} = 20\log_{10}\left(\frac{v_1}{v_2}\sqrt{\frac{Z_{02}}{Z_{01}}}\right) \tag{4-25}$$

The ratio of the root-mean-square voltage values (Sec. 4-2) is the same as that of the maximum values.

The conversion factor between nepers and decibels may be derived as follows.

$$\log_{10} u = (\log_{10} e)(\ln u)$$

$$= 0.4343 \ln u$$

$$20 \log_{10} u = 8.686 \ln u \tag{4-26}$$

Hence an attenuation constant in nepers per unit length of line may be converted into decibels per unit of length by multiplying it by 8.686.

The decibel unit is commonly used in the electrical communications industry in many parts of the world as a measure of line attenuation, amplifier gain, etc. It probably gained early acceptance largely because base-10 logarithms seemed more familiar than natural logarithms. For the basic study of transmission-line phenomena, however, the use of natural logarithms, nepers, and radians is recommended because those units are directly related to the operations of calculus and the solution of differential equations.

(2) *Phase Velocity.* The concept of velocity of propagation (Eq. 2-29) was introduced in Sec. 2-2b for the general traveling-wave functions f_A and f_B on a lossless line, by noting the change δz that should accompany a change δt to keep the argument of the function constant. The same concept may be adapted to the attenuating sinusoidal wave as the velocity at which any point on the voltage or current wave, defined as a stated fraction of the maximum value at each location on the line, appears to move. For $v_1(z, t)$, Eq. 4-7, the constant-argument constraint is

$$\omega\, \delta t = \beta\, \delta z \tag{4-27}$$

As for the lossless line, let

$$\nu = \left| \frac{\delta z}{\delta t} \right| \tag{2-29}$$

Substitution of Eq. 4-27 in 2-29 yields

$$\nu = \frac{\omega}{\beta} \tag{4-28}^*$$

* By comparison with Eq. 2-30 one may note that

$$\beta = \omega\sqrt{lc} \quad \text{(lossless line)} \tag{4-29}$$

For the lossy line, β, which is given by the imaginary part of Eq. 4-16, is not directly proportional to ω, hence the velocity is a function of frequency.

Because the mode of variation with respect to time has, for v_1 in Eq. 4-7, been expressly constrained to be sinusoidal, and because the velocity defined by Eq. 4-28 is a function of frequency on a lossy line, the expression *phase velocity* has been adopted as a more specific designator for ν in this application. The same results are applicable to v_2 as defined in Eq. 4-18. (A system in which the phase velocity is a function of frequency is said to be *dispersive*. Hollow metallic waveguides, such as those discussed in Chapters 12 and 13, have this property even if the system is assumed to be lossless.)

d. Voltage-Current Relationship—Complex Characteristic Impedance

Expressions should be found for the current functions which accompany voltages $v_1(z,t)$ and $v_2(z,t)$. An expression for $i_1(z,t)$ may be found by differentiating $v_1(z,t)$, Eq. 4-14, with respect to t and substituting in Eq. 4-4. Integration of this result with respect to z yields

$$i_1(z,t) = \mathrm{Re}\left[V_{1M}\frac{g + j\omega c}{\alpha + j\beta}\epsilon^{j\omega t}\epsilon^{-(\alpha + j\beta)z}\right]$$

Substituting of Eq. 4-16 for the denominator term $\alpha + j\beta$ yields

$$i_1(z,t) = \mathrm{Re}\left[V_{1M}\sqrt{\frac{g + j\omega c}{r + j\omega l}}\,\epsilon^{j\omega t}\epsilon^{-(\alpha + j\beta)z}\right] \tag{4-30}$$

The quantity $\sqrt{(g + j\omega c)/(r + j\omega l)}$ has the dimensions of admittance; its reciprocal will be designated by the complex quantity

$$Z_0 = |Z_0|\exp(j\theta_0) = \sqrt{\frac{r + j\omega l}{g + j\omega c}} \tag{4-31}$$

Hence

$$i_1(z,t) = \mathrm{Re}\left[\frac{V_{1M}}{|Z_0|}\exp[j(\omega t - \theta_0) - (\alpha + j\beta)z]\right] \tag{4-32}$$

$$i_1(z,t) = \frac{V_{1M}}{|Z_0|}\epsilon^{-\alpha z}\cos(\omega t - \theta_0 - \beta z) \tag{4-33}$$

It appears from comparison of Eq. 4-33 with Eq. 4-7 that in a lossy line the current and voltage traveling waves are not in time phase at any value of z. The relationship between current and voltage waves of a single frequency may be expressed by a complex quantity Z_0, Eq. 4-31, which, by analogy with the lossless line (see Eqs. 2-32, 2-34, and 2-35), will be called the *characteristic impedance*.

Similarly, the current accompanying $v_2(z,t)$ may be shown to be

$$i_2(z,t) = -\frac{V_{2M}}{|Z_0|}\epsilon^{\alpha z}\cos(\omega t - \theta_0 + \beta z) \tag{4-34}$$

In general the magnitude $|Z_0|$ and the angle θ_0 are both functions of frequency.

e. Direction of Travel: Consistency with Attenuation Function

The functions $v_1(z,t)$ and $i_1(z,t)$ describe sinusoidal waves which travel in the positive direction with increasing t, as indicated in Fig. 4-1. The maximum value which each function can attain at any particular z is

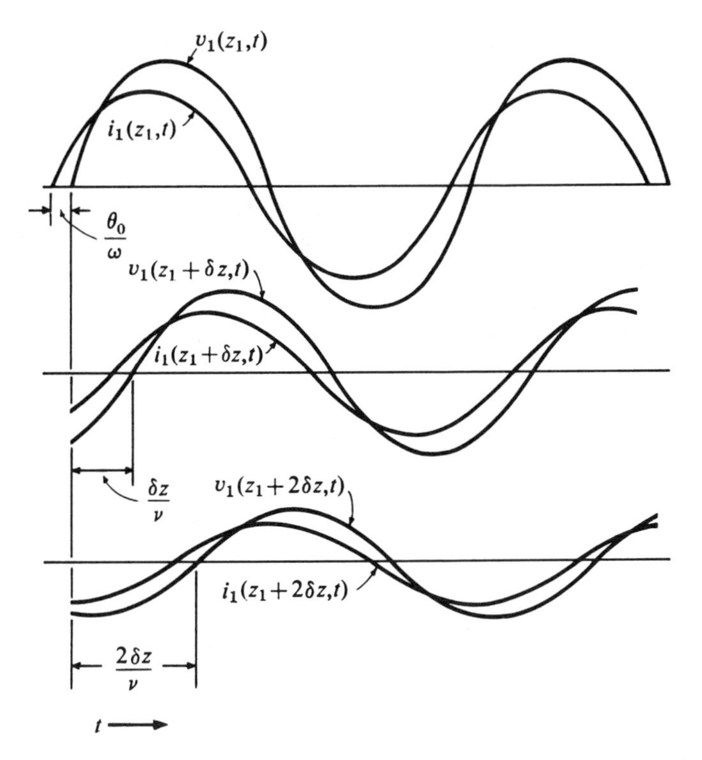

Figure 4-1. Sinusoidal traveling waves observed as functions of time at three locations on a transmission line with losses.

progressively reduced as the location considered is moved in the positive z direction, because of the attenuation function. This is to be expected physically, because the flow of conduction current through the conductor resistance and of leakage current through the shunt conductance converts a part of the electromagnetic-field energy associated with the current and voltage waves into heat.

The functions $v_2(z,t)$ and $i_2(z,t)$, on the other hand, represent sinusoidal waves traveling in the negative direction of z. Since an increase in time is accompanied by a *decrease* in z for any reference point on the moving wave, the factor $\epsilon^{\alpha z}$ becomes progressively smaller for any such traveling point, corresponding to reduction in the amount of electromagnetic-field energy remaining in the wave system.

Thus the three aspects of direction of wave travel, (1) the direction of movement of a constant phase point with increase of time, (2) the direction of average power flow (Sec. 2-2d), and (3) the direction in which reduction in wave-pattern magnitude takes place, are consistent for both cases considered here.

4-2 PHASOR NOTATION FOR TRAVELING-WAVE FUNCTIONS

Up to this point the current and voltage functions have been written out in "longhand," with the functional relationship to both z and t explicitly stated. This was to make the operations of differentiation and integration as obvious as possible and free of any unwritten "understandings." A more abbreviated form of notation may be used advantageously if one is concerned simply with the sinusoidal steady-state behavior of a system.* The traveling-wave voltage $v_1(z,t)$ as given in Eq. 4-14, which is mathematically complete, may be represented in "shorthand" by

$$V_1(z) = V_{1M}\epsilon^{-(\alpha+j\beta)z} \tag{4-35}$$

The function $V_1(z)$, known as a *phasor*, is related to $v_1(z,t)$ in the following manner:

$$v_1(z,t) = \mathrm{Re}\left[V_1(z)\epsilon^{j\omega t}\right] \tag{4-36}$$

*If a given voltage or current function is the sum of two or more sinusoidal functions of different frequencies, the phasor method of analysis must be applied to each frequency component separately. (Use appropriate subscripts to distinguish one from another.) The separate-frequency solutions obtained for another voltage or current in the system would have to be converted to the explicit form before they could be added together to form a single equation. Likewise a separate phasor diagram would be associated with each frequency component; these would rotate at different speeds.

Equation 4-35 may be further shortened by substitution of Eq. 4-17:

$$V_1(z) = V_{1M}\epsilon^{-\gamma z} \tag{4-37}$$

It will be understood that the subscript M attached to V or I designates the maximum value of a sinusoidal wave; this value is a positive real number. The capital letters V and I without the subscript M will designate phasors; V_1 in Eq. 4-37 is an example. In transmission-line problems the voltage and current phasors are functions of line location z; thus $V(z)$; $I(z)$.

Voltmeters and ammeters are conventionally scaled in terms of root-mean-square voltage or current. The corresponding numerical values (or functions of z in the case of transmission-line theory) are real and positive and will be designated by the subscript "rms." For example, in terms of $v_1(z,t)$ as given in Eq. 4-7,

$$\begin{aligned}[V_1(z)_{\text{rms}}]^2 &= \frac{\omega}{2\pi}\int_0^{2\pi/\omega}[v_1(z,t)]^2\,dt \\ &= \frac{\omega}{2\pi}\int_0^{2\pi/\omega}[V_{1M}\epsilon^{-\alpha z}\cos(\omega t - \beta z)]^2\,dt \\ &= \frac{(V_{1M}\epsilon^{-\alpha z})^2}{2}\end{aligned} \tag{4-38}$$

The magnitude of the phasor $V_1(z)$ in Eq. 4-35 is

$$|V_1(z)| = V_{1M}\epsilon^{-\alpha z} \tag{4-39}$$

Hence

$$V_1(z)_{\text{rms}} = \frac{|V_1(z)|}{\sqrt{2}} \tag{4-40}$$

The voltage ratios in Eqs. 4-20 and 4-21 in terms of maximum values, for example, would be the same as $V_1(z_1)_{\text{rms}}/V_1(z_2)_{\text{rms}}$, or the ratio of voltmeter readings at z_1 and z_2.

The capital letter Z will be understood to represent an impedance as a complex quantity; thus

$$Z_0 = |Z_0|\exp(j\theta_0) \tag{4-41}$$*

* For convenience in routine calculations, the complex exponential is often abbreviated by the notation

$$|Z_0|\exp(j\theta_0) = |Z_0|\underline{/\theta_0} \tag{4-42}$$

This may be illustrated in expressions for $i_1(z, t)$:

$$i_1(z,t) = \frac{V_{1M}}{|Z_0|}\epsilon^{-\alpha z}\cos(\omega t - \theta_0 - \beta z) \tag{4-33}$$

$$I_1(z) = \frac{V_{1M}}{Z_0}\epsilon^{-(\alpha + j\beta)z} \tag{4-43}$$

The relationship between $i_1(z, t)$ and $I_1(z)$ parallels Eq. 4-36 for voltage.

$$i_1(z,t) = \mathrm{Re}\left[I_1(z)\epsilon^{j\omega t}\right] \tag{4-44}$$

From Eqs. 4-17 and 4-37,

$$I_1(z) = \frac{V_{1M}}{Z_0}\epsilon^{-\gamma z} \tag{4-45}$$

$$I_1(z) = \frac{V_1(z)}{Z_0} \tag{4-46}$$

Both forms of notation, phasor and explicit, will be used in the remainder of this text, the phasor form where mathematical manipulation will be expedited thereby, and the explicit form on occasion for the display and examination of results.

4-3 PROPAGATION CHARACTERISTICS AS FUNCTIONS OF FREQUENCY

The expressions for the propagation function and the characteristic impedance in Eqs. 4-16 and 4-31 show that those quantities vary with frequency, but the dominant trends of those variations are not apparent at a casual glance. Power series expansions are oftentimes helpful in situations of this sort.

Within the audio-frequency range (about 100 to 3600 Hz, for satisfactory voice communication[7]) on telephone lines, r is commonly comparable to or larger than ωl. But in the carrier-frequency range,* r is much smaller than ωl.

---

*Owing to skin effect, r is a function of frequency. This was mentioned in Sec. 1-2a(1), and is discussed at length in Chap. 14.

Loss caused by an alternating electric field in solid insulation is approximately proportional to frequency. This is the principal component of the voltage-dependent losses in cables; those losses are accounted for in the shunt conductance parameter, g. Some insulation materials which are satisfactory for audio-frequency applications are excessively lossy at radio and microwave frequencies.

a. Attenuation and Phase Velocity

The propagation function may be expanded in a Taylor series to show the asymptotic behavior of the attenuation function and the phase velocity.

(1) *High-Frequency Range.* As already mentioned, for investigation of the high-frequency region, r will be assumed to be small compared with ωl and g small compared with ωc. Equation 4-16 may be rearranged as

$$\gamma = \sqrt{j\omega l\left(1 + \frac{r}{j\omega l}\right)j\omega c\left(1 + \frac{g}{j\omega c}\right)}$$

The terms $1 + (r/j\omega l)$ and $1 + (g/j\omega c)$ are each replaced by the series expansion

$$\sqrt{1 \pm u} = 1 \pm \frac{u}{2} - \cdots \qquad |u| < 1$$

$$\gamma \approx j\omega\sqrt{lc}\left[1 + \frac{1}{2}\frac{r}{j\omega l} - \cdots\right] \times \left[1 + \frac{1}{2}\frac{g}{j\omega c} - \cdots\right] \tag{4-47}$$

One precaution to be observed in multiplying series together is that all terms involving small quantities of a given order should be carried if the approximation is to be valid to that order. Here terms of the first order were written for both component series; the product of those expressions will be a valid approximation for terms up to and including the first order but not for higher-order terms:

$$\gamma \approx j\omega\sqrt{lc}\left[1 + \frac{1}{2}\left(\frac{r}{j\omega l} + \frac{g}{j\omega c}\right)\right]$$

$$\approx \frac{r}{2}\sqrt{\frac{c}{l}} + \frac{g}{2}\sqrt{\frac{l}{c}} + j\omega\sqrt{lc} \qquad r \ll \omega l;\ g \ll \omega c \tag{4-48}$$

The imaginary part of this result, $\omega\sqrt{lc}$, when substituted in Eq. 4-28, yields

$$\nu = 1/\sqrt{lc} \quad \text{(high-frequency asymptote)} \tag{4-49}$$

This is the same as Eq. 2-30 for the lossless line.

For a well-designed line, g/c is much smaller than r/l; then the attenuation function, represented by the real part of Eq. 4-48, reduces to

$$\alpha \approx \frac{r}{2}\sqrt{\frac{c}{l}} \tag{4-50}$$

In accordance with Eq. 2-35, $\sqrt{l/c}$ is the characteristic impedance of a lossless line, and it also proves to be the leading term in the series expansion for the characteristic impedance of a low-loss line (Eqs. 4-55 through 4-57). Thus if two lines have the same resistance per unit length but different characteristic impedances, the one with the higher characteristic impedance will have the lower attenuation. A physical explanation of this effect is as follows: The higher characteristic impedance will cause a signal of a given power level to be transmitted at a higher voltage and lower current than will the lower characteristic impedance. With satisfactory insulation, the increase in shunt leakage loss (v^2g) will be negligible compared with the reduction in series resistance loss (i^2r). Thus the power loss per unit length of time for a given transmitted power will be lower, as will the attenuation. (This is the theoretical basis for *loading*, or ferromagnetically increasing the inductance of a line to reduce attenuation.*)

(2) *Unloaded Cable at Low Frequencies.* The distributed inductance of a cable (unless it is loaded, as described in the footnote) is low compared with that of an open-wire line because of the close spacing of the conductors, and the following approximations may be valid over most

*Two general methods have been used: (1) *Continuous loading*. Here the center conductor of a coaxial cable is wrapped with high-permeability magnetic tape; this form of construction was used in some submarine telegraph cables. (2) *Lumped loading*.[1-5, 7] In this method ferromagnetic-cored coils are inserted in series with each conductor of a two-wire line or cable at regularly spaced intervals. Such a line is inherently a *low-pass filter*. Lumped loading was used on many long-distance open-wire lines from about 1900 until shortly after the development of vacuum-tube amplifiers (1914), and subsequently on cable systems intended for audio-frequency operation.

of the audio-frequency band: $r \gg \omega l$ and $g \ll \omega c$. Equation 4-16 reduces to

$$\gamma \approx \sqrt{j\omega cr}$$

$$\approx \sqrt{\frac{\omega cr}{2}} + j\sqrt{\frac{\omega cr}{2}} \qquad (4\text{-}51)$$

The corresponding velocity of propagation may be found with the aid of Eq. 4-28:

$$\nu \approx \sqrt{\frac{2\omega}{cr}} \qquad (4\text{-}52)$$

It is apparent that the attenuation and phase velocity of an unloaded cable vary widely over the audio-frequency range.

The distributed inductance of an actual line or cable cannot be zero.* (The maximum velocity of propagation of energy in free space, 3.00×10^8 m/s, fixes, in accordance with Eq. 2-30 ($\nu = 1/\sqrt{lc}$), a minimum limit for the product lc.)

b. Characteristic Impedance

The limit approached by Z_0 as frequency approaches infinity is found from Eq. 4-31 to be

$$Z_0(\infty) = \sqrt{\frac{l}{c}} \qquad (4\text{-}53)$$

This result is applicable at radio frequencies as an excellent approximation for Z_0. For the usual open-wire line, it is a good approximation at audio frequencies. It is sometimes referred to as the *surge impedance*.

---

*The phrase "ideal cable" is sometimes used in connection with approximate analyses in which l and g are neglected; there the adjective "ideal" refers to the resulting mathematical simplification, not to electrical transmission properties. The effect of setting l equal to zero in the analysis of step-function response is to ignore the physical property of a finite velocity of propagation.

For frequencies approaching zero, Z_0 approaches the limit

$$Z_0(0) = \sqrt{\frac{r}{g}} \tag{4-54}$$

A series expansion for Z_0, valid for a low-loss line, may be obtained if Eq. 4-31 is factored as

$$Z_0 = \sqrt{\frac{l}{c}\left(1 + \frac{r}{j\omega l}\right)\left(1 + \frac{g}{j\omega c}\right)^{-1}} \tag{4-55}$$

The following series expansion may be used in conjunction with Eq. 4-47:

$$\frac{1}{\sqrt{1 \pm w}} = 1 \mp \frac{w}{2} + \cdots \qquad |w| < 1 \tag{4-56}$$

Substitution of Eqs. 4-47 and 4-56 in 4-55 yields the following (as in Eq. 4-48, only the first-order terms are used):

$$Z_0 \approx \sqrt{\frac{l}{c}}\left[1 - \frac{j}{2\omega}\left(\frac{r}{l} - \frac{g}{c}\right)\right] \tag{4-57}$$

For the usual line in which $(r/l) > (g/c)$, the reactive component of the characteristic impedance is capacitive.

If $r \gg \omega l$ and $g \ll \omega c$, a simple approximation for the characteristic impedance results:

$$Z_0 \approx \sqrt{\frac{r}{j\omega c}}$$

$$\approx \sqrt{\frac{r}{\omega c}}\,\epsilon^{-j(\pi/4)} \tag{4-58}$$

c. Distortion

Changes in signal waveshape with distance traveled along a linear transmission line may be classified as *attenuation distortion* and *phase distortion*. Attenuation distortion is the result of variation of the attenuation with frequency, and phase distortion is the result of variation of phase velocity with frequency. Dependence of the characteristic impedance on frequency is not referred to as distortion, although (a) it results in a difference between a voltage wave and the accompanying current wave which, if components of more than one frequency are present, is more pronounced than a mere phase shift, and (b) it aggravates the problem of obtaining matched terminations. These three effects are necessarily all present if any one is present; only if the line is lossless are they absent.

Attenuation-distortion and phase-distortion effects are cumulative with distance traveled, but they may be offset by means of terminal networks known as equalizers.*[1, 4–7] An instrumental musical note, for example, includes harmonic frequency components (overtones), among which the proportions differ for various instruments. For high *quality* audio transmission, both of music and speech (output sounding the same to the human ear as the input), both attenuation distortion and phase distortion should be kept to practical minima. The utilitarian aspect of *intelligibility* of speech (syllables, words, and phrases readily and correctly recognized) seems to be little affected by phase distortion.[1] *Video images* lose definition (become "fuzzy") under phase distortion.

4-4 CONCLUSIONS

A uniform lossy transmission line will propagate, in either direction, at a uniform velocity, and with no distortion, a sinusoidally varying voltage-and-current wave set. These waves are reduced in magnitude in proportion to the exponential of the product of the distance traveled and an attenuation factor.

Phasor notation is derivable from the complex exponential form of the cosine function. It may be adapted to the traveling-wave function readily, and the functional variation with respect to distance becomes $\epsilon^{-\gamma z}$ or $\epsilon^{\gamma z}$, depending on the direction of travel. Here γ is the *propagation function*, a complex quantity of which the real part is the *attenuation function* α (nepers per unit of distance) and the imaginary part is the phase function

*An *attenuation equalizer* has an attenuation characteristic which, ideally, complements that of the transmission line; it attenuates as little as possible those frequencies which are heavily attenuated by the line, but attenuates more heavily the frequencies which are transmitted with little attenuation. The overall attenuation of the line and equalizer is almost independent of frequency over a selected band. *Phase* or *delay equalization* is based on a similar concept; it retards the high-velocity components with respect to the lower-velocity ones so as to give very nearly the same resultant delay time (line plus equalizer) for all frequencies within the given band.

TABLE 4-1. Transmission-Line Parameters

Type of Line	Identifying Feature	Characteristic Impedance	Attenuation Function	Phase Velocity
Lossless	$r = 0$, $g = 0$	$\sqrt{\frac{l}{c}}$	0	$\frac{1}{\sqrt{lc}}$
Low-loss	$r \ll \omega l$, $g = 0$	$\sqrt{\frac{l}{c}}\,\epsilon^{-j(r/2\omega l)}$	$\frac{r}{2}\sqrt{\frac{c}{l}}$	$\frac{1}{\sqrt{lc}}$
Unloaded cable $f \to 0$	$\omega l \ll r$, $g = 0$	$\sqrt{\frac{r}{\omega c}}\,\epsilon^{-j(\pi/4)}$	$\sqrt{\frac{r\omega c}{2}}$	$\sqrt{\frac{2\omega}{rc}}$

β (radians per unit of distance):

$$\gamma = \sqrt{(r + j\omega l)(g + j\omega c)} \tag{4-16}$$

A value of attenuation in nepers may be converted to *decibels* by multiplying it by 8.686 (Eq. 4-26).

The phase function is related to the *phase velocity* ν in the following manner:

$$\nu = \frac{\omega}{\beta} \tag{4-28}$$

The *characteristic impedance* Z_0 relates the voltage and current phasors of a traveling-wave pair:

$$Z_0 = \sqrt{\frac{r + j\omega l}{g + j\omega c}} \tag{4-31}$$

At any given location, corresponding voltage and current sinusoidal traveling waves are, in general, not in time phase with each other.

The attenuation function, the phase velocity, and the characteristic impedance, are all functions of frequency unless the line is lossless. Some useful first-order approximations are summarized in Table 4-1.

PROBLEMS

4-1. Given the following: $V_1(z) = 50.0\underline{/30.0^\circ}\epsilon^{-\gamma z}$ V, $Z_0 = 500\underline{/-6.0^\circ}$ Ω, $\gamma = (0.95 + j9.00) \times 10^{-3}$ m^{-1}, $\omega = 2.55 \times 10^6$ rad/s. Write expressions for $I_1(z)$, $i_1(z,t)$, and $v_1(z,t)$, with numerical substitutions for all quantities except z and t.

4-2. Given the following traveling-wave voltage function:

$$v_A(z,t) = 100\exp(-2.00 \times 10^{-4} z)$$

$$\cdot \cos(3.00 \times 10^5 t + 30.0^\circ - 1.25 \times 10^{-3} z) \text{ V}$$

Here t is in seconds and z is in meters. Write the corresponding phasor, $V_A(z)$. If $Z_0 = 50.0\angle{-8.0^\circ}\ \Omega$, write the phasor of the accompanying current wave, $I_A(z)$. What are the numerical values of (a) the attenuation function, (b) the phase function, (c) the frequency, and (d) the phase velocity?

4-3. Given a 48-km length of transmission line which is terminated with an impedance equal to the characteristic impedance of the line. (As is shown in Sec. 5-2a, only a single traveling-wave pair approaching the termination could then exist.) The input and terminal voltages and currents are

$$V_{\text{in}} = 10.0\angle{0^\circ} \text{ V}$$

$$V_{\text{out}} = 6.0\angle{-150^\circ} \text{ V}$$

$$I_{\text{in}} = 0.20\angle{7.5^\circ} \text{ A}$$

$$I_{\text{out}} = 0.120\angle{-142.5^\circ} \text{ A}$$

Find the characteristic impedance and evaluate the attenuation function and the propagation function.

4-4. The following data have been calculated from impedance measurements at 2.50×10^5 Hz on a section of coaxial cable:

$$Z_0 = 93.0\epsilon^{-j1.7^\circ}\ \Omega$$

$$\alpha = 4.60 \times 10^{-4} \text{ Np/m}$$

$$\beta = 6.28 \times 10^{-3} \text{ rad/m}$$

Find the distributed circuit parameters r, l, g and c, and the phase velocity.

4-5. A given transmission line has the following parameters:

$$Z_0 = 600\angle{-6^\circ}\ \Omega$$

$$\alpha = 2.0 \times 10^{-5} \text{ dB/m}$$

$$\nu = 2.97 \times 10^8 \text{ m/s}$$

$$f = 10^3 \text{ Hz}$$

(a) Write the phasor functions $V(z)$ and $I(z)$, and the instantaneous functions $v(z, t)$ and $i(z, t)$ for a wave set traveling in the $+z$ direction, if the maximum value of the current wave at $z = 0$ is 3.0×10^{-2} A and the current wave at $z = 0$ is at its maximum positive value with respect to time at $t = 0$.

(b) Sketch the current and voltage at $z = 0$ as functions of time.

4-6. Given the following set of traveling waves moving in opposite directions on a transmission line, for which $Z_0 = 650\underline{/-5^\circ}\ \Omega$:

$$V(z) = 20.0 \exp\left[-(0.572 + j3.26) \times 10^{-5} z\right]$$

$$+ 6.0 \exp\left[(0.572 + j3.26) \times 10^{-5} z + j35.0^\circ\right] \text{ V}$$

Here $f = 1500$ Hz and z is in meters. Write expressions for (a) $I(z)$, (b) $v(z, t)$, and (c) $i(z, t)$. Sketch a phasor diagram for the voltage and current components at $z = 0$.

4-7. Your study of traveling waves on transmission lines is interrupted by a call for help from a friend, an engineering student who is majoring in a branch other than electrical engineering. Your friend is taking a "service" course for nonelectrical students; he was told in lecture today that the differential equation for a resistance-inductance series AC circuit is

$$Ri + L\frac{di}{dt} = V_{\max} \cos(\omega t + \theta)$$

He was also told that the solution to this equation is

$$I_{\max}\underline{/\phi} = \frac{V_{\max}\underline{/\theta}}{R + j\omega L}$$

$$= \frac{V_{\max}\underline{/\theta - \tan^{-1}(\omega L/R)}}{\sqrt{R^2 + (\omega L)^2}}$$

The gist of his plea, after deleting the expletives, is: "Time, t, is the independent variable of the differential equation, yet the purported solution isn't even a function of t. What's missing? Those $\underline{/\phi}$ and $\underline{/\theta}$ symbols weren't used in any mathematics course that I took. ... The instructor drew some wavy graphs and triangles on the blackboard, but he erased them before I could copy them. ... Is this supposed to be what they called a 'complete solution' in the differential equations short course, or is it just an approximation?"

As a student who has just mastered the application of the phasor method to traveling wave problems, you should be well prepared to apply Euler's identity, convert the given phasor solution into an explicit function of time, sketch waves of voltage and current as functions of time, and otherwise resolve your friend's confusion. Prepare a concise summary which he may take with him.

REFERENCES

1. Hamsher, D. H., Ed., *Communication Systems Engineering Handbook*, McGraw-Hill, New York, 1967.
2. Creamer, W. J., *Communication Networks and Lines*, Harper & Brothers, New York, 1951, 132.
3. Everitt, W. L. and Anner, G. E., *Communication Engineering*, McGraw-Hill, New York, 1956.
4. Freeman, R. L., *Reference Manual for Telecommunications Engineering*, Wiley, New York, 1985.
5. Freeman, R. L., *Telecommunication Transmission Handbook*, 2nd ed., Wiley, New York, 1982.
6. Rounds, P. W. and Larkin, G. L., Equalization of cables for local television transmission, *Bell Syst. Tech. J.*, 34, 713, 1955.
7. Technical Staff, Bell Telephone Laboratories, *Transmission Systems for Communications*, Graybar Electric Company, Inc., New York, 1964.

CHAPTER 5

Wave Distortion on Lossy Lines—Step-Function Source

A study of the step-function response of a lossy transmission line will complement the analysis of the preceding chapters. The Laplace transform provides a general operational technique which will accommodate problems of this complexity.[2, 17] The inverse transforms of irrational and transcendental functions will be used; these will not be derived, but reference will be made to analyses and tabulated results.[1, 12]

5-1 TRANSFORMS OF VOLTAGE AND CURRENT ON A LOSSY LINE

The partial differential equations previously derived for the lossless line (Eqs. 2-6 and 2-8) will be converted to Laplace transforms with respect to time, and the parameters modified to accommodate frequency-dependent loss. The transform equations may then be solved as ordinary differential equations in z.

a. The Operational Equation of Telegraphy

Let $V(z,s)$ be the Laplace transform of $v(z,t)$ with respect to t, in accordance with the standard definition:

$$V(z,s) = \int_0^\infty v(z,t)\epsilon^{-st}\,dt = \mathscr{L}[v(z,t)] \tag{5-1}$$

Capital letters, with functional dependence on s explicitly stated, will be used to designate the Laplace transforms of functions of time. The operation of taking the Laplace transform of a given function of time will be indicated by the symbol $\mathscr{L}$; correspondingly, the inverse-transform operation will be indicated by $\mathscr{L}^{-1}$.

The transform of the time derivative of a function for which the transform has been defined as $V(z, s)$ is

$$\mathscr{L}\left[\frac{\partial v(z,t)}{\partial t}\right] = sV(z,s) - v(z,0) \tag{5-2}$$

The initial-value term $v(z,0)$ describes the voltage distribution along the line (as a function of z) existing at the instant of zero reference time.

The transform integral, Eq. 5-1, is unaffected by differentiation with respect to variables other than t; hence the partial derivative of $v(z,t)$ with respect to z (as in Eq. 2-8) has the transform

$$\mathscr{L}\left[\frac{\partial v(z,t)}{\partial z}\right] = \frac{\partial V(z,s)}{\partial z} \tag{5-3}$$

If the initial-value term $v(z,0)$ is zero, the Laplace transform of Eq. 2-8 is

$$\frac{\partial V(z,s)}{\partial z} = -lsI(z,s) \tag{5-4}$$

Correspondingly, the Laplace transform of Eq. 2-6 is

$$\frac{\partial I(z,s)}{\partial z} = -csV(z,s) \tag{5-5}$$

The term ls (l independent of frequency) is the *distributed operational impedance* of the lossless line. It is analogous to the distributed complex impedance which was mentioned after Eq. 4-16. For the lossy line, ls will be replaced by the general expression $\mathscr{z}(s)$,* for which a specific form

---

*An important constraint on the nature of every operational immitance or transfer function is that it be analytic (satisfy the Cauchy–Riemann equations) in the right half-plane. All actual voltages and currents, when stated in explicit forms (as distinguished from phasors, for example), are necessarily real functions of time, and each has a Laplace transform (not necessarily one in closed form) which is analytic in the right half-plane. The ratio of two analytic functions will also be analytic. Those functions of time which may be encountered in physical systems, including, as limiting forms, the step function and the impulse function, satisfy the relevant convergence requirements for the Laplace transform.

Interrelated with this proposition are (1) the principle of *causality* (in elementary terms, if a physical system is initially quiescent, response to some activating source cannot become nonzero before that source becomes nonzero) and (2) the *Hilbert transform*, which mutually relates the real and imaginary parts of an analytic function. The analogy in field theory, relating the dissipative and dielectric properties of a medium, is known as the *Kronig–Kramers relationship*.[4, 6, 10, 11, 15]

will be derived from the analysis of skin effect. Likewise for generality, cs will be replaced by $\mathscr{y}(s)$ as the *distributed operational admittance.*

$$\frac{\partial V(z,s)}{\partial z} = -\mathscr{z}(s)I(z,s) \tag{5-6}$$

$$\frac{\partial I(z,s)}{\partial z} = -\mathscr{y}(s)V(z,s) \tag{5-7}$$

By differentiating Eq. 5-6 with respect to z and substituting 5-7, the current transform may be eliminated from the set, which gives

$$\frac{\partial^2 V(z,s)}{\partial z^2} = \mathscr{z}(s)\mathscr{y}(s)V(z,s) \tag{5-8}$$

This may be called the *operational equation of telegraphy.*

b. Traveling-Wave Solutions

Equation 5-8 may be solved with respect to z by the following substitution, in which $V_A(s)$ and m are undetermined functions which are postulated to be independent of z. Let

$$V(z,s) = V_A(s)\epsilon^{mz} \tag{5-9}$$

Differentiation of Eq. 5-9 twice with respect to z, substitution in Eq. 5-8, cancellation of common terms, and solution for m yields

$$m = \pm\sqrt{\mathscr{z}(s)\mathscr{y}(s)} \tag{5-10}$$

Because m has two values, a second arbitrary function $V_B(s)$ should be introduced:

$$V(z,s) = V_A(s)\exp\left[-z\sqrt{\mathscr{z}(s)\mathscr{y}(s)}\right] + V_B(s)\exp\left[z\sqrt{\mathscr{z}(s)\mathscr{y}(s)}\right] \tag{5-11}$$

The current transform may be found from the voltage transform by substituting Eq. 5-11 into Eq. 5-6:

$$I(z,s) = \sqrt{\frac{\mathscr{y}(s)}{\mathscr{z}(s)}}\, V_A(s)\exp\left[-z\sqrt{\mathscr{z}(s)\mathscr{y}(s)}\right]$$

$$-\sqrt{\frac{\mathscr{y}(s)}{\mathscr{z}(s)}}\, V_B(s)\exp\left[z\sqrt{\mathscr{z}(s)\mathscr{y}(s)}\right] \tag{5-12}$$

Let

$$Z_0(s) = \sqrt{\frac{\mathscr{z}(s)}{\mathscr{y}(s)}} \tag{5-13}$$

and

$$\gamma(s) = \sqrt{\mathscr{z}(s)\mathscr{y}(s)} \tag{5-14}$$

Here it may be observed that the characteristic impedance of the lossless line, Eq. 2-35, and that of the lossy line in AC steady state, Eq. 4-31, have an analogy in the *operational characteristic impedance*, Eq. 5-13. The function defined in Eq. 5-14 is analogous to the propagation function of the lossy line in AC steady state, and will be designated as the *operational propagation function*.

Substitution of Eqs. 5-13 and 5-14 in Eqs. 5-11 and 5-12 permits the latter to be written in condensed form:

$$V(z,s) = V_A(s)\epsilon^{-z\gamma(s)} + V_B(s)\epsilon^{z\gamma(s)} \tag{5-15}$$

$$I(z,s) = \frac{V_A(s)}{Z_0(s)}\epsilon^{-z\gamma(s)} - \frac{V_B(s)}{Z_0(s)}\epsilon^{z\gamma(s)} \tag{5-16}$$

The terms involving the exponentials of $-z\gamma(s)$ prove to give rise to waves traveling in the $+z$ direction, whereas the terms involving the exponentials of $+z\gamma(s)$ give rise to waves traveling in the $-z$ direction.

c. Boundary Conditions for Transform Functions

The boundary-condition relationships developed in Sec. 3-1 for the lossless line, Eqs. 3-5 through 3-10 (see Fig. 3-1), are applicable here with minor modifications.

If a lossy line is terminated by a lumped impedance $Z_T(s)$ at distance z_r from the source, the following constraint applies:

$$V(z_r, s) = Z_T(s)I(z_r, s) \tag{5-17}$$

Equations 5-15 and 5-16 may be substituted with $z = z_r$, and the resulting equation solved for $V_B(s)$:

$$V_B(s)\exp[z_r\gamma(s)][Z_T(s) + Z_0(s)]$$
$$= V_A(s)\exp[-z_r\gamma(s)][Z_T(s) - Z_0(s)]$$

An *operational reflection coefficient* may be defined as

$$\rho_T(s) = \frac{Z_T(s) - Z_0(s)}{Z_T(s) + Z_0(s)} \tag{5-18}$$

The following result is obtained:

$$V_B(s) = V_A(s)\rho_T(s)\exp[-2z_r\gamma(s)] \tag{5-19}$$

At the source, let $Z_G(s)$ represent the operational Thévénin impedance of the generator:

$$V(0,s) = V_g(s) - Z_G(s)I(0,s) \tag{5-20}$$

Let

$$\rho_G(s) = \frac{Z_G(s) - Z_0(s)}{Z_G(s) + Z_0(s)} \tag{5-21}$$

If Eqs. 5-15 and 5-16 are substituted in Eq. 5-20 with $z = 0$, and Eq. 5-21 is also substituted, the following results:

$$V_A(s) = \frac{V_g(s)Z_0(s)}{Z_G(s) + Z_0(s)} + V_B(s)\rho_G(s) \tag{5-22}$$

Equations 5-15, 5-16, 5-19, and 5-22, together with the definitions of the operational reflection coefficients in Eqs. 5-18 and 5-21, provide the means with which to assemble the transforms of successive outgoing and returning traveling waves on a single section of lossy line.

5-2 OPERATIONAL PARAMETERS—FIRST-ORDER SKIN-EFFECT APPROXIMATION

Let the loss in the insulation be assumed zero; this will reduce the distributed operational admittance to

$$\mathscr{y}(s) = cs \tag{5-23}$$

The variation of distributed complex impedance with frequency for a lossy coaxial line is investigated in Chap. 14. If that analysis were repeated in terms of Laplace transforms with the complex variable s (and all initial-value terms were assumed to be zero), the corresponding operational distributed impedance $\mathcal{z}(s)$ would have the same form as the distributed complex impedance, provided each $j\omega$ were replaced by s.

Inverse transforms in closed form (results in terms of functions which have been defined, studied, and tabulated) have not been developed for voltage and current transforms based on the exact functions for conductor impedance given in Eqs. 14-36 and 14-37, although such expressions might be inverse-transformed numerically. Here a high-frequency approximation for the impedance function will be used, one which gives results which agree with oscilloscopic observations of the initial portion of the response to step-function excitation.

a. Approximation of Skin-Effect Complex Impedance

For a coaxial line consisting of (1) a uniform, solid circular inner conductor and (2) a uniform annular outer conductor, the following approximation, consisting of the leading terms of the asymptotic series* for the complex distributed impedance, will serve as a starting point†:

$$\mathfrak{r} + j\omega l \approx \frac{j\omega\mu_i}{2\pi}\ln\left(\frac{r_b}{r_a}\right) + \sqrt{j\omega}\left(\frac{1}{2\pi r_a}\sqrt{\frac{\mu_a}{\sigma_a}} + \frac{1}{2\pi r_b}\sqrt{\frac{\mu_b}{\sigma_b}}\right)$$

$$+ \frac{1}{4\pi}\left(\frac{1}{r_a^2\sigma_a} - \frac{1}{r_b^2\sigma_b}\right)$$

$$r_a\sqrt{\omega\sigma_a\mu_a} \gg 1 \quad \text{and} \quad r_b\sqrt{\omega\sigma_b\mu_b} \gg 1 \qquad \text{(14-51)}$$

---

*Asymptotic series approach the given function more and more closely as the *magnitude of the argument*, is increased; a good introductory discussion is given by von Kármán and Biot,[16] pp. 51–54.

†A composite conductor in which a solid circular wire of one metal is plated or overlaid with a different metal will have a more complicated characteristic, particularly in the vicinity of the frequency for which the skin depth of the outer metal ($\sqrt{2/\omega\mu\sigma}$; see Eq. 10-79) is equal to the layer thickness. The inner conductors of some flexible coaxial cables are composed of copper strands which are coated with tin or silver. Likewise the strands of braided conductors are sometimes coated. The step-function response of such lines has been investigated analytically,[7] but the mathematical results are formidable.

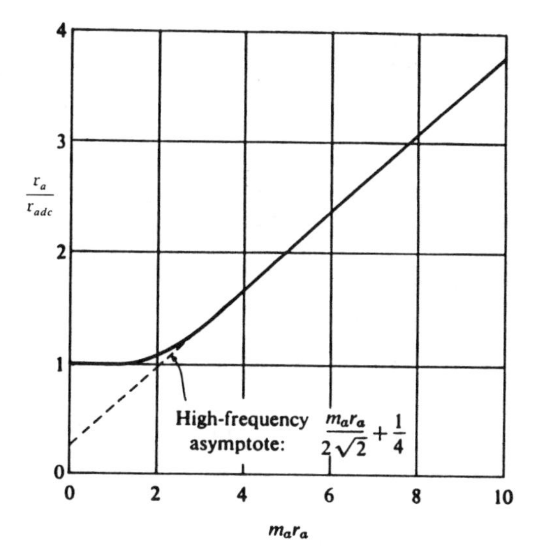

Figure 5-1. Normalized resistance of solid, uniform circular conductor $m_a = \sqrt{\omega \mu_a \sigma_a}$; r_a is the radius. (From Eqs. 14-39 and 14-45.)

Here r_a is the radius (in meters) of the inner conductor and r_b is the inner radius of the outer conductor; the σ's are the conductivities (siemens per meter) of the respective conductors, and μ_i, μ_a, and μ_b are the absolute permeabilities of the insulating medium and of the conducting materials. (Air, solid dielectrics generally, and nonferrous metals all have absolute permeabilities of essentially $4\pi \times 10^{-7}$ H/m.)

Figure 5-1, based on Eq. 14-39, displays the resistance of a solid, uniform circular conductor, normalized with respect to its zero-frequency value, versus a dimensionless variable which is proportional to the square root of the frequency. The asymptotic approximation taken from Eq. 14-45 is also shown. The high-frequency behavior of the resistance of the outer conductor is similar.

b. Operational Distributed Impedance

The first term on the right-hand side of Eq. 14-51 is the component of impedance caused by time-varying flux between the conductors; the corresponding component of inductance will be designated by l':

$$l' = \frac{\mu_i}{2\pi} \ln\left(\frac{r_b}{r_a}\right) \qquad (14\text{-}34)$$

The last term in Eq. 14-51 is small compared with the other terms in the frequency range of interest, and may be deleted. As a further algebraic simplification, let

$$\xi = \frac{1}{2\pi}\left(\frac{1}{r_a}\sqrt{\frac{\mu_a}{\sigma_a}} + \frac{1}{r_b}\sqrt{\frac{\mu_b}{\sigma_b}}\right) \tag{5-24}$$

Equation 14-51 then reduces to

$$r + j\omega l \approx j\omega l' + \xi\sqrt{j\omega} \tag{5-25}$$

The corresponding operational distributed impedance has the same form:

$$\mathscr{z}(s) \approx l's + \xi\sqrt{s} \tag{5-26}$$

When substituting this result in Eqs. 5-13 and 5-14 for the operational characteristic impedance and propagation function, the square root of $\mathscr{z}(s)$ is needed. Equation 5-26 may be restated as

$$\mathscr{z}(s) \approx l's\left(1 + \frac{\xi}{l'\sqrt{s}}\right) \tag{5-27}$$

The first two terms of the Taylor series for the square root indicated in Eq. 5-14 are

$$\sqrt{1 + \frac{\xi}{l'\sqrt{s}}} \approx 1 + \frac{\xi}{2l'\sqrt{s}} \qquad \frac{\xi}{l'\sqrt{s}} \ll 1 \tag{5-28}$$

Hence

$$\sqrt{\mathscr{z}(s)} \approx \sqrt{l's}\left(1 + \frac{\xi}{2l'\sqrt{s}}\right) \tag{5-29}$$

$$\approx \sqrt{l's} + \frac{\xi}{2\sqrt{l'}} \tag{5-30}$$

Equation 5-30 (with ξ defined in Eq. 5-24) leads to transforms for current and voltage for which closed-form inverse transforms are known. The resulting solution may be referred to as a *first-order-skin-effect approximation*.

c. Propagation Function

Substitution of Eqs. 5-23 and 5-30 in Eq. 5-14 yields

$$\gamma(s) \approx s\sqrt{l'c} + \frac{\xi}{2}\sqrt{\frac{cs}{l'}} \tag{5-31}$$

Letting

$$a = \frac{\xi}{2}\sqrt{\frac{c}{l'}} \tag{5-32}$$

$$\gamma(s) \approx s\sqrt{l'c} + a\sqrt{s} \tag{5-33}$$

For physical interpretation of the constant a in Eq. 5-33, consider the corresponding complex propagation function, the real part of which is the attenuation function:

$$\gamma(j\omega) \approx j\omega\sqrt{l'c} + \frac{a}{\sqrt{2}}(1+j)\sqrt{\omega} \tag{5-34}$$

$$\alpha(j\omega) \approx (a/\sqrt{2})\sqrt{\omega} \tag{5-35}$$

Thus $a/\sqrt{2}$ is the *slope of the high-frequency asymptote of a plot of attenuation* (nepers per unit of length) *versus the square root of the radian frequency*.

d. Characteristic Impedance

Similarly, substitution of Eqs. 5-23 and 5-29 in Eq. 5-13 yields

$$Z_0(s) \approx \sqrt{\frac{l'}{c}}\left(1 + \frac{\xi}{2l'\sqrt{s}}\right)$$

Let

$$b = \frac{\xi}{2l'} = \frac{a}{\sqrt{l'c}} \tag{5-36}$$

The constant value approached by Z_0 as frequency is increased may be designated by the symbol $\tilde{Z}_0$; thus

$$\tilde{Z}_0 = \sqrt{\frac{l'}{c}} \tag{5-37}$$

Hence

$$\tilde{Z}_0(s) \approx \tilde{Z}_0\left(1 + \frac{b}{\sqrt{s}}\right) \tag{5-38}$$

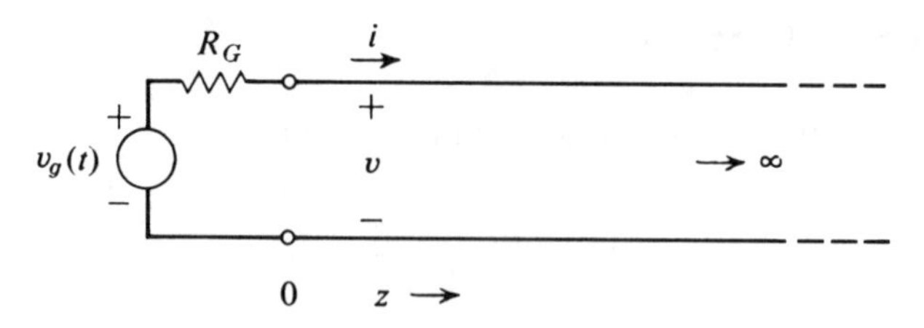

Figure 5-2. Semi-infinite line.

5-3 RESPONSE ON SEMI-INFINITE LINE WITH ZERO SOURCE IMPEDANCE

The propagation of voltage on a skin-effect-lossy line caused by a step-function voltage source connected directly to the line was first analyzed by Pleijel*[9] in 1918, and the same result (Eq. 5-50 or 5-58) was independently derived by other investigators during succeeding years. Wigington and Nahman[18] presented extensive experimental data along with analyses for impulse and ramp-function sources.

a. Outgoing-Wave Transforms

The immediate analysis will be concerned with the initial outgoing wave set alone; the circuit shown in Fig. 5-2, with $R_G = 0$, is applicable. In terms of the operational form in Eq. 5-22,

$$Z_G(s) = 0 \tag{5-39}$$

$$V_g(s) = \frac{V_0}{s} \tag{5-40}$$

The mathematical concept of a semi-infinite line ($0 < z < \infty$) eliminates the possibility of returning waves. Hence

$$V_B(s) = 0 \tag{5-41}$$

---

*A 1939 paper by Gates[3] seems to be the earliest English-language reference in which Pleijel's result was applied.

Substitution of Eqs. 5-22, 5-39, 5-40 and 5-41 in Eqs. 5-15 and 5-16 yields

$$V(z,s) = \frac{V_0}{s}\epsilon^{-z\gamma(s)} \tag{5-42}$$

$$I(z,s) = \frac{V_0}{sZ_0(s)}\epsilon^{-z\gamma(s)} \tag{5-43}$$

Substitution of Eqs. 5-33 and 5-38 in Eqs. 5-42 and 5-43 yields

$$V(z,s) = \frac{V_0}{s}\exp(-zs\sqrt{l'c} - az\sqrt{s}) \tag{5-44}$$

$$I(z,s) = \frac{V_0}{\tilde{Z}_0}\cdot\frac{1}{\sqrt{s}\,(\sqrt{s}+b)}\exp(-zs\sqrt{l'c} - az\sqrt{s}) \tag{5-45}$$

b. Inverse Transforms; The erfc Function

The exponential of $-zs\sqrt{l'c}$ is transformable in accordance with the shifting theorem; it results in a delay of $z\sqrt{l'c}$ seconds in every aspect of the response. Let

$$t_z = z\sqrt{l'c} \tag{5-46}$$

The following transform pairs, derivable by contour integration, are applicable to the remaining portions of $V(z, s)$ and $I(z, s)$[12, 17]:

$$\mathscr{L}^{-1}\left[\frac{\epsilon^{-az\sqrt{s}}}{s}\right] = \operatorname{erfc}\left(\frac{az}{2\sqrt{t}}\right)U(t) \tag{5-47}$$

$$\mathscr{L}^{-1}\left[\frac{\epsilon^{-az\sqrt{s}}}{\sqrt{s}\,(\sqrt{s}+b)}\right] = \exp(b^2t + abz)\operatorname{erfc}\left(b\sqrt{t} + \frac{az}{2\sqrt{t}}\right)U(t) \tag{5-48}$$

The abbreviation "erfc" designates the *complementary error function*, a tabulated function[1, 8, 13] which is defined by the integral

$$\operatorname{erfc}(p) = \frac{2}{\sqrt{\pi}}\int_p^{\infty}\exp(-u^2)\,du \tag{5-49}$$

Application of the transforms in Eqs. 5-47 and 5-48 to Eqs. 5-44 and 5-45 yields

$$v(z,t) = V_0 \operatorname{erfc}\left(\frac{az}{2\sqrt{t - t_z}}\right) U(t - t_z) \tag{5-50}$$

$$i(z,t) = \frac{V_0}{\tilde{Z}_0} \exp\left[b^2(t - t_z) + abz\right] \cdot \operatorname{erfc}\left(b\sqrt{t - t_z} + \frac{az}{2\sqrt{t - t_z}}\right) U(t - t_z) \tag{5-51}$$

From Eqs. 5-36 and 5-46 the following identity is obtained:

$$b^2 t_z = abz \tag{5-52}$$

This may be substituted in Eq. 5-51:

$$i(z,t) = \frac{V_0}{\tilde{Z}_0} \cdot \exp(b^2 t) \operatorname{erfc}\left(b\sqrt{t - t_z} + \frac{az}{2\sqrt{t - t_z}}\right) U(t - t_z) \tag{5-53}$$

This result was derived by Bohn.[2]

The asymptotic approximation for erfc,[1] valid for large values of the argument, is particularly useful in examining the voltage and current functions at small values of time, since t appears in the denominator of the argument in those instances, and also for the current function at large values of time:

$$\operatorname{erfc}(p) \approx \frac{\exp(-p^2)}{p\sqrt{\pi}} \left[1 - \frac{1}{2p^2} + \frac{3}{(2p^2)^2} - \cdots\right] \qquad p \gg 1 \tag{5-54}$$

The first terms of the power series for erfc are useful when the argument is small; this is the case for the voltage function at large values of time:

$$\operatorname{erfc}(p) = 1 - \frac{2}{\sqrt{\pi}} \left(p - \frac{p^3}{3} + \cdots\right) \qquad p \ll 1 \tag{5-55}$$

Empirical approximations have been derived by Hastings;* see also Abramowitz and Stegun.[1]

---

*Cecil Hastings, Jr., *Approximations for Digital Computers*. Princeton, N.J.: Princeton University Press, 1955.

TABLE 5-1. The Complementary Error Function

u	erfc(u)
0	1.000
0.0889	0.900
0.4769	0.500
1.1631	0.100
∞	0

Table 5-1 lists some selected values of erfc and its argument; they indicate its bounds, and are useful in calculating rise times of v in Eqs. 5-50 or 5-58. (See Prob. 5-4.)

c. Dimensionless Variables

Some economy in the plotting of curves may be achieved by substituting dimensionless variables. Let

$$T = b^2 t \tag{5-56}$$

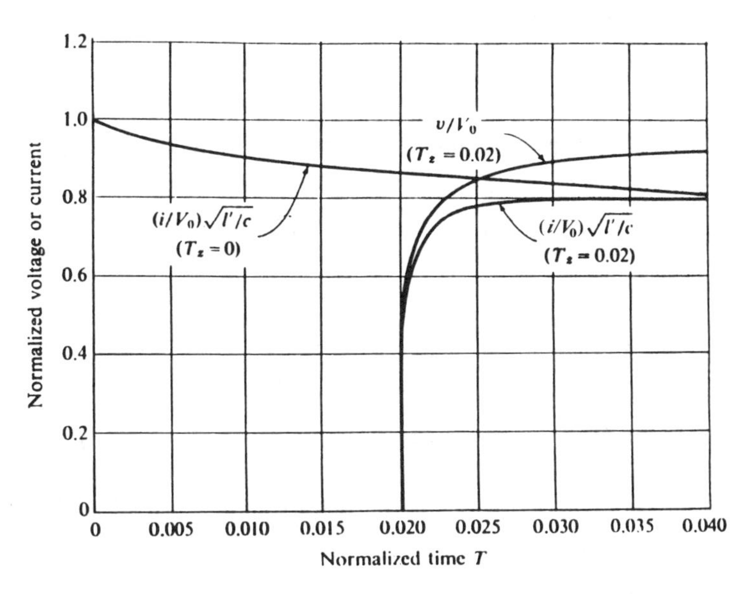

Figure 5-3. Influence of skin effect on current and voltage wave fronts; step-function voltage source connected directly to semi-infinite line.

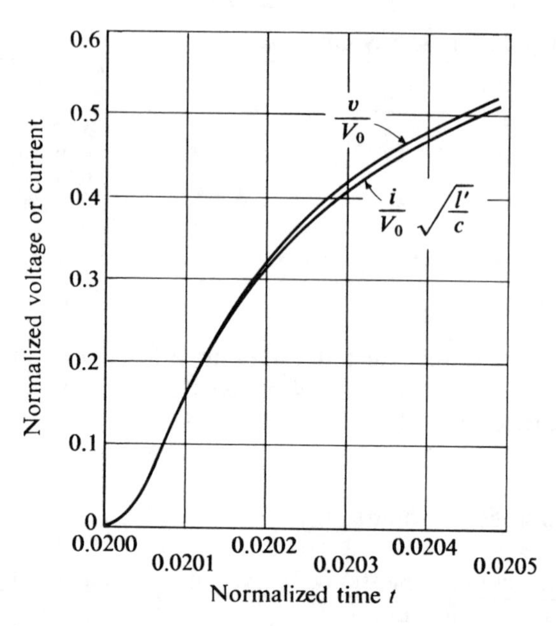

Figure 5-4. Influence of skin effect on current and voltage wave fronts; expanded time scale plot of part of Fig. 5-3. Normalized delay time $T_z = 0.02$.

Correspondingly (see Eq. 5-52), let

$$T_z = b^2 t_z = abz \tag{5-57}$$

Substitution of Eqs. 5-56 and 5-57 in Eqs. 5-50 and 5-53 will give

$$v(z, T) = V_0 \operatorname{erfc}\left(\frac{T_z}{2\sqrt{T - T_z}} \right) U(T - T_z) \tag{5-58}$$

$$i(z, T) = \frac{V_0}{\tilde{Z}_0} \epsilon^T \operatorname{erfc}\left(\sqrt{T - T_z} + \frac{T_z}{2\sqrt{T - T_z}} \right) U(T - T_z) \tag{5-59}$$

These functions are represented in Figs. 5-3 and 5-4, and by the curves marked "$R_G = 0$" in Figs. 5-5 and 5-6.

Example—To illustrate the foregoing in physical terms, consider a coaxial cable with the following characteristics:

$$\nu = 2.0 \times 10^8 \text{ m/s (at radio frequencies)}$$

$$\alpha = 2.50 \times 10^{-3} \text{ N/m at 10.0 MHz}$$

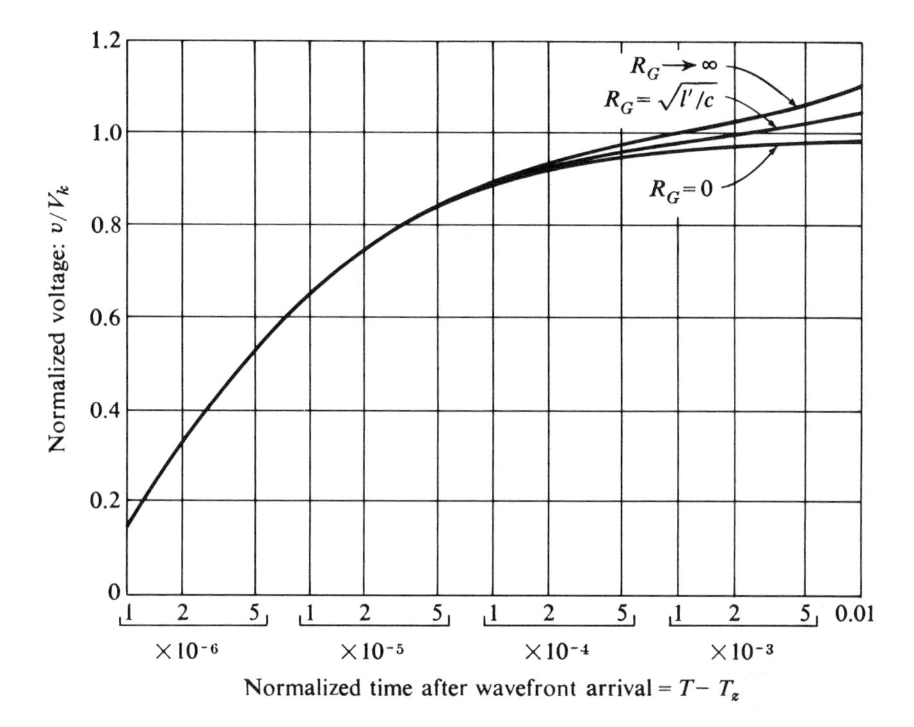

Figure 5-5. Computed normalized voltages on skin-effect lossy transmission line at normalized distance $T_z = 0.002$. Step-function source: $V_k(R_G + \tilde{Z}_0)/\tilde{Z}_0$.

Find the various parameters which are used in the analysis of step-function response, and relate the normalized quantities T and T_z in Figs. 5-3, 5-4, and 5-6 to t and z.

From Eq. 5-35,

$$\begin{aligned} a &= \alpha_1\sqrt{2/\omega_1} \\ &= 2.50 \times 10^{-3}\sqrt{2/(2\pi \times 1.00 \times 10^7)} \\ &= 4.46 \times 10^{-7}\ \mathrm{s}^{1/2}/\mathrm{m} \end{aligned}$$

From Eq. 5-36,

$$\begin{aligned} b &= a/\sqrt{l'c} = av \\ &= 4.46 \times 10^{-7} \times 2.0 \times 10^8 \\ &= 8.92 \times 10^1\ \mathrm{s}^{-1/2} \end{aligned}$$

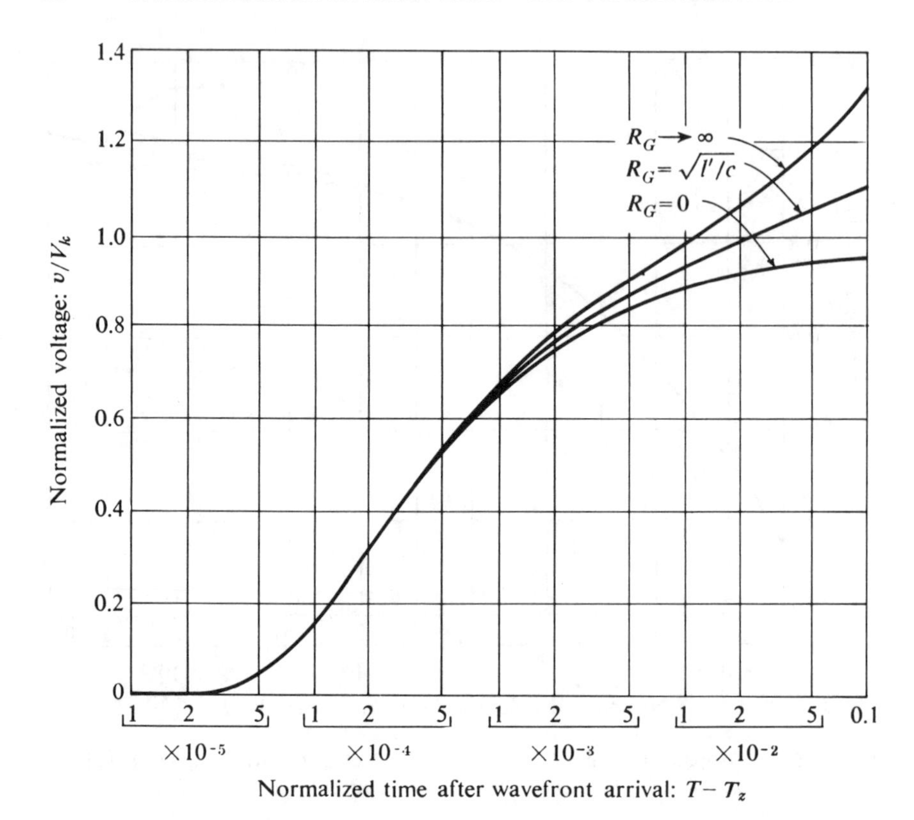

Figure 5-6. Computed normalized voltages on skin-effect-lossy transmission line at normalized distance $T_z = 0.02$. Step-function source: $V_k(R_G + \tilde{Z}_0)/\tilde{Z}_0$.

From Eq. 5-57,

$$\begin{aligned} T_z &= b^2 t_z \\ &= (8.92 \times 10^1)^2 t_z \\ &= 7.96 \times 10^3 t_z \end{aligned}$$

For the value of $T_z = 0.02$, which was used in Figs. 5-3, 5-4, and 5-6,

$$\begin{aligned} t_z &= 0.02/7.96 \times 10^3 \\ &= 2.51 \times 10^{-6}\ \text{s} \\ z &= t_z v \\ &= 2.51 \times 10^{-6} \times 2.0 \times 10^8 \\ &= 5.02 \times 10^2\ \text{m} \end{aligned}$$

For this cable, values of T on the abscissa scales of those graphs may be converted to time in seconds by multiplying by $(7.96 \times 10^3)^{-1}$ or 1.25×10^{-4} s.

d. Analysis of Response Functions

Some features of these results which are readily amenable to mathematical examinations are (1) the principal modes of variation of $v(z,T)$ and $i(z,T)$ for values of T only slightly larger than T_z; (2) the time at which $v(z,T)$ is increasing most rapidly, and the corresponding value of v; and (3) the principal modes of variation of $v(z,T)$ and $i(z,T)$ as T tends toward infinity.

(1) ***Initial-Rise Portions of Wave Fronts.*** As $T - T_z$ approaches zero, the arguments of the complementary error functions in Eqs. 5-58 and 5-59 approach infinity; in this range the asymptotic approximation (Eq. 5-54) is valid. Substitution of the leading term in Eq. 5-58 yields

$$v(z,T) \approx \frac{2V_0\sqrt{T - T_z}}{T_z\sqrt{\pi}} \exp\left[\frac{-T_z^2}{4(T - T_z)}\right] U(T - T_z) \qquad (T - T_z) \ll T_z^2 \tag{5-60}$$

Squaring the argument of erfc in Eq. 5-59 for current yields

$$\left(\sqrt{T - T_z} + \frac{T_z}{2\sqrt{T - T_z}}\right)^2 = (T - T_z) + T_z + \frac{T_z^2}{4(T - T_z)}$$

$$= T + \frac{T_z^2}{4(T - T_z)} \tag{5-61}$$

Substitution of this result in Eqs. 5-54 and 5-59 yields

$$i(z,T) \approx \frac{V_0}{\tilde{Z}_0} \cdot \frac{\exp\left[-T_z^2/4(T - T_z)\right]}{\left[\sqrt{T - T_z} + \left(T_z/2\sqrt{T - T_z}\right)\right]\sqrt{\pi}} U(T - T_z)$$

$$(T - T_z) \ll T_z^2 \quad \text{or} \quad (T - T_z) \gg 1 \tag{5-62}$$

Under the constraint that $(T - T_z) \ll T_z^2$, the inverse term in the denominator of Eq. 5-62 dominates, and the relationship may be further approximated:

$$i(z,t) \approx \frac{2V_0\sqrt{T - T_z}}{\tilde{Z}_0\sqrt{\pi}} \cdot \exp\left[\frac{-T_z^2}{4(T - T_z)}\right]U(T - T_z)$$

$$(T - T_z \ll T_z^2) \quad \text{and} \quad (T - T_z \ll T_z) \qquad (5\text{-}63)$$

If one divides Eq. 5-60 by Eq. 5-63, the following results:

$$\frac{v(z,T)}{i(z,T)} \approx \tilde{Z}_0 \qquad T - T_z \ll T_z^2 \qquad (5\text{-}64)$$

Observations which may be made from these results and from Fig. 5-4 are that immediately after $T = T_z$, (a) v and i are both zero and (b) the ratio between v and i begins as $\tilde{Z}_0$, but then increases. A further property, discernible after some differentiations, is that all infinite-ordered derivatives of v and i are zero at $T = T_z$, the instant at which the wave front would arrive if the line were lossless.* Let

$$\tau = \frac{4(T - T_z)}{T_z^2} \qquad (5\text{-}72)$$

---

*These results may be compared with those predicted if one were to assume that the distributed resistance and inductance are frequency-independent (skin effect would be ignored). Designating the frequency-independent parameters by r_{dc} and l_{dc} yields

$$z(s) = r_{\text{dc}} + l_{\text{dc}}s \qquad (5\text{-}65)$$

As before

$$y(s) = cs \qquad (5\text{-}23)$$

Substitution in Eqs. 5-13 and 5-14 for characteristic impedance and propagation function leads to the following transform for the current response to a step-function source:

$$I(z,s) = \frac{V_0}{s}\sqrt{\frac{cs}{r_{\text{dc}} + l_{\text{dc}}s}}\,\exp\left[-z\sqrt{cs(r_{\text{dc}} + l_{\text{dc}}s)}\right] \qquad (5\text{-}66)$$

The inverse transform is[1, 2, 12, 17]

$$i(z,t) = V_0\sqrt{\frac{c}{l_{\text{dc}}}}\,\exp\left(-\frac{r_{\text{dc}}t_z}{2l_{\text{dc}}}\right)I_0\left[\frac{r_{\text{dc}}}{2l_{\text{dc}}}\sqrt{t^2 - t_z^2}\right]U(t - t_z) \qquad (5\text{-}67)$$

and
$$f(\tau) = \sqrt{\tau}\epsilon^{-1/\tau} \tag{5-73}$$

Then
$$v(z,t) \approx \frac{V_0}{\sqrt{\pi}} f(\tau) \qquad \tau \ll 1 \tag{5-74}$$

$$i(z,t) \approx \frac{V_0}{Z_{\hat{0}}\sqrt{\pi}} f(\tau) \qquad \tau \ll 1$$

$$\frac{df}{d\tau} = \epsilon^{-1/\tau}\left(\frac{\tau^{-1/2}}{2} + \tau^{-3/2}\right) \tag{5-75}$$

Here $I_0(\ )$ represents the modified Bessel function of the first kind, zero order (Sec. B-2):

$$I_0(0) = 1.0$$

At $t = t_{z+}$ (an instant after the elapse of travel time to location z) this reduces to

$$i(z, t_{z+}) = V_0\sqrt{\frac{c}{l_{dc}}} \exp\left(-\frac{r_{dc}t_z}{2l_{dc}}\right)U(t_{z+} - t_z) \tag{5.68}$$

The inverse transform for voltage is more complicated:

$$V(z,s) = \frac{V_0}{s}\exp\left[-z\sqrt{cs(r_{dc} + l_{dc}s)}\right] \tag{5-69}$$

$$v(z,t) = V_0\left\{\exp\left[-0.5zr_{dc}\sqrt{c/l_{dc}}\right] + 0.5zr_{dc}\sqrt{\frac{c}{l_{dc}}}\right.$$

$$\left.\times\int_{t_z}^{t}\exp\left[-\frac{r_{dc}\tau}{2l_{dc}}\right]\frac{I_1\left[(r_{dc}/2l_{dc})\sqrt{\tau^2 - t_z^2}\right]}{\sqrt{\tau^2 - t_z^2}}\,d\tau\right\}U(t - t_z) \tag{5-70}$$

The corresponding wave-front value is

$$v(z, t_{z+}) = V_0 \exp\left(-\frac{r_{dc}t_z}{2l_{dc}}\right)U(t_{z+} - t_z) \tag{5-71}$$

Equations 5-68 and 5-71 predict step-function changes in current and voltage at the wave-arrival time t_z at any location z, a feature which is contrary to experimental observation. Hence the assumption of frequency-independent distributed resistance and inductance (Eq. 5-65) does not lead to an adequate formulation of the lossy-line step-function-response problem.

Each higher-order derivative will consist of a sum of terms of the type $\epsilon^{-1/\tau}\tau^{-(n+1/2)}$, where n is an integer. Each such term will be indeterminate in the form 0/0, but it will be shown that all those terms approach zero as τ approaches zero. To demonstrate the fact, l'Hôpital's rule (indicated by the first arrow in the following equation) may be applied to the first term on the right-hand side of Eq. 5-75:

$$\frac{\tau^{-1/2}}{2\epsilon^{1/\tau}} \to \frac{-\frac{1}{2}\tau^{-3/2}}{-2\epsilon^{1/\tau}\tau^{-2}} = \frac{1}{4}\tau^{1/2}\epsilon^{-1/\tau} \to 0 \qquad \tau \to 0 \tag{5-76}$$

Any term with a value of n greater than zero may be reduced by l'Hôpital's rule to a term of the same form with the next smaller value for n:

$$\frac{\tau^{-(n+1/2)}}{\epsilon^{1/\tau}} \to \frac{-\left(n+\frac{1}{2}\right)\tau^{-(n+3/2)}}{-\epsilon^{1/\tau}\tau^{-2}} = \frac{\left(n+\frac{1}{2}\right)\tau^{-(n+1/2)}}{\epsilon^{1/\tau}} \qquad \tau \to 0 \tag{5-77}$$

Thus the functions $v(z,t)$ and $i(z,t)$ cannot be expanded in Taylor's series for the instant $T = T_z$.

(2) *Maximum Slope of Voltage Wave Front.* Equation 5-58 may be differentiated twice with respect to T, and the result equated to zero to fix the normalized time at which the slope is a maximum. The following rule for differentiation with respect to the lower limit of a definite integral such as the complementary error function, Eq. 5-49, should be noted:

$$\frac{d}{dp}\int_p^{\infty} f(u)\,du = -f(p) \tag{5-78}$$

$$\frac{\partial v(z,t)}{\partial T} = \frac{V_0 T_z}{2\sqrt{\pi}\,(T-T_z)^{3/2}}\exp\left[-T_z^2/4(T-T_z)\right] \tag{5-79}$$

$$\frac{\partial^2 v(z,t)}{\partial T^2} = \frac{\partial v(z,t)}{\partial T}\left[\frac{T_z^2}{4(T-T_z)^2} - \frac{3}{2(T-T_z)}\right] \tag{5-80}$$

Let $T = T_1$ when $\partial^2 v/\partial T^2 = 0$:

$$T_1 - T_z = \frac{T_z^2}{6} \tag{5-81}$$

Substitution of this result in Eq. 5-58 yields

$$v(z, T_1) = V_0 \operatorname{erfc}(\sqrt{1.5}) = 0.0832V_0 \qquad (5\text{-}82)$$

From the ordinate-at-maximum-slope result given in Eq. 5-82, and from inspection of Figs. 5-3 and 5-4, it appears that the wave front has a small "steep rise" soon after the elapse of time t_z, but that the rate of rise decreases drastically thereafter. When voltage waves at two different values of z are compared, it may be seen from Eq. 5-50 or 5-58 that the times after t_z at which the voltages reach some prescribed value are proportional to the square of the distance from the source. Correspondingly, the rates of change of the voltages with time, at a given voltage, decrease as the square of distance from the source.

Example (continued)—The result given in Eq. 5-81 may be applied to the numerical example in Sec. 5-3c by substituting $T_z = 0.02$; for that condition the maximum rate of increase of voltage occurs when $T - T_z = 6.67 \times 10^{-5}$, or if this is multiplied by 1.25×10^{-4}, when $t - t_z = 8.33 \times 10^{-9}$ s.

(3) *Behavior as Time Approaches Infinity.* The asymptotic approximation given in Eq. 5-62 for $i(z, T)$ is also valid for $T - T_z$ much greater than unity, as may be seen by inspection of Eq. 5-59:

$$i(z, T) \approx \frac{V_0}{\tilde{Z}_0\sqrt{\pi}} \cdot \frac{1}{\sqrt{T - T_z}\,[1 + (T_z/2(T - T_z))]} \cdot \exp\left[-\frac{T_z^2}{4(T - T_z)}\right] U(T - T_z) \qquad (T - T_z) \gg 1 \qquad (5\text{-}83)$$

If, in addition, $T - T_z$ is much greater than T_z^2, the exponential function approaches unity and the denominator may be simplified to yield

$$i(z, T) \approx \frac{V_0}{\tilde{Z}_0} \cdot \frac{1}{\sqrt{\pi(T - T_z)}} \qquad T - T_z \gg 1;\ T - T_z \gg T_z^2 \qquad (5\text{-}84)$$

For the voltage function in Eq. 5-58, time approaching infinity corresponds to the argument of the complementary error function approaching zero, so the power series of that function (Eq. 5-55) should be used:

$$v(z,T) \approx V_0\left[1 - \frac{T_z}{\sqrt{\pi(T - T_z)}}\right] \qquad (T - T_z) \gg T_z^2 \tag{5-85}$$

Hence the ultimate mode of time variation of both current and voltage as they approach their respective asymptotes of zero and V_0 is that of the inverse square root of time.*

It should be appreciated that decrease as the inverse square root of time is very different from an exponential decrease with time. The proportionate effect of a unit decrease in the argument of a negative exponential is the same whether the initial value of the argument is small or large:

$$\frac{\epsilon^{-2}}{\epsilon^{-1}} = 0.368$$

$$\frac{\epsilon^{-1001}}{\epsilon^{-1000}} = 0.368$$

A corresponding comparison for the inverse-square-root function yields

$$\frac{(1/\sqrt{2})}{(1/\sqrt{1})} = 0.707$$

$$\frac{(1/\sqrt{1001})}{(1/\sqrt{1000})} = 0.9995$$

To achieve the same proportionate reduction when starting with an

---

*One may apply the *Final Value Theorem* to Eqs. 5-44 and 5-45 to verify the asymptotes noted here.

$$i(z,\infty) \to [sI(z,s)]_{\lim s\to 0} \tag{5-86}$$

$$\to \left[\frac{V_0\sqrt{s}}{\tilde{Z}_0\sqrt{s} + b}\exp(-zs\sqrt{l'c} - az\sqrt{s})\right]_{\lim s\to 0}$$

$$\to 0 \tag{5-87}$$

Similarly $$v(z,\infty) \to [sV(z,s)]_{\lim s\to 0}$$

$$\to V_0 \tag{5-88}$$

argument of 1000, as was achieved by changing the argument from 1 to 2, the new argument would have to be 2000.

5-4 SEMI-INFINITE LINE, NONZERO SOURCE RESISTANCE*

The problem which was just worked may be readily generalized to include a lumped frequency-independent resistance in series with the voltage source, as shown in Fig. 5-2. As noted in Sec. 3-1a(1), this equivalent source gives a good approximation to the behavior of an electronic pulse generator. Let

$$Z_G(s) = R_G \tag{5-89}$$

a. Current and Voltage Transforms

From Eqs. 5-16, 5-22, 5-40 and 5-41, the transform of the current at $z = 0$ is

$$I(0, s) = \frac{V_0}{s[R_G + Z_0(s)]} \tag{5-90}$$

Substitution of Eq. 5-38 for $Z_0(s)$ gives

$$I(0, s) = \frac{V_0}{s\left[R_G + \tilde{Z}_0 + b\tilde{Z}_0/\sqrt{s}\right]} \tag{5-91}$$

Let

$$k = \frac{\tilde{Z}_0}{R_G + \tilde{Z}_0} \tag{5-92}$$

The denominator of Eq. 5-91 may be rearranged by substitution of Eq. 5-92:

$$I(0, s) = \frac{V_0}{s(R_G + \tilde{Z}_0)\left(1 + \dfrac{bk}{\sqrt{s}}\right)}$$

$$= \frac{V_0}{\sqrt{s}\,(R_G + \tilde{Z}_0)(\sqrt{s} + bk)} \tag{5-93}$$

---

*This section is based on the IEEE copyrighted 1968 paper by P. C. Magnusson, Transient wavefronts on lossy transmission lines—effect of source resistance, which was published in the *IEEE Trans. Circuit Theory*, CT-15, no. 3, 290, September 1968. The substance is reprinted by permission.

The transform for current at any other location on the line may be found by multiplying Eq. 5-93 by the exponential of $-\gamma(s)z$ and substituting Eqs. 5-33 and 5-46:

$$I(z,s) = \frac{V_0}{\sqrt{s}\left(R_G + \tilde{Z}_0\right)\left(\sqrt{s} + bk\right)} \exp\left(-st_z - az\sqrt{s}\right) \tag{5-94}$$

The transform for voltage at the input to the line is

$$V(0,s) = \frac{V_0}{s} - R_G I(0,s) \tag{5-95}$$

The transform for voltage anywhere on the semi-infinite line may be found by multiplying this expression by the exponential of $-\gamma(s)z$. It may be noted that

$$R_G I(0,s)\exp[-\gamma(s)z] = R_G I(z,s)$$

Hence

$$V(z,s) = \frac{V_0}{s}\exp\left(-st_z - az\sqrt{s}\right) - R_G I(z,s) \tag{5-96}$$

b. Inverse Transforms; Dimensionless Variables

The inverse transform of $I(z,s)$, Eq. 5-94, may be found by substitution in Eq. 5-48, if b in the latter equation is replaced by bk. Equation 5-92 will be substituted for $R_G + \tilde{Z}_0$:

$$i(z,t) = \frac{kV_0}{\tilde{Z}_0}\exp\left[b^2k^2(t-t_z) + abkz\right] \cdot \operatorname{erfc}\left(bk\sqrt{t-t_z} + \frac{az}{2\sqrt{t-t_z}}\right)U(t-t_z) \tag{5-97}$$

The inverse transform of the first term for $V(z,s)$, Eq. 5-96, may be found by substitution in Eq. 5-47; that of the second term is simply $R_G i(z,t)$:

$$v(z,t) = V_0 \operatorname{erfc}\left(\frac{az}{2\sqrt{t-t_z}}\right)U(t-t_z) - R_G i(z,t) \tag{5-98}$$

Equations 5-97 and 5-98 may be restated in terms of dimensionless variables by substituting Eqs. 5-56 and 5-57:

$$i(z,T) = \frac{kV_0}{\tilde{Z}_0}\exp\left[k^2(T - T_z) + kT_z\right] \cdot \operatorname{erfc}\left(k\sqrt{T - T_z} + \frac{T_z}{2\sqrt{T - T_z}}\right)U(T - T_z) \quad (5\text{-}99)$$

$$v(z,T) = V_0 \operatorname{erfc}\left(\frac{T_z}{2\sqrt{T - T_z}}\right)U(T - T_z) - R_G i(z,T) \quad (5\text{-}100)$$

c. Asymptotic Behavior; Normalizing for Plotting

The analysis of Sec. 5-3c(1) by substituting the leading term of the asymptotic expansion for the complementary error function may be repeated for Eqs. 5-99 and 5-100 (Prob. 5-6) with the following results:

$$i(z,T) \approx \frac{2kV_0\sqrt{T - T_z}}{\tilde{Z}_0 T_z\sqrt{\pi}}\exp\left[\frac{-T_z^2}{4(T - T_z)}\right]U(T - T_z) \qquad T - T_z \ll T_z^2;\ T - T_z \ll T_z \quad (5\text{-}101)$$

$$v(z,T) \approx \frac{2V_0\sqrt{T - T_z}}{T_z\sqrt{\pi}}\exp\left[\frac{-T_z^2}{4(T - T_z)}\right]U(T - T_z) - R_G i(z,T) \quad (5\text{-}102)$$

This reduces to

$$v(z,T) \approx k\frac{2V_0\sqrt{T - T_z}}{T_z\sqrt{\pi}}\exp\left[\frac{-T_z^2}{4(T - T_z)}\right]U(T - T_z) \qquad T - T_z \ll T_z^2 \quad (5\text{-}103)$$

Here, as in Eqs. 5-60 and 5-63, $v(z,T)/i(z,T)$ is initially $\approx \tilde{Z}_0$. Comparison between Eqs. 5-103 and 5-60 shows that the results differ by the factor k. Accordingly curves of v and i may be normalized with respect to their initial-rise behavior by assuming that the voltage source is

$$v_g(t) = V_0 U(t)$$

$$= \frac{V_k}{k} U(t) \tag{5-104}$$

Here V_k is a constant.

The curves in Figs. 5-5 and 5-6 for $R_G = \tilde{Z}_0$ have been plotted on this basis.

At t increases without limit, $i(z,T)$ approaches zero and $v(z,T)$ approaches V_0.

5-5 CONCLUSIONS

The response of a lossy transmission line to a step-function voltage source has been derived on the basis that the distributed resistance and the component of inductive reactance due to magnetic flux within the conductors are assumed to vary as the square root of frequency, because of skin effect. The capacitance is assumed to be frequency-independent, and the shunt conductance, zero.

The following operational propagation function was used:

$$\gamma(s) \approx s\sqrt{l'c} + a\sqrt{s} \tag{5-33}$$

Here l' is the component of distributed inductance associated with magnetic flux external to the conductors, and $a/\sqrt{2}$ is the slope of the high-frequency asymptote of a plot of attenuation (nepers per unit of length) versus the square root of the radian frequency.

The corresponding operational characteristic impedance is

$$\tilde{Z}_0(s) \approx \tilde{Z}_0\left(1 + \frac{b}{\sqrt{s}}\right) \tag{5-38}$$

where

$$\tilde{Z}_0 = \sqrt{\frac{l'}{c}} \tag{5-37}$$

and

$$b = \frac{a}{\sqrt{l'c}} \tag{5-36}$$

The following results were obtained for step-function excitation, when the source impedance is zero:

$$v(z,t) = V_0 \operatorname{erfc}\left(\frac{az}{2\sqrt{t - t_z}}\right)U(t - t_z) \tag{5-50}$$

$$i(z,t) = \frac{V_0}{\tilde{Z}_0}\exp(b^2 t)\operatorname{erfc}\left(b\sqrt{t - t_z} + \frac{az}{2\sqrt{t - t_z}}\right)U(t - t_z) \tag{5-53}$$

$$t_z = z\sqrt{l'c} \tag{5-46}$$

Here erfc designates the complementary error function, which is defined by the integral

$$\operatorname{erfc}(p) = \frac{2}{\sqrt{\pi}}\int_p^\infty \exp(-u^2)\,du \tag{5-49}$$

The following results are obtained if the source impedance is a frequency-independent resistance R_G:

$$i(z,t) = \frac{kV_0}{\tilde{Z}_0}\exp\left[b^2k^2(t - t_z) + abkz\right]$$

$$\cdot \operatorname{erfc}\left(bk\sqrt{t - t_z} + \frac{az}{2\sqrt{t - t_z}}\right)U(t - t_z) \tag{5-97}$$

where

$$k = \frac{\tilde{Z}_0}{R_G + \tilde{Z}_0} \tag{5-92}$$

$$v(z,t) = V_0 \operatorname{erfc}\left(\frac{az}{2\sqrt{t - t_z}}\right)U(t - t_z) - R_G i(z,t) \tag{5-98}$$

Analysis of the equations just given for $v(z,t)$ and $i(z,t)$ discloses the following:

1. a delay of $t_z = z_1\sqrt{l'c}$ before the voltage and current at any location z_1 become nonzero.
2. Immediately after lapse of the corresponding delay time the values of voltage and current and of all their finite-ordered derivatives with respect to time are zero.
3. An infinitesimal interval of time later the voltage and current are increasing at accelerating rates.
4. Voltage and current soon experience their maximum rates of increase, after which the voltage continues to rise at an ever slower rate, and approaches the value generated by the source.

5. The current rises gradually to a maximum and then declines slowly toward zero.

These features are in agreement with oscillographic observations on actual lines.

A lumped resistance in series with the voltage source causes voltage division with respect to the high-frequency characteristic impedance $\tilde{Z}_0$ during the initial-rise period. As input current decreases, the source resistance has progressively less effect on voltage and current on the line.

PROBLEMS

Note: Problems 5-1 through 5-3 are intended primarily for review in Laplace transform techniques.

5-1. Given the circuit shown in Fig. 5-7. The line is initially uncharged and the initial current in L_T is zero. Find the functions $V_A(s)$ and $V_B(s)$ and reduce Eqs. 5-15 and 5-16. Find the inverse transforms and sketch v and i as functions of time at two or more locations on the line. *Partial answer:*

$$v(z,t) = \frac{V_0}{2}U(t - t_z)$$

$$- \frac{V_0}{2}\left\{1 - 2\exp\left[\left(-\sqrt{l/c}\,/L_T\right)(t - 2t_r + t_z)\right]\right\}$$

$$\times U(t - 2t_r + t_z)$$

5-2. Repeat Prob. 5-1, letting the terminating impedance consist of L_T in series with a resistor $R_T = \sqrt{l/c}$.

5-3. Repeat Prob. 5-1, letting the terminating impedance consist of an initially uncharged capacitor C_T instead of L_T.

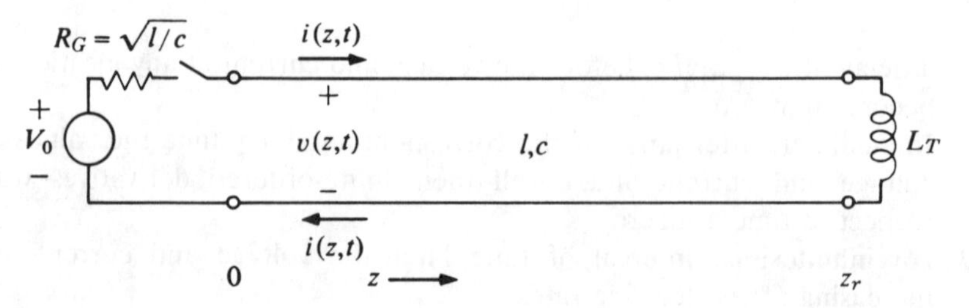

Figure 5-7. Lossless transmission line with purely inductive termination and reflectionless source.

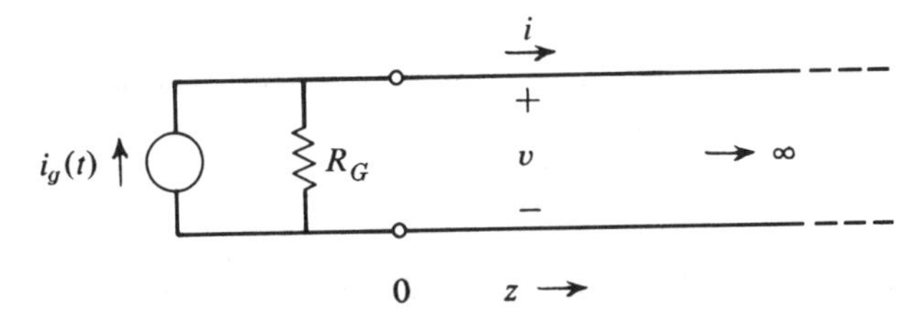

Figure 5-8. Semi-infinite line with equivalent current source.

5-4. Given the line described in the example at the close of Sec. 5-3c; for $T_z = 0.02$, find these rise times for v: (a) from zero to 0.5 V_0 and (b) from 0.1 V_0 to 0.9 V_0. (Use Table 5-1.)

5-5. Given the circuit shown in Fig. 5-2, in which the line is a skin-effect-lossy polyethylene insulated cable ($\epsilon = 2.25\epsilon_0$; $\mu = \mu_0$), and R_G is assumed to be zero. Let two locations on the line be specified as follows: $z_1 = 80$ m; $z_2 = 150$ m; $\nu \approx 1/\sqrt{\mu\epsilon}$.
   (a) Find time t_z for locations z_1 and z_2.
   (b) Let it be assumed that if the source impresses a step-function voltage V_0 at the input to the line, the rise time $(t - t_z)$ to $0.5V_0$ at location z_1 is 1.4×10^{-9} s. What is the corresponding rise time at location z_2?

5-6. Derive Eqs. 5-101 and 5-103 for the initial-rise behavior with nonzero source resistance from Eqs. 5-99 and 5-100.

5-7. Examine the initial-rise behavior of the current and voltage at the input ($z = 0$) of a skin-effect-lossy line with a nonzero source resistance R_G.

5-8. Find approximations for the current and voltage on a skin-effect-lossy line with a nonzero source resistance, Eqs. 5-99 and 5-100, for time approaching infinity, and compare results with Eqs. 5-84 and 5-85.

5-9. Given the circuit shown in Fig. 5-8 with a current source and lumped shunt resistance. Let $i_g(t) = I_0 U(t)$. Write the equation for $I(0, s)$ in terms of $I_g(s)$ and relevant parameters. Find the inverse transforms for $I(z, s)$ and $V(z, s)$ for the case in which R_G is infinite. *Note:* the following transform pair[1] may be useful:

$$\mathcal{L}^{-1}\left[\frac{\exp(-h\sqrt{s})}{s^{3/2}}\right] = \left[2\sqrt{\frac{t}{\pi}}\exp\left(\frac{h^2}{4t}\right) - h\,\mathrm{erfc}\left(\frac{h}{2\sqrt{t}}\right)\right]U(t) \quad (5\text{-}105)$$

REFERENCES

1. Abramowitz, M. and Stegun, I. A., *Handbook of Mathematical Functions*, Dover, New York, 1965.
2. Bohn, Erik. V., *The Transform Analysis of Linear Systems*, pp. 281–95, 308–15, 336–39. Reading, Mass.: Addison-Wesley Publishing Company, Inc., 1963.

3. Gates, B. G., The attenuation and distortion of traveling waves on overhead power transmission systems at voltages below the corona limit, *J. IEE*, 84, 711, June 1939.
4. Guillemin, E. A., *Theory of Linear Physical Systems*, Wiley, New York, 1963, 552.
5. Magnusson, P. C., Transient wavefronts on lossy transmission lines—Effect of source resistance, *IEEE Trans. Circuit Theory*, CT-15, no. 3, 290, September 1968.
6. Murakami, T. and Corrington, M. S., Relation between amplitude and phase in electrical networks, *RCA Review*, 9, 602, 1948. (Excellent historical summary.)
7. Nahman, N. S., A discussion on the transient analysis of coaxial cables considering high-frequency losses, *IRE Trans. Circuit Theory*, CT-9, no. 2, 144, June 1962.
8. Peirce, B. O., *A Short Table of Integrals*, 3rd rev. ed., Ginn and Company, Boston, 1929, 116.
9. Pleijel, H., Transient waves and their deformation under propagation along lines, *Tek. Tidskr. Elektrotek.* 48, no. 11, 129, November 1918. (In Swedish.)
10. Ramo, S., Whinnery, J. R., and Van Duzer, T. *Fields and Waves in Communication Electronics*, Wiley, New York, 1965, 333 and 623.
11. Richards, P. I., General impedance-function theory, *Quart. Appl. Math.*, 6, no. 1, 21, April 1948.
12. Roberts, G. E. and Kaufman, H., *Table of Laplace Transforms*, W. B. Saunders Company, Philadelphia, 1966, 246.
13. U.S. Department of Commerce, *Tables of the Error Function and Its Derivative*, 2nd ed., Applied Mathematics Series 41, National Bureau of Standards, Washington, DC, 1954.
14. Thomas, R. J., Choosing coaxial cable for fast pulse response, *Microwaves*, 7, no. 11, 56, November 1968.
15. Tuttle, D. F., Jr., *Network Synthesis*, vol. 1, Wiley, New York, 1958, 107 and 381.
16. von Kármán, T. and Biot, M. A., *Mathematical Methods in Engineering*, McGraw-Hill Book, New York, 1940.
17. Weber, E., *Linear Transient Analysis*, vol. II, Wiley, New York, 1956, 318 and 372.
18. Wigington, R. L. and Nahman, N. S., Transient analysis of coaxial cable considering skin effect, *Proc. IRE*, 45, no. 2, 166, 1957.

CHAPTER 6

Sinusoidal Voltages and Currents on a Terminated Line — Standing Waves

If a line is terminated in other than its characteristic impedance, the steady-state component of the response to a sinusoidal driving function will not be a single traveling-wave pair of voltage and current. It will be a more complicated function, which may be resolved mathematically into the sum of two oppositely moving sinusoidal traveling-wave pairs. The boundary conditions relating those wave functions at the terminal are the same in principle as those introduced in Sec. 3-1 for a resistive termination to a lossless line. Through the use of phasors, impedances other than purely resistive ones may be accommodated. As noted in Sec. 4-3b, a lossy line has a frequency-dependent, complex characteristic impedance.

6-1 INCIDENT AND REFLECTED WAVE FUNCTIONS

In the analysis of the steady-state response of a transmission line and passive load, or an assembly of line sections and lumped networks, it is usually best to begin at the load, where the relationship between the incident and reflected steady-state traveling-wave sets is fixed by the ratio of the load impedance to the characteristic impedance, and work back toward the source. The elementary situation illustrated in Fig. 6-1, that of a uniform line terminated with a lumped impedance, will be examined first. Mathematical descriptions of the traveling-wave functions at the termination are less complicated if the zero point for measuring distance along the line is placed at the termination. The coordinate-direction plan of Fig. 2-4 will be followed so that locations on the line are designated with positive values of z'.

The following expressions for postulated voltage and current may be written. Waves traveling to the right and impinging on the terminating

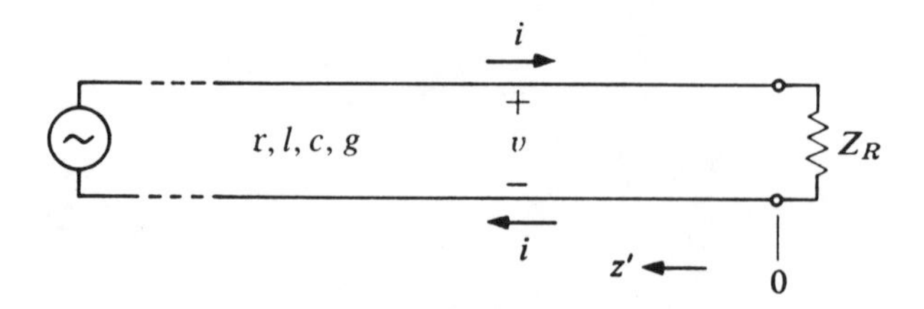

Figure 6-1. Transmission line terminated with lumped impedance.

impedance will be known as *incident* waves; those leaving the terminated end and traveling to the left will be called *reflected* waves. The voltage-and-current polarity pairings of Eqs. 2-45 through 2-48 are applicable.

Let
$$v_{\text{inc}}(z', t) = V_{IM}\epsilon^{\alpha z'} \cos(\omega t + \beta z') \tag{6-1}$$

or
$$V_{\text{inc}}(z') = V_{IM}\epsilon^{\gamma z'} \tag{6-2}$$

Then
$$I_{\text{inc}}(z') = \frac{V_{IM}}{Z_0}\epsilon^{\gamma z'} \tag{6-3}$$

Likewise let

$$v_{\text{ref}}(z', t) = V_{RM}\epsilon^{-\alpha z'} \cos(\omega t + \theta_K - \beta z') \tag{6-4}$$

or
$$V_{\text{ref}}(z') = V_{RM} \exp(j\theta_K - \gamma z') \tag{6-5}$$

$$I_{\text{ref}}(z') = -\frac{V_{RM}}{Z_0} \exp(j\theta_K - \gamma z') \tag{6-6}$$

The angle θ_K allows for a time-phase difference between the incident and reflected components of voltage at the terminated end where z' equals zero.

a. Relationships at Termination

The analysis of the relationships at the terminating impedance Z_R closely parallels that in Sec. 3-1, except that the voltages and currents are stated as phasors, and Z_0 and Z_R are not restricted to pure resistances. The resultant voltage at the termination ($z' = 0$) will be designated by

V_{ter}, and the resultant current, by I_{ter}. For continuity of voltage,

$$V_{\text{ter}} = V_{\text{inc}}(0) + V_{\text{ref}}(0) \tag{6-7}$$

Equations 6-2 and 6-5 are substituted in Eq. 6-7:

$$V_{\text{ter}} = V_{IM} + V_{RM} \exp(j\theta_K) \tag{6-8}$$

For continuity of current,

$$I_{\text{ter}} = I_{\text{inc}}(0) + I_{\text{ref}}(0) \tag{6-9}$$

Equations 6-3 and 6-4 are substituted in Eq. 6-9:

$$I_{\text{ter}} = \frac{V_{IM}}{Z_0} - \frac{V_{RM}}{Z_0} \exp(j\theta_K) \tag{6-10}$$

In accordance with Ohm's law,

$$V_{\text{ter}} = Z_R I_{\text{ter}} \tag{6-11}$$

Equations 6-8 and 6-10 may be substituted into Eq. 6-11 and the resulting expression reduced to

$$V_{RM} \exp(j\theta_K)(Z_R + Z_0) = V_{1M}(Z_R - Z_0)$$

Let*

$$K = |K| \exp(j\theta_K) = \frac{Z_R - Z_0}{Z_R + Z_0} \tag{6-12}$$

Substitution of Eq. 6-12 into the expression preceding it, and the further substitution of Eqs. 6-2 and 6-5 with $z' = 0$, yields

$$V_{\text{ret}}(0) = KV_{\text{inc}}(0) \tag{6-14}$$

The complex quantity K is shown as the *reflection coefficient*. It may be

---

*The numerator and denominator of Eq. 6-12 may be divided by Z_0 to yield

$$K = \frac{(Z_R/Z_0) - 1}{(Z_R/Z_0) + 1} \tag{6-13}$$

Thus K is basically a function of the ratio Z_R/Z_0, or the load impedance normalized with respect to the characteristic impedance. This will be developed further in Sec. 7-5b.

compared with ρ_T in Eq. 3-6, which is a real number, and with $\rho_T(s)$ in Eq. 5-18, which is a function of the complex variable s. The following example will serve to illustrate the application of K. Let

$$Z_0 = 600\underline{/-20.0^\circ}\ \Omega$$

$$Z_R = 500\underline{/35.0^\circ}\ \Omega$$

Substitution in Eq. 6-12 yields

$$K = \frac{410 + j287 - 564 + j205}{410 + j287 + 564 - j205}$$

$$= \frac{-154 + j492}{974 + j82}$$

$$= \frac{516\underline{/107.4^\circ}}{977\underline{/4.8^\circ}}$$

$$= 0.528\underline{/102.6^\circ}$$

A phasor diagram relating the voltages and currents at the terminal of the line is sketched in Fig. 6-2. In drawing such a diagram it is convenient

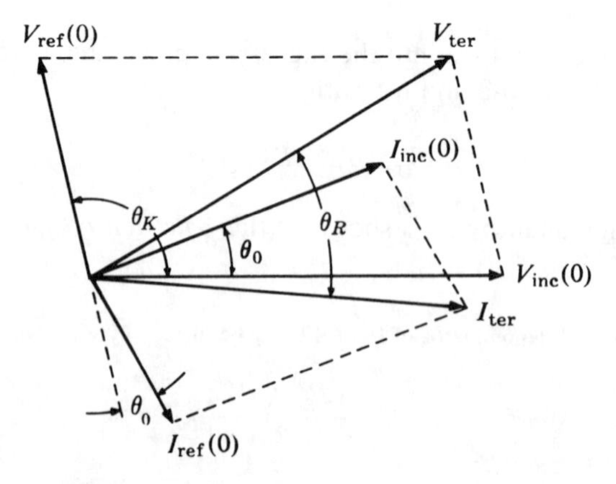

Figure 6-2. Phasor diagram at termination of transmission line: $Z_0 = 600\underline{/-20.0^\circ}\ \Omega$, $Z_R = 500\underline{/35.0^\circ}\ \Omega$; ($K = 0.528\underline{/102.6^\circ}$).

to start with $V_{inc}(0)$. The reflected component of voltage leads it by angle θ_K and is in proportion to $|K|$. The incident current phasor is related to the incident voltage by angle θ_0; here it leads by 20°. Reflected current leads the negative of reflected voltage by the same amount. The two voltage components add to form V_{ter}, and the two current components add to form I_{ter}. Load current should lag the load voltage by the angle of Z_R, in this instance 35°.

b. Resultant Voltage and Current away from Termination

A phasor diagram relating voltages and currents at the termination may be generalized to show those quantities at any desired point on the line by adding traveling-wave phasor loci as shown in Fig. 6-3. It was assumed in the construction of that diagram that the ratio α/β is 0.1 The phasor

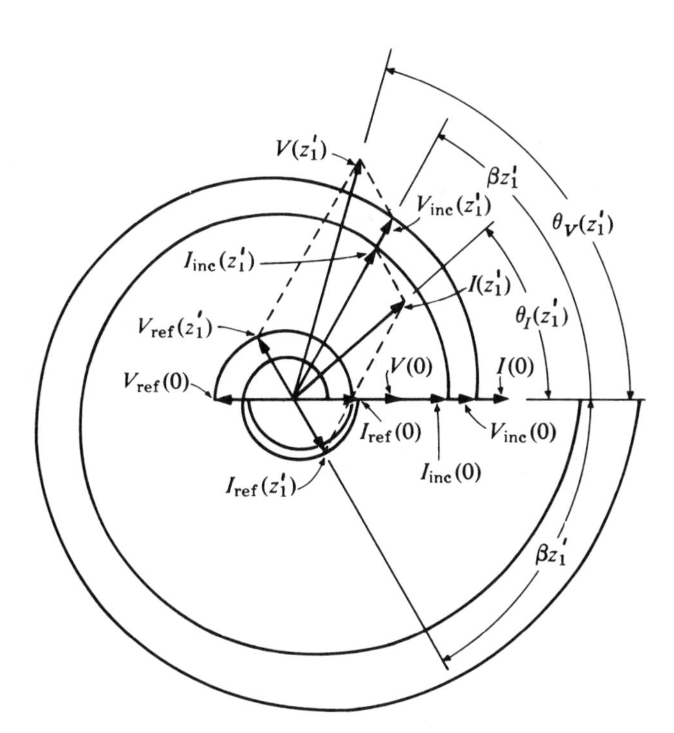

Figure 6-3. Phasor loci for lossy transmission line with resistive termination: $Z_0 = 600\underline{/0^\circ}$ Ω, $Z_R = 250\underline{/0^\circ}$ Ω, $\alpha/\beta = 0.1$; ($K = 0.412\underline{/180.0^\circ}$).

for the incident component of voltage at distance z'_1 (meters, miles, etc.) from the termination is positioned on the diagram by drawing a line radially from the origin at an angle of $\beta z'_1$ radians *counterclockwise* from the phasor of incident voltage at the termination. The phasor for the reflected component of voltage is positioned in direction by drawing a radial line at angle $\beta z'_1$ radians *clockwise* for the phasor of reflected voltage at the termination. The lengths of both phasors are determined by spirals that start at the respective terminal phasors and progress outward and counterclockwise in the case of incident voltage and inward and clockwise for the reflected voltage. After one full revolution the spiral intersects the radial line from which it started, and the ratio of its larger intercept to its smaller one is $\epsilon^{2\pi\alpha/\beta}$—in this case $\epsilon^{0.628}$, or 1.873. The same principle applies for current components. Resultant voltage and current, including their phase angles relative to the reference voltage, may be found by phasor addition of the respective incident and reflected components drawn for the appropriate $\beta z'_1$.

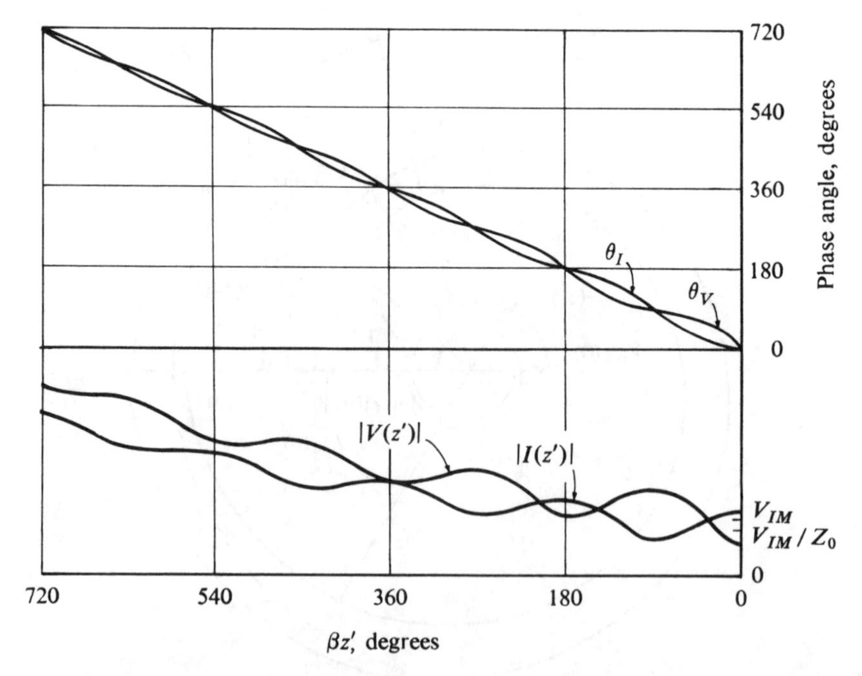

Figure 6-4. Magnitudes and relative phase angles of voltage and current on lossy transmission line with resistive termination: $Z_0 = 600\underline{/0°}\ \Omega$, $Z_R = 250\underline{/0°}\ \Omega$, $\alpha/\beta = 0.1$; ($K = 0.412\underline{/180.0°}$).

Plots of magnitudes of resultant voltage and current phasors with respect to distance from the termination, and also of their phase angles relative to the terminal voltage, are instructive and are given in Fig. 6-4 for the situation studied in Fig. 6-3.

Mathematical expressions paralleling the procedures followed in Fig. 6-3 may be found in Eqs. 6-2, 6-3, 6-5, 6-6, and 6-14:

$$V(z') = V_{\text{inc}}(z') + V_{\text{ref}}(z')$$

$$= V_{IM}(\epsilon^{\gamma z'} + K\epsilon^{-\gamma z'}) \tag{6-15}$$

$$I(z') = I_{\text{inc}}(z') + I_{\text{ref}}(z')$$

$$= \frac{V_{IM}}{Z_0}(\epsilon^{\gamma z'} - K\epsilon^{-\gamma z'}) \tag{6-16}$$

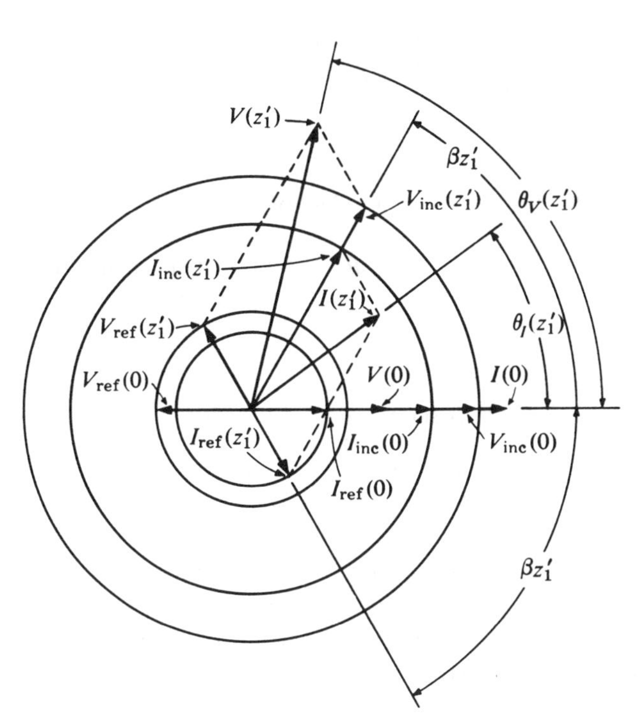

Figure 6-5. Phasor loci for lossless transmission line with resistive termination: $Z_0 = 600\angle 0°\ \Omega$, $Z_R = 250\angle 0°\ \Omega$; ($K = 0.412\angle 180.0°$).

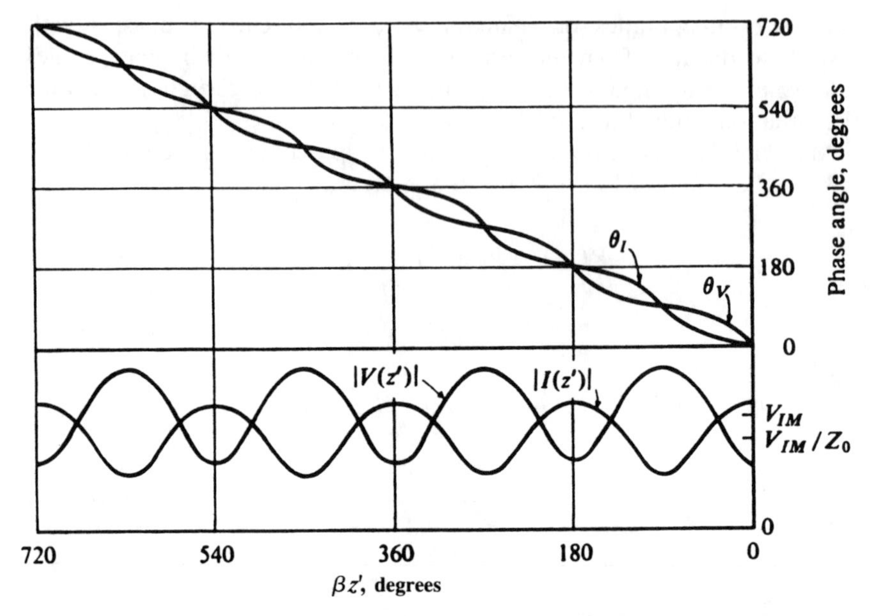

Figure 6-6. Magnitudes and relative phase angles of voltage and current on lossless transmission line with resistive termination: $Z_0 = 600\angle 0° \ \Omega$, $Z_R = 250\angle 0° \ \Omega$; ($K = 0.412\angle 180.0°$).

These functions may be stated in instantaneous form as

$$v(z', t) = V_{IM}[\epsilon^{\alpha z'} \cos(\omega t + \beta z') + |K|\epsilon^{-\alpha z'} \cos(\omega t + \theta_K - \beta z')] \tag{6-17}$$

$$i(z', t) = \frac{V_{IM}}{|Z_0|}[\epsilon^{\alpha z'} \cos(\omega t - \theta_0 + \beta z') - |K|\epsilon^{-\alpha z'} \cos(\omega t - \theta_0 + \theta_K - \beta z')] \tag{6-18}$$

If the line is lossless, the loci of V_{inc}, V_{ref}, I_{inc}, and I_{ref} in the phasor diagram (Fig. 6-5) become circles, and, as shown in Fig. 6-6, plots of the magnitudes of resultant voltage and current phasors (or rms voltage and current, in accordance with Eq. 4-40) then repeat every 180° in $\beta z'$. Such variations with distance are described as *standing waves*; the concept is most meaningful for a lossless line or one with very low losses, such that each undulation differs but little from those adjacent to it.

c. Electrical Distance and Wavelength

In the study of standing wave functions it is convenient to refer to distances along the line in terms of *electrical distance*, the argument $\beta z'$ in the sinusoidal wave equations. This may be stated in radians or degrees; the latter is used for the abscissa variable in the graphs of standing waves in Fig. 6-4 and following plots.

An electrical distance may also be stated in terms of the ratio z'/λ, where λ is the *wavelength*; it corresponds to a span of 2π in $\beta z'$. Hence

$$\lambda = \frac{2\pi}{\beta} \tag{6-19}$$

6-2 LIMITING-CASE REFLECTION SITUATIONS

Certain special cases yield much simpler results than the general examples just given.

a. Terminating Impedance Equal to Characteristic Impedance

Should the terminating impedance be equal to the characteristic impedance, the numerator of Eq. 6-12 would vanish and the reflection coefficient would be zero. Root-mean-square voltage and current would be uniform along a lossless line under these circumstances; they would increase exponentially with distance from the termination toward the source on a line having losses. Relative phase angles would increase linearly in the leading direction with distance from the termination.

b. Open-Circuited End: Lossless Line

Letting Z_R approach infinity will reduce the reflection coefficient to unity. In terms of instantaneous quantities, taken from Eqs. 6-17 and 6-18, with α and θ_0 both set equal to zero,

$$\begin{aligned} v(z',t) &= V_{IM}[\cos(\omega t + \beta z') + \cos(\omega t - \beta z')] \\ &= 2V_{IM} \cos \omega t \, \cos \beta z' \end{aligned} \tag{6-20}$$

$$\begin{aligned} i(z',t) &= \frac{V_{IM}}{Z_0}[\cos(\omega t + \beta z') - \cos(\omega t - \beta z')] \\ &= -2\frac{V_{IM}}{Z_0} \sin \omega t \sin \beta z' \end{aligned} \tag{6-21}$$

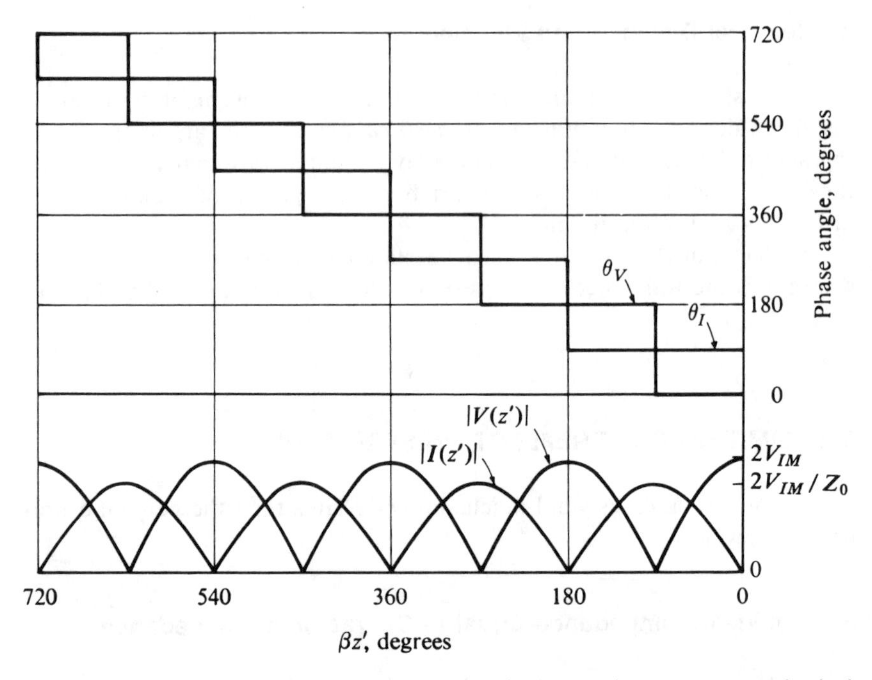

Figure 6-7. Magnitudes and relative phase angles of voltage and current on open-circuited lossless transmission line.

In terms of phasors, from Eqs. 6-15 and 6-16,

$$V(z') = V_{IM}(\epsilon^{j\beta z'} + \epsilon^{-j\beta z'})$$

$$= 2V_{IM} \cos \beta z' \tag{6-22}$$

$$I(z') = \frac{V_{IM}}{Z_0}(\epsilon^{j\beta z'} - \epsilon^{-j\beta z'})$$

$$= j\frac{2V_{IM}}{Z_0} \sin \beta z' \tag{6-23}$$

Plots of phasor magnitudes and relative phase angles are given in Fig. 6-7. It should be noted that resultant voltage and current are in time quadrature with each other at all locations; in the quarter-wavelength of line nearest the open-circuited end the current leads the voltage, and in the next quarter-wavelength the current lags the voltage. This pattern is repeated every half-wavelength. It should also be noted that the resultant current and resultant voltage each dip to zero every half-wavelength, but the standing waves are displaced a quarter-wavelength such that the resultant current is a maximum at those locations at which voltage is zero, and vice versa.

c. Open-Circuited End: Line with Loss

When losses are present, less dramatic simplifications result from substituting the open-circuit reflection coefficient in Eqs. 6-14 and 6-15:

$$V(z') = V_{IM}(\epsilon^{\gamma z'} + \epsilon^{-\gamma z'})$$

$$I(z') = \frac{V_{IM}}{Z_0}(\epsilon^{\gamma z'} - \epsilon^{-\gamma z'}) \tag{6-24}$$

These may be reduced to rectangular complex form.

$$\begin{aligned} V(z') &= V_{IM}(\epsilon^{\alpha z'} e^{j\beta z'} + \epsilon^{-\alpha z'}\epsilon^{-j\beta z'}) \\ &= V_{IM}[\epsilon^{\alpha z'}(\cos\beta z' + j\sin\beta z') + \epsilon^{-\alpha z'}(\cos\beta z' - j\sin\beta z')] \\ &= V_{IM}[(\epsilon^{\alpha z'} + \epsilon^{-\alpha z'})\cos\beta z' + j(\epsilon^{\alpha z'} - \epsilon^{-\alpha z'})\sin\beta z'] \end{aligned} \tag{6-25}$$

Similarly

$$I(z') = \frac{V_{IM}}{Z_0}[(\epsilon^{\alpha z'} - \epsilon^{-\alpha z'})\cos\beta z' + j(\epsilon^{\alpha z'} + \epsilon^{-\alpha' z'})\sin\beta z'] \tag{6-26}$$

Plots of the phasor magnitudes and relative phase angles of the resultant voltage and current on an open-circuited lossy line, computed from Eqs. 6-25 and 6-26, are given in Fig. 6-8. Significant differences between these curves and those in Fig. 6-7 are (1) the standing-wave patterns are not recurrent with distance along the line; (2) the curves of magnitude no longer touch the axis at voltage or current minima, and they are rounded at those locations rather than forming sharp cusps; and (3) the plots of phase angles are no longer discontinuous at half-wavelength intervals and uniform in between; rather they change continuously at varying, but always finite, nonzero rates.

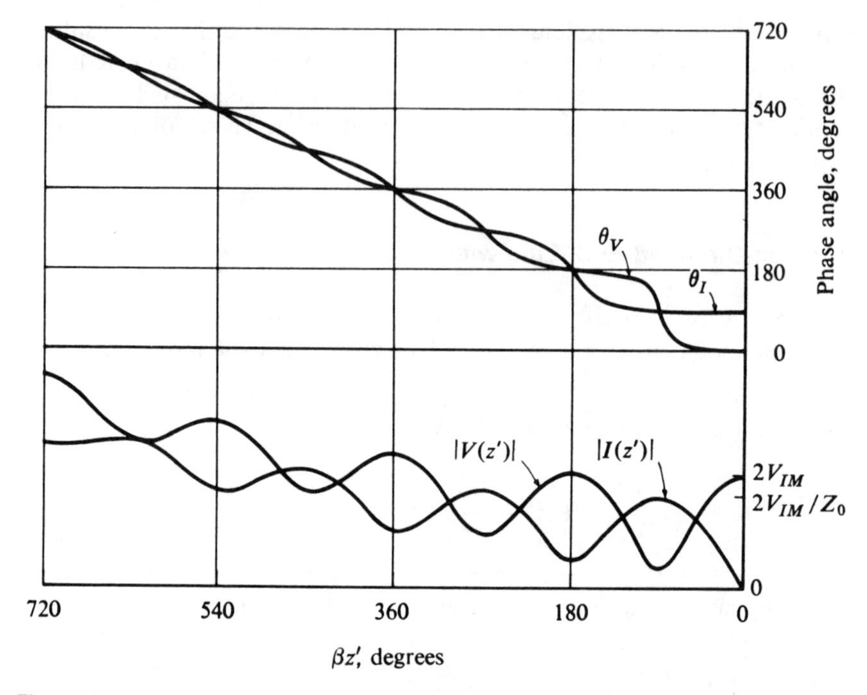

Figure 6-8. Magnitudes and relative phase angles of voltage and current on open-circuited lossy transmission line: $\alpha/\beta = 0.1$.

d. Short-Circuited End: Lossless Line

With Z_R set equal to zero, the reflection coefficient becomes -1. If the same development as for Eqs. 6-22 and 6-23 is used, the following results may be written:

$$V(z') = j2V_{IM} \sin \beta z'$$

$$I(z') = 2\frac{V_{IM}}{Z_0} \cos \beta z' \tag{6-27}$$

Plots of phasor magnitude and phase angle of voltage and current are given in Fig. 6-9. Comparison of these with Fig. 6-7 indicates that the patterns are identical if a shift of a quarter-wavelength in $\beta z'$ is made.

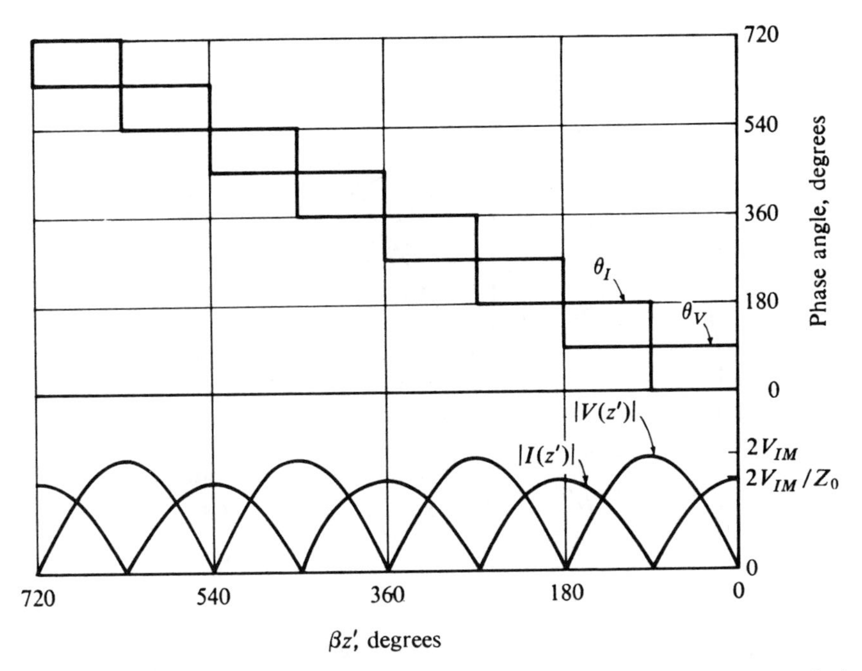

Figure 6-9. Magnitudes and relative phase angles of voltage and current on short-circuited lossless transmission line.

e. Pure Reactance as Termination: Lossless Line

A lossless line will necessarily have a purely resistive characteristic impedance, and that, combined with a purely reactive Z_R will invariably give a reflection coefficient whose magnitude is unity. The angle θ_K will depend on the relative magnitudes of Z_0 and Z_R, namely, R_0 and X_R. From Eq. 6-12,

$$K = \frac{jX_R - R_0}{jX_R + R_0}$$

$$\begin{aligned}\theta_K &= \tan^{-1}\left(\frac{X_R}{-R_0}\right) - \tan^{-1}\left(\frac{X_R}{R_0}\right) \\ &= \frac{\pi}{2} + \tan^{-1}\left(\frac{R_0}{X_R}\right) - \frac{\pi}{2} + \tan^{-1}\left(\frac{R_0}{X_R}\right) \\ &= 2\tan^{-1}\left(\frac{R_0}{X_R}\right) \qquad (6\text{-}28)\end{aligned}$$

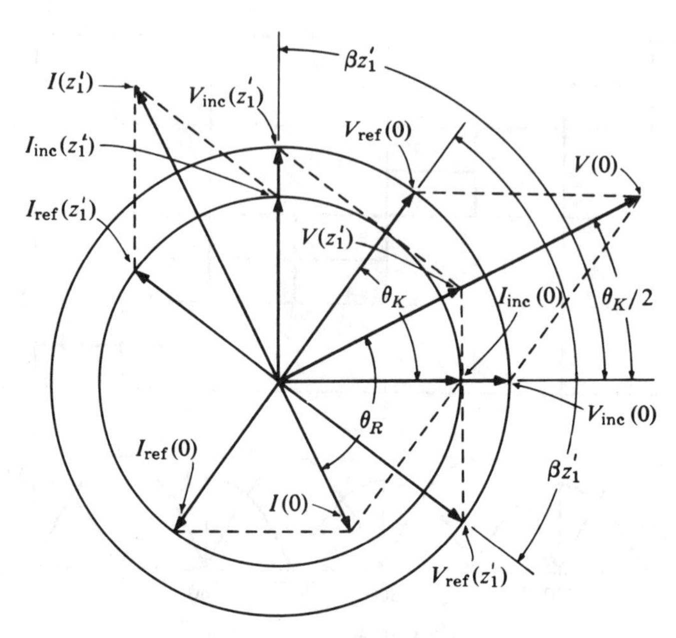

Figure 6-10. Phasor loci for lossless transmission line with purely reactive termination: $Z_0 = 600\underline{/0°}\ \Omega$, $Z_R = 1200\underline{/90.0°}\ \Omega$; ($K = 1.0\underline{/53.2°}$).

In an example illustrated in Figs. 6-10 and 6-11, Z_R equals $0 + j1200\ \Omega$ and Z_0 equals $600 + j0\ \Omega$. Thus the reflection coefficient is $1.0\underline{/53.2°}$. The general equations for voltage and current may be reduced as follows:

$$V(z') = V_{IM}\{\exp(j\beta z') + \exp[j(\theta_K - \beta z')]\}$$

$$= V_{IM}\exp(j\theta_K/2)\{\exp[j(\beta z' - \theta_K/2)] + \exp[-j(\beta z' - \theta_K/2)]\}$$

$$V(z') = 2V_{IM}\exp(j\theta_K/2)\cos\left(\beta z' - \frac{\theta_K}{2}\right) \tag{6-29}$$

$$I(z') = 2\frac{V_{IM}}{Z_0}j\exp(j\theta_K/2)\sin\left(\beta z' - \frac{\theta_K}{2}\right) \tag{6-30}$$

Comparison of Fig. 6-11 with Fig. 6-7 indicates that the difference between the standing-wave pattern with a purely reactive termination and that with an open-circuit termination is a shift of $\theta_K/2$ in $\beta z'$.

The open-circuit termination, the short-circuit termination, and the purely reactive termination have an important physical property in common—they are lossless. And as has been noted, when used with a lossless line they all yield reflection coefficients with magnitudes of unity. For conciseness of expression, the term "lossless termination" will be used

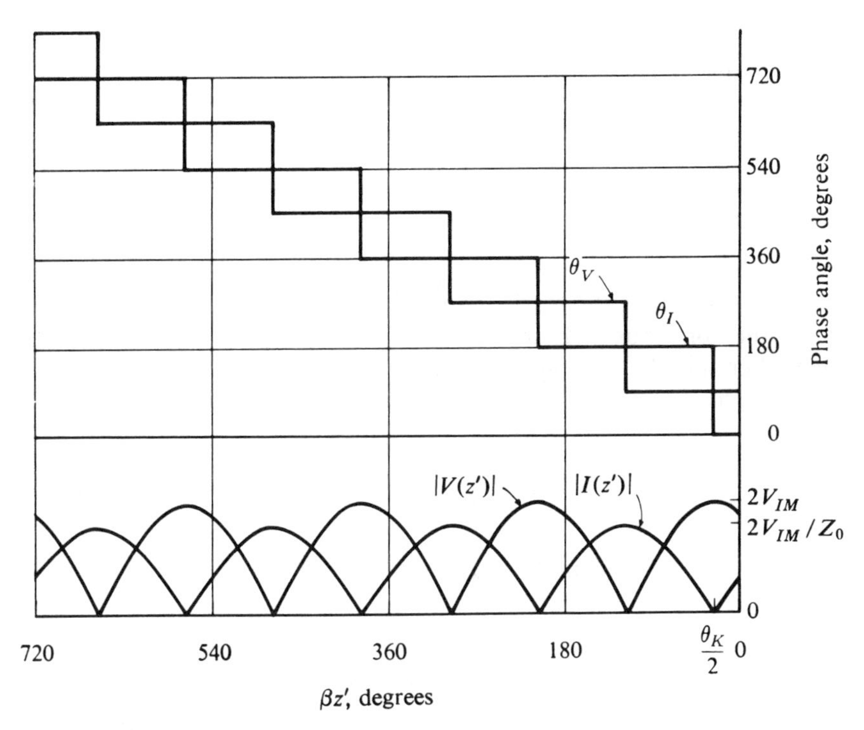

Figure 6-11. Magnitudes and relative phase angles of voltage and current on lossless transmission line with purely reactive termination: $Z_0 = 600\angle 0°\ \Omega$, $Z_R = 1200\angle 90.0°\ \Omega$; $K = 1.0\angle 53.2°$).

hereafter in statements which are applicable interchangeably to open-circuit, short-circuit, and purely reactive terminations.

6-3 STANDING-WAVE RATIO

Standing waves result from the simultaneous presence of waves traveling in both directions on a line. The *standing-wave* ratio is a quantity which is easily found from measured voltage or current data, and it is functionally related to the magnitude of the reflection coefficient.

The ratio of the maximum resultant voltage (phasor magnitude or rms) to the minimum resultant voltage along a transmission line is called the *voltage-standing-wave ratio*, abbreviated VSWR. The corresponding ratio for currents is called the *current-standing-wave ratio*, ISWR. On a lossless line, all rms voltage maxima are equal, and so are all rms voltage minima, rms current maxima, and rms current minima. Thus the ratio between any voltage maximum and any voltage minimum would yield the same numerical value, and the same is true of the ratio between any current maximum and any current minimum:

$$\text{VSWR} = \frac{|V(z')|_{\max}}{|V(z')|_{\min}} \tag{6-31}$$

$$\text{ISWR} = \frac{|I(z')|_{\max}}{|I(z')|_{\min}} \tag{6-32}$$

On a lossless line, the maximum resultant rms voltage or current occurs where the corresponding incident and reflected waves are in phase with each other, and the minimum resultant rms voltage or current occurs when they are phase-opposed. Thus

$$|V(z')|_{\max} = V_{IM}(1 + |K|) \tag{6-33}$$

$$|V(z')|_{\min} = V_{IM}(1 - |K|) \tag{6-34}$$

$$|I(z')|_{\max} = \frac{V_{IM}}{Z_0}(1 + |K|) \tag{6-35}$$

$$|I(z')|_{\min} = \frac{V_{IM}}{Z_0}(1 - |K|) \tag{6-36}$$

From Fig. 6-5, or from the corresponding diagram for a general termination, one may see that the voltage maxima occur at the same values of $\beta z'$ (location on the line) as do the current minima, and vice versa.

If Eqs. 6-33 through 6-36 are substituted in Eqs. 6-31 and 6-32, one finds that the standing-wave ratio of a lossless line is the same for both voltage and current and that it is related to the magnitude of the reflection coefficient in the following manner:

$$\text{SWR} = \frac{1 + |K|}{1 - |K|} \tag{6-37}$$

The lossless line with a lossless termination will have a reflection coefficient of magnitude unity; hence the standing-wave ratio will approach infinity. At the other extreme, a lossless line terminated with a resistor equal to the characteristic impedance will have a reflection coefficient of zero and a standing-wave ratio of unity; such a line is said to be "flat."

When losses are present, as indicated in Fig. 6-8, the minima located at successively greater distances from the terminated end become progressively greater. If the line is sufficiently long, a point will be found beyond which the slopes of the current and voltage magnitude plots are continuously upward, although with varying steepness, and no minima exist.

Clearly the standing-wave ratio as a function lacks a unique value on a high-loss line, and the quantity must be interpreted with some discretion on any practical line.

6-4 RESULTANT STANDING-WAVE PHASORS — HYPERBOLIC FUNCTIONS

Equations may be derived for the resultant input voltage and current phasors of a line or network in terms of the resultant and output phasors. Those equations involve sums and differences of $\exp(\gamma z')$ and $\exp(-\gamma z')$; substitution of hyperbolic functions of $\gamma z'$ for such combinations is possible and is sometimes advantageous, although physical interpretation in terms of traveling-wave components is obscured thereby.

The input-vs.-output-phasor relationships may be readily organized in matrix form (*ABCD* matrix).

a. Hyperbolic Functions of a Complex Argument

The hyperbolic sine, cosine, and tangent are best regarded for our purposes simply as algebraic combinations of exponentials with real arguments, just as trigonometric functions may be regarded as algebraic combinations of exponentials with imaginary arguments:

$$\sinh \alpha z' = \frac{\epsilon^{\alpha z'} - \epsilon^{-\alpha z'}}{2} \tag{6-38}$$

$$\cosh \alpha z' = \frac{\epsilon^{\alpha z'} + \epsilon^{-\alpha z'}}{2} \tag{6-39}$$

$$\tanh \alpha z' = \frac{\sinh \alpha z'}{\cosh \alpha z'} \tag{6-40}$$

$$\sin \beta z' = \frac{\epsilon^{j\beta z'} - \epsilon^{-j\beta z'}}{2j} \tag{6-41}$$

$$\cos \beta z' = \frac{\epsilon^{j\beta z'} + \epsilon^{-j\beta z'}}{2} \tag{6-42}$$

$$\tan \beta z' = \frac{\sin \beta z'}{\cos \beta z'} \tag{6-43}$$

The definitions given in Eqs. 6-38, 6-39, and 6-40 may be extended to include a complex argument:

$$\sinh \gamma z' = \frac{\epsilon^{\gamma z'} - \epsilon^{-\gamma z'}}{2} \tag{6-44}$$

$$\cosh \gamma z' = \frac{\epsilon^{\gamma z'} + \epsilon^{-\gamma z'}}{2} \tag{6-45}$$

$$\tanh \gamma z' = \frac{\sinh \gamma z'}{\cosh \gamma z'} \tag{6-46}$$

Also

$$\coth \gamma z' = 1/\tanh \gamma z' \tag{6-47}$$

$$\operatorname{csch} \gamma z' = 1/\sinh \gamma z' \tag{6-48}$$

If $\gamma z'$ is purely imaginary, Eqs. 6-44 to 6-46 reduce to

$$\sinh j\beta z' = j \sin \beta z' \tag{6-49}$$

$$\cosh j\beta z' = \cos \beta z' \tag{6-50}$$

$$\tanh j\beta z' = j \tan \beta z' \tag{6-51}$$

Numerical values of the hyperbolic functions of a real argument may be assembled in compact single-entry tables, but for complex arguments a double-entry situation arises. This is because the real part and the imaginary part of a general complex argument may be chosen independently. Fortunately the hyperbolic functions of a complex argument may be resolved into combinations of hyperbolic functions of the real part of the argument and trigonometric functions of the imaginary part, as follows.

$$\begin{aligned} \sinh(\alpha z' + j\beta z') &= \sinh \alpha z' \cosh j\beta z' + \cosh \alpha z' \sinh j\beta z' \\ &= \sinh \alpha z' \cos \beta z' + j \cosh \alpha z' \sin \beta z' \end{aligned} \tag{6-52}$$

$$\begin{aligned} \cosh(\alpha z' + j\beta z') &= \cosh \alpha z' \cosh j\beta z' + \sinh \alpha z' \sinh j\beta z' \\ &= \cosh \alpha z' \cos \beta z' + j \sinh \alpha z' \sin \beta z' \end{aligned} \tag{6-53}$$

$$\begin{aligned} \tanh(\alpha z' + j\beta z') &= \frac{\tanh \alpha z' + \tanh j\beta z'}{1 + \tanh \alpha z' \tanh j\beta z'} \\ &= \frac{\tanh \alpha z' + j \tan \beta z'}{1 + j \tanh \alpha z' \tan \beta z'} \end{aligned} \tag{6-54}$$

The following power series and asymptotes for hyperbolic functions with real arguments may be noted:

$$\sinh u = u + \frac{u^3}{3!} + \frac{u^5}{5!} + \frac{u^7}{7!} + \cdots \tag{6-55}$$

$$\sinh u \to 0.5\epsilon^u \qquad u \to \infty \tag{6-56}$$

$$\cosh u = 1 + \frac{u^2}{2} + \frac{u^4}{4!} + \frac{u^6}{6!} + \cdots \tag{6-57}$$

$$\cosh u \to 0.5\epsilon^u \qquad u \to \infty \tag{6-58}$$

$$\tanh u \approx u - \frac{u^3}{3} \qquad u \ll 1 \tag{6-59}$$

$$\tanh u \to 1 - 2\epsilon^{-2u} \qquad u \to \infty \tag{6-60}$$

Some expressions, such as Eqs. 6-24 for voltage and current on an open-circuited lossy line, may be made more concise by substituting Eqs. 6-44 and 6-45:

$$V(z')_{\text{oc}} = 2V_{IM} \cosh \gamma z' \tag{6-61}$$

$$I(z')_{\text{oc}} = 2\frac{V_{IM}}{Z_0} \sinh \gamma z' \tag{6-62}$$

It should be recognized that an algebraic substitution, such as the replacing of a combination of exponential functions with an equivalent combination of hyperbolic functions, does not alter any physical laws or behavior.

b. Standing Waves in Terms of Terminal Voltage and Current

Expressions for voltage and current along the line in terms of receiving-end voltage or current and hyperbolic functions of $\gamma z'$ may also be derived.

From Eqs. 6-2, 6-7, 6-12, and 6-14,

$$V_{\text{ter}} = \frac{2V_{IM}Z_R}{Z_R + Z_0} \tag{6-63}$$

If this result is solved for V_{IM} and substituted in Eq. 6-15, and Eq. 6-12 is also substituted in Eq. 5-14, the following may be obtained:

$$V(z') = \frac{V_{\text{ter}}}{2Z_R}\left[(Z_R + Z_0)\epsilon^{\gamma z'} + (Z_R - Z_0)\epsilon^{-\gamma z'}\right]$$

Next, Eqs. 6-44 and 6-45 may be substituted:

$$V(z') = \frac{V_{\text{ter}}}{Z_R}(Z_R \cosh \gamma z' + Z_0 \sinh \gamma z') \tag{6-64}$$

Similarly,

$$I(z') = \frac{V_{\text{ter}}}{Z_R Z_0}(Z_R \sinh \gamma z' + Z_0 \cosh \gamma z') \tag{6-65}$$

If V_{ter}/Z_R is replaced by I_{ter}, the following may be obtained:

$$V(z') = V_{\text{ter}} \cosh \gamma z' + Z_0 I_{\text{ter}} \sinh \gamma z' \tag{6-66}$$

$$I(z') = \frac{V_{\text{ter}}}{Z_0} \sinh \gamma z' + I_{\text{ter}} \cosh \gamma z' \tag{6-67}$$

For a lossless line, substitution of Eqs. 6-49 and 6-50 will reduce these to the following:

$$V(z') = V_{\text{ter}} \cos \beta z' + jZ_0 I_{\text{ter}} \sin \beta z' \tag{6-68}$$

$$I(z') = \frac{jV_{\text{ter}}}{Z_0} \sin \beta z' + I_{\text{ter}} \cos \beta z' \tag{6-69}$$

c. The *ABCD* matrix

The *ABCD* matrix organizes line or network transfer functions in a manner adaptable to tandem or cascade arrangements. The basic form is illustrated in Fig. 6-12. Here

$$\begin{bmatrix} V_{IN} \\ I_{IN} \end{bmatrix} = \begin{bmatrix} A & B \\ C & D \end{bmatrix} = \begin{bmatrix} V_{OUT} \\ I_{OUT} \end{bmatrix} \tag{6-70}$$

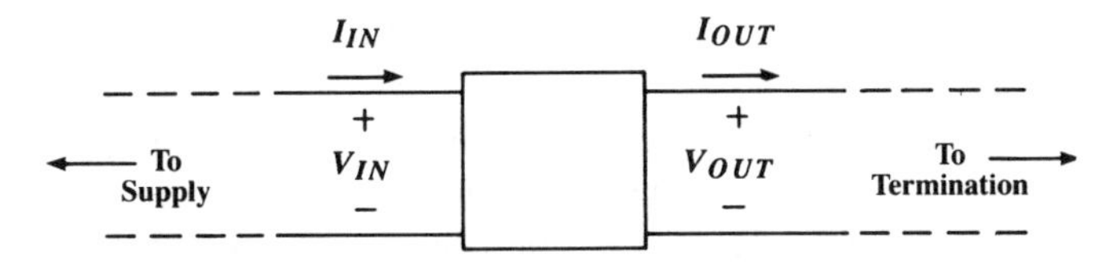

Figure 6-12. Two-terminal-pair network: polarity convention for *ABCD* matrix.

Equations 6-66 and 6-67 for a section of transmission line may be restated in matrix form thus:

$$\begin{bmatrix} V(z') \\ I(z') \end{bmatrix} = \begin{bmatrix} A & B \\ C & D \end{bmatrix} = \begin{bmatrix} V_{\text{ter}} \\ I_{\text{ter}} \end{bmatrix} \tag{6-71}$$

where

$$\begin{bmatrix} A & B \\ C & D \end{bmatrix} = \begin{bmatrix} \cosh \gamma z' & Z_0 \sinh \gamma z' \\ (1/Z_0)\sinh \gamma z' & \cosh \gamma z' \end{bmatrix} \tag{6-72}$$

For the lossless line this reduces to

$$\begin{bmatrix} A & B \\ C & D \end{bmatrix} = \begin{bmatrix} \cos \beta z' & jZ_0 \sin \beta z' \\ (j/Z_0)\sin \beta z' & \cos \beta z' \end{bmatrix} \tag{6-73}$$

If the network is composed of bilateral elements (voltage versus current relationship the same for both directions of current) the following is valid:

$$\begin{bmatrix} V_{OUT} \\ I_{OUT} \end{bmatrix} = \begin{bmatrix} A & B \\ C & D \end{bmatrix}^{-1} \begin{bmatrix} V_{IN} \\ I_{IN} \end{bmatrix} \tag{6.74}$$

where

$$\begin{bmatrix} A & B \\ C & D \end{bmatrix}^{-1} = \begin{bmatrix} D & -B \\ -C & A \end{bmatrix} (AD - BC) \tag{6-75}$$

Substitution of the elements in Eqs. 6-72 or 6-73 yields

$$AD - BC = 1 \tag{6-76}$$

This relationship proves to be true in general.[1] Hence

$$\begin{bmatrix} V_{OUT} \\ I_{OUT} \end{bmatrix} = \begin{bmatrix} D & -B \\ -C & A \end{bmatrix} \begin{bmatrix} V_{IN} \\ I_{IN} \end{bmatrix} \tag{6-77}$$

Matrix formulation has an advantage in that it is readily adaptable to computer programming.

6-5 LUMPED-IMPEDANCE DISCONTINUITIES

Thus far only a simple type of discontinuity, that of a lumped-impedance termination, has been considered. More general situations include the following: (1) the tandem connection of two lines of differing characteristic impedances, (2) a lumped impedance connected from one conductor to the other midway along a lone, (3) lumped impedances inserted in series with each conductor midway along a line, and (4) three or more lines converging at a junction. Discontinuities may be treated more directly in terms of impedances or admittances (as will be done in Chap. 7), rather than the voltage and current functions. For the sake of additional insight, two elementary examples will be examined here from the voltage-and-current point of view.

a. Shunt Discontinuity

As an example of a line with a shunt discontinuity, consider the circuit shown in Fig. 6-13. Here a lossless line is terminated with impedance Z_R, and a shunt resistor R_C is connected between the conductors one-sixth of a wavelength from the termination. Let $Z_0 = 600\ \Omega$, $Z_R = 250\ \Omega$, and $R_C = 400\ \Omega$. $z'_{\mathrm{IC}} = \pi/(3\beta)$. Find the voltage and current at $z'_{\mathrm{II}} = 0$.

From Eqs. 6-68 and 6-69

$$V_{\mathrm{I}}(z'_{\mathrm{IC}}) = V_{\mathrm{I}}(0)\left(\cos 60.0^\circ + j\frac{600}{250}\sin 60.0^\circ\right)$$

$$= V_{\mathrm{I}}(0)(0.500 + j2.078)$$

$$I_{\mathrm{I}}(z'_{\mathrm{IC}}) = V_{\mathrm{I}}(0)\left(\frac{j\sin 60.0^\circ}{600} + \frac{\cos 60.0^\circ}{250}\right)$$

$$= V_{\mathrm{I}}(0)(2.00 + j1.44) \times 10^{-3}$$

$$I_C = \frac{V_{\mathrm{I}}(z'_{\mathrm{IC}})}{R_C}$$

$$= V_{\mathrm{I}}(0)\frac{0.500 + j2.078}{400}$$

$$= V_{\mathrm{I}}(0)(1.25 + j5.20) \times 10^{-3}$$

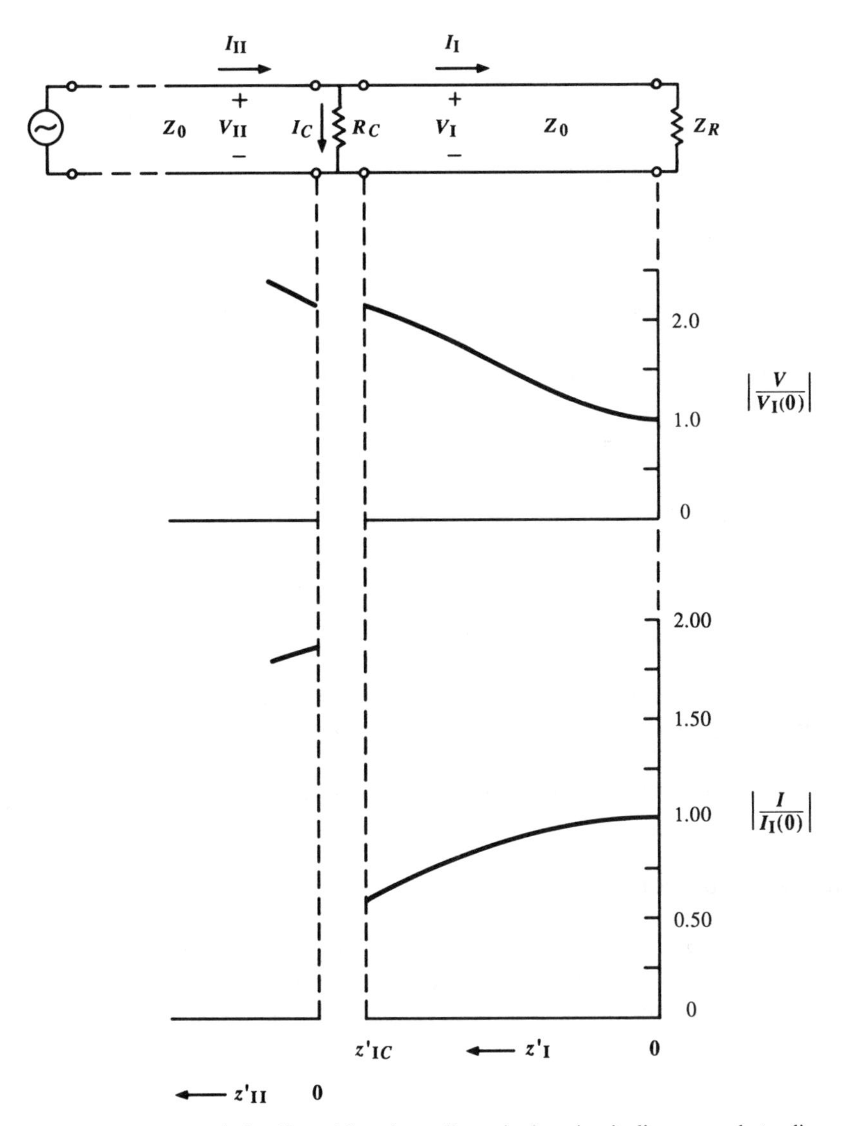

Figure 6-13. Transmission line with a shunt discontinuity; circuit diagram and standing-wave patterns ($Z_R = 0.417Z_0$; $R_C = 0.667Z_0$).

At the left-hand side of the junction,

$$V_{\mathrm{II}}(0) = V_{\mathrm{I}}(z'_{\mathrm{IC}})$$

$$= 2.14\underline{/76.5^\circ}\, V_{\mathrm{I}}(0)$$

$$I_{\mathrm{II}}(0) = I_{\mathrm{I}}(z'_{\mathrm{IC}}) + I_C$$

$$= V_{\mathrm{I}}(0)(2.00 + j1.44 + 1.25 + j5.20) \times 10^{-3}$$

$$= V_{\mathrm{I}}(0)(3.25 + j6.64) \times 10^{-3}$$

$$= 7.39 \times 10^{-3}\underline{/63.9^\circ}\, V_{\mathrm{I}}(0)$$

In matrix form

$$\begin{bmatrix} V_{\mathrm{II}}(0) \\ I_{\mathrm{II}}(0) \end{bmatrix} = \begin{bmatrix} 1 & 0 \\ 1/R_C & 1 \end{bmatrix} \begin{bmatrix} V_{\mathrm{I}}(z'_{\mathrm{IC}}) \\ I_{\mathrm{I}}(z'_{\mathrm{IC}}) \end{bmatrix} \tag{6-78}$$

Voltage and current at points to the left of the discontinuity may be found by applying Eqs. 6-71 and 6-73, with $V_{\mathrm{II}}(0)$ and $I_{\mathrm{II}}(0)$ as V_{OUT} and I_{OUT}, respectively. Problem 6-8 calls for some calculations of this type. Standing waves of $|V|$ and $|I|$ have been plotted in Fig. 6-13.

b. Series Discontinuity

In Fig. 6-14 series inductors have been inserted in each conductor of a lossless transmission line. For simplicity the terminal impedance has been set equal to the characteristic impedance of the line, 600 Ω. Let $\omega L_A =$

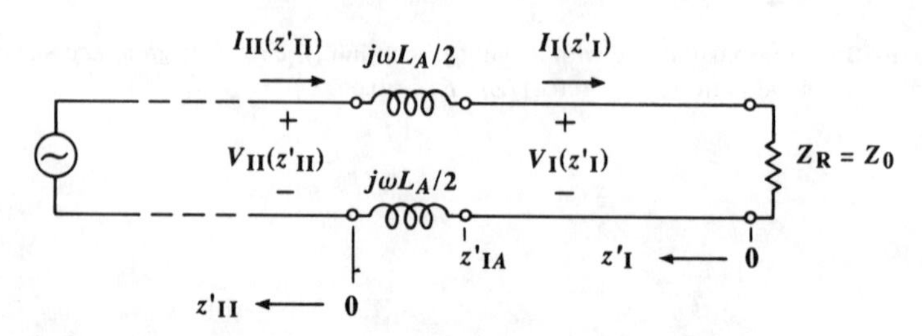

Figure 6-14. Terminated transmission line with series discontinuity.

400 Ω and $z'_{IA} = (\pi/3\beta)$. Substitution in Eqs. 6-68 and 6-69 yields

$$V_I(z'_{IA}) = V_I(0)\cos 60.0° + jZ_0 I_I(0)\sin 60.0°$$

$$I_I(0) = V_I(0)/Z_R$$

$$= 1.667 \times 10^{-3} V_I(0)$$

$$V_I(z'_{IA}) = V_I(0)(\cos 60.0° + j\sin 60.0°)$$

Similarly

$$I_I(z'_{IA}) = 1.667 \times 10^{-3} V_I(0)(\cos 60.0° + j\sin 60.0°)$$

In matrix form

$$\begin{bmatrix} V_{II}(0) \\ I_{II}(0) \end{bmatrix} = \begin{bmatrix} 1 & j\omega L_A \\ 0 & 1 \end{bmatrix} \begin{bmatrix} V_I(z'_{IA}) \\ I_I(z'_{IA}) \end{bmatrix} \tag{6.79}$$

Problem 6-9 calls for some additional calculations and a sketch of the standing-wave pattern.

6-6 CONCLUSIONS

The steady-state voltage and current functions existing on a transmission line with any terminating impedance may be expressed in terms of two oppositely moving traveling-wave component functions. The traveling-wave functions approaching and departing from the termination in question are known as the incident and reflected components, respectively. The ratio of the reflected-voltage phasor to the incident-voltage phasor, computed at the terminus, is known as the *reflection coefficient*. It is the following function of the load impedance Z_R and the characteristic impedance Z_0:

$$K = \frac{Z_R - Z_0}{Z_R + Z_0} \tag{6-12}$$

(In general Z_0, Z_R, and K are all complex quantities.)

At points away from the terminus, toward the source of the incident wave, the incident voltage and current phasors are increased in magnitude by the factor $\epsilon^{\alpha z'}$ and lead the corresponding phasors at the termination by the angle $\beta z'$. Reflected-wave phasors, on the other hand, are reduced in magnitude by the factor $\epsilon^{-\alpha z'}$ and lag behind the corresponding phasors at the termination by the same angle $\beta z'$.

In the event that the line is lossless and the termination too is lossless (open circuit, short circuit, or pure reactance), the reflection coefficient has a magnitude of unity. The rms values of resultant voltage and current vary with distance as sinusoidal half-wave loops, the voltage nulls occurring at the same locations as the current maxima, and vice versa. Within each half-wave span between nulls, the voltages at all points are in time phase with one another. Voltages at points with one null between them are in phase opposition. The same principle applies to the phase alignment among currents at various locations, but voltage and current are in time quadrature with each other at every location.

Losses in the line or in the load change the situation, in that the minima of rms voltage and current do not reach zero, and the phase angle of resultant voltage or current with respect to a given reference varies continuously, although in general not uniformly, with distance.

The ratio of maximum resultant rms voltage to minimum resultant rms voltage along a lossless line is equal to the corresponding ratio between resultant rms currents, and it is known as the *standing-wave ratio*, SWR. It is related to the magnitude of the reflection coefficient

$$\text{SWR} = \frac{1 + |K|}{1 - |K|} \tag{6-37}$$

Hyperbolic functions of $\gamma z'$ are defined by combinations of exponential functions of $\gamma z'$ and $-\gamma z'$, which parallel the definitions of the trigonometric functions in terms of exponential functions with imaginary arguments. Hyperbolic functions of a complex argument may be resolved into combinations of hyperbolic functions of the real part of the argument and trigonometric functions of the imaginary part. Relevant formulas are listed in Eqs. 6-38 through 6-54.

The *ABCD* transmission matrix is convenient for relating line or network input voltage and current to output voltage and current:

$$\begin{bmatrix} V_{IN} \\ I_{IN} \end{bmatrix} = \begin{bmatrix} A & B \\ C & D \end{bmatrix} \begin{bmatrix} V_{OUT} \\ I_{OUT} \end{bmatrix} \tag{6-70}$$

The transmission matrix for a section of transmission line has the following elements:

$$\begin{bmatrix} A & B \\ C & D \end{bmatrix} = \begin{bmatrix} \cosh \gamma z' & Z_0 \sinh \gamma z' \\ (1/Z_0) \sinh \gamma z' & \cosh \gamma z' \end{bmatrix} \tag{6-72}$$

PROBLEMS

6-1. A lossless transmission line is under steady-state AC excitation as in Fig. 6-1. $\gamma = 0 + j2.50 \times 10^{-3}\ \text{m}^{-1}$.

$$V_{\text{inc}}(0) = 20.0\epsilon^{j0.0°}\ \text{V}$$

$$V_{\text{ref}}(0) = 11.7\epsilon^{-j78.0°}\ \text{V}$$

$$I_{\text{inc}}(0) = 4.00 \times 10^{-2}\epsilon^{j0.0°}\ \text{A}$$

$$I_{\text{ref}}(0) = 2.34 \times 10^{-2}\epsilon^{j102.0°}\ \text{A}$$

Find (a) the complex reflection coefficient, (b) the characteristic impedance of the line, (c) the resultant voltage and current $V(0)$ and $I(0)$, (d) the terminating impedance, (e) the distance to the nearest maximum of the resultant voltage, and (f) the corresponding resultant voltage and current. Draw a phasor diagram for $z' = 0$, and for the location of the maximum resultant voltage.

6-2. See Fig. 6-2. Assume that the line described in the caption has the following properties: $\gamma = (0.684 + j1.865) \times 10^{-3}\ \text{m}^{-1}$; $V_{\text{inc}}(0) = 20\underline{/0°}$ V. At what distances from the termination will V_{inc} and V_{ref} be (1) in phase with each other or (2) phase opposed? What are their magnitudes at the location of each type closest to the termination? Draw phasor diagrams showing the incident, reflected, and resultant voltages and currents at those locations.

6-3. From Fig. 6-7, sketch $v(t)$ and $i(t)$ at the locations corresponding to $\beta z'$ equal to (1) $\pi/4$, (2) $\pi/2$, and (3) $3\pi/4$. Use a common abscissa scale.

6-4. Construct a phasor diagram similar to Fig. 6-10 for a lossless line with a short-circuit termination at $z' = 0$, showing phasors for (1) $\beta z' = 0$ and (2) $\beta z' = 75°$.

6-5. Given a lossy transmission line as shown in Fig. 6-1. Let

$$v(z', t) = 3.00\epsilon^{4.00\times10^{-4}z'}\cos(8.00 \times 10^5 t + 25.0° + 3.20 \times 10^{-3}z')$$

$$+ 1.80\epsilon^{-4.00\times10^{-4}z'}\cos(8.00 \times 10^5 t - 15.0° - 3.20 \times 10^{-3}z')\ \text{V}$$

(a) Write the corresponding phasor $V(z')$.
(b) Find the reflection coefficient.
(c) Find the phase velocity.
(d) If the characteristic impedance is $500\underline{/-7.0°}\ \Omega$, write the corresponding current phasor $I(z')$.

6-6. A lossless transmission line ($Z_0 = 600\angle 0°\ \Omega$) is terminated at $z' = 0$ with a pure capacitance of 2.0×10^{-8} F.

$$v_{\text{inc}}(z', t) = 50 \cos(6.28 \times 10^4 t + 2.09 \times 10^{-4} z')$$
$$+ 40 \cos(1.256 \times 10^5 t + 4.18 \times 10^{-4} z') \text{ V}$$

(Here z' is in meters.) Draw phasor diagram(s) showing the incident, reflected, and resultant voltages and currents at $z' = 0$, and write an expression for the voltage across the capacitor as a function of time.

6-7. A lossless line with an electrical length of more than 2π radians has a characteristic impedance of $700\angle 0°\ \Omega$, and is terminated with an impedance $Z_R = 300 - j900\ \Omega$. What value of standing-wave ratio would be measured on the line?

6-8. Refer to the example in Sec. 6-5a and Fig. 6-13 involving a shunt discontinuity on a transmission line. Using the data in that example, calculate, with the *ABCD* matrix, the resultant voltage and current at $z'_{\text{II}} = \pi/(4\beta)$. Draw a phasor diagram which shows the voltage and current at $z'_{\text{I}} = 0$, $z'_{\text{I}} = z'_{\text{I}C}$, $z'_{\text{II}} = 0$, and $z'_{\text{II}} = \pi/(4\beta)$.

6-9. Refer to the example in Sec. 6-5b and Fig. 6-14 involving a series discontinuity on a transmission line. Using the data in that example, calculate the resultant voltage and current at $z'_{\text{II}} = 0$ and at $z'_{\text{II}} = \pi/(4\beta)$. Draw a phasor digram which shows the voltage and current at those locations and at $z'_{\text{I}} = 0$ and at $z'_{\text{I}} = z'_{\text{I}A}$. Sketch the standing wave pattern.

6-10. Assume that the circuit shown in Fig. 3-10 (two lines joined by a resistive network) is in steady state. Find the elements of the *ABCD* matrix for the tandem-linking network.

6-11. If a transmission line for which the characteristic impedance has a nonzero angle is to be terminated with a pure resistance, but reflections are to be kept to a minimum, what value of terminating resistance should be chosen? Find the corresponding reflection coefficient, and sketch a phasor diagram. *Answer*: $R_T = |Z_0|$.

REFERENCES

1. Van Valkenberg, M. E., *Network Analysis*, 3rd ed., Prentice-Hall, Inc., Englewood Cliffs, NJ, 1974, 331.

CHAPTER 7

IMPEDANCE, ADMITTANCE, AND THE SMITH* CHART

A most useful property of sinusoidal functions of the same frequency is that the relationship between any two functions may be expressed in terms of a single complex quantity. That quantity has (1) a magnitude, which has dimensions appropriate to the physical interpretation of the two functions, and (2) a phase-displacement angle.

Impedance is one such quantity; it relates voltage and current in linear circuit analysis, and its units are ohms. Admittance is simply the reciprocal of impedance, and has units of siemens. Calculations related to network transformations, such as the paralleling of branches or the changing from a "T" into an equivalent delta, may be made more concisely in terms of impedances or admittances than in terms of branch currents and voltage drops. This simplification is welcome, but it should be recognized that complex impedance or admittance is, in a sense, one step farther removed from physical reality than voltage or current phasors.

In transmission-line calculations, the impedance concept provides an expedient means of handling most discontinuities. Direct analysis in terms of voltage and current functions, as outlined in Sec. 6-5, is usually more cumbersome.

7-1 INPUT IMPEDANCE: GENERAL

The complex ratio of the voltage phasor at any location on a transmission line to the current phasor at the same location is dimensionally an impedance. Physically, on a line with an energy source at only one end,

*The word "Smith", used in connection with the chart which is reproduced in Fig. 7-9, is a trademark of Analog Instruments Co., New Providence, NJ.

that ratio is the impedance which would be observed if one opened both conductors at the given point and connected an impedance-measuring instrument (oscillator and bridge, for instance) to the pair of conductors leading to the terminated end. This impedance-dimensioned ratio may be readily envisioned as a continuous function of distance from the termination; it is commonly spoken of as the *input impedance* of the line at the given location:

$$Z(z') = \frac{V(z')}{I(z')} \tag{7-1}$$

Equations 6-15 and 6-16, or 6-64 and 6-65, may be substituted (these are all based on the polarity-and-direction convention of Fig. 6-1):

$$Z(z') = Z_0 \frac{\epsilon^{\gamma z'} + K\epsilon^{-\gamma z'}}{\epsilon^{\gamma z'} - K\epsilon^{-\gamma z'}}$$

$$= Z_0 \frac{1 + K\epsilon^{-2\gamma z'}}{1 - K\epsilon^{-2\gamma z'}} \tag{7-2}$$

or

$$Z(z') = Z_0 \frac{Z_R \cosh \gamma z' + Z_0 \sinh \gamma z'}{Z_0 \cosh \gamma z' + Z_R \sinh \gamma z'} \tag{7-3}$$

Division of numerator and denominator by $\cosh \gamma z'$ yields

$$Z(z') = Z_0 \frac{Z_R + Z_0 \tanh \gamma z'}{Z_0 + Z_R \tanh \gamma z'} \tag{7-4}$$

If the line is lossless, Eqs. 7-2, 7-3, and 7-4 reduce to

$$Z(z') = Z_0 \frac{1 + K \exp(-j2\beta z')}{1 - K \exp(-j2\beta z')} \tag{7-5}$$

$$Z(z') = Z_0 \frac{Z_R \cos \beta z' + jZ_0 \sin \beta z'}{Z_0 \cos \beta z' + jZ_R \sin \beta z'} \tag{7-6}$$

$$Z(z') = Z_0 \frac{Z_R + jZ_0 \tan \beta z'}{Z_0 + jZ_R \tan \beta z'} \tag{7-7}$$

Equations for input admittance may be found by replacing each impedance in the foregoing equations by the reciprocal of its admittance and simplifying the resulting fractions. Thus from Eqs. 7-4 and 7-7,

$$Y(z') = Y_0 \frac{Y_R + Y_0 \tanh \gamma z'}{Y_0 + Y_R \tanh \gamma z'} \tag{7-8}$$

$$Y(z') = Y_0 \frac{Y_R + jY_0 \tan \beta z'}{Y_0 + jY_R \tan \beta z'} \quad \text{(lossless line)} \tag{7-9}$$

Variation of impedance or admittance on a lossless line is readily adaptable to a nomographic chart (Sec. 7-5).

It should be noted in passing that the quotient of an instantaneous voltage divided by an instantaneous current has no obvious physical meaning (unless they are in time phase with each other); in particular, one should realize that such a ratio is not a complex impedance. (As an illustration, divide Eq. 6-20 by 6-21, and note that the result cannot be equated to that found by dividing Eq. 6-22 by Eq. 6-23.)

7-2 LIMITING-CASE INPUT IMPEDANCES

Some of the limiting cases treated in Chap. 6 in terms of voltage and current functions may be reexamined here in terms of input impedance.

a. Terminating Impedance Equal to Characteristic Impedance

If Z_R is equal to Z_0, Eq. 7-3 will immediately reduce to

$$Z(z') = Z_0 \tag{7-10}$$

Thus any line which is terminated in its characteristic impedance will have that same value as its input impedance, regardless of how long the line is. This is a reflectionless termination, as noted in Sec. 6-2a.

b. Open-Circuited End

If Z_R approaches infinity, Eq. 7-4 will reduce to

$$Z_{oc}(z') = \frac{Z_0}{\tanh \gamma z'} \tag{7-11}$$

As an illustration of results, plots of $|Z_{oc}|$ and θ_Z are given in Fig. 7-1 for a

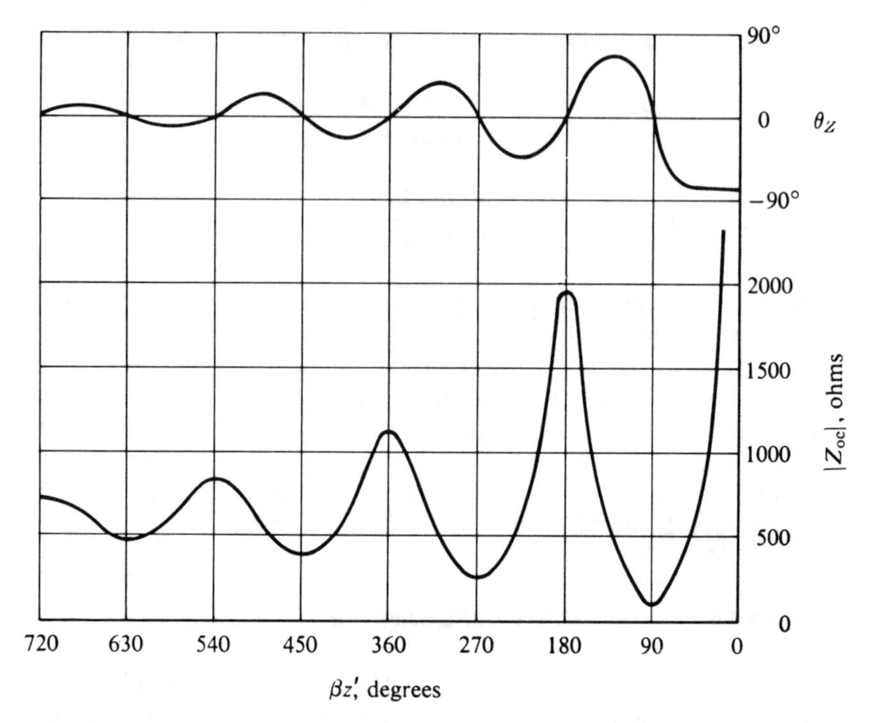

Figure 7-1. Open-circuit impedance of lossy transmission line: $\alpha/\beta = 0.1$, $Z_0 = 600\ \Omega$ (see Fig. 6-8).

lossy line, the same one for which voltage and current plots were given in Fig. 6-8. More tractable functions result when γ is either purely real or purely imaginary.

Direct-current excitation of a line with losses, which yields a purely real γ, is the subject of Prob. 7-1.

Should the line be lossless, but the frequency other than zero, γ is purely imaginary:

$$Z_{oc}(z') = -jZ_0 \cot \beta z' \tag{7-12}$$

The cotangent function is infinite at $\beta z'$ equal to zero, and becomes infinite at 180° intervals from there. This is in accordance with the results previously noted, in which voltage was finite but current zero at those locations. The cotangent is zero when its argument is 90°, and at recurrent intervals of 180°. Correspondingly the voltage is zero, but the current finite, at those locations.

The j factor in Eq. 7-12 indicates that for the lossless line with open-circuit termination, the input impedance is a pure reactance and that the voltage and current are in time quadrature with each other at all points on the line. The cotangent is alternately positive and negative in consecutive 90° intervals; hence the input impedance changes abruptly from a capacitive reactance to an inductive reactance and vice versa. This is in agreement with the results shown in Fig. 6-7.

Equation 7-12 could have been obtained by dividing Eq. 6-22 by Eq. 6-23, expressions for $V(z')$ and $I(z')$, respectively, for this line.

c. Short-Circuited End

Setting Z_R equal to zero reduces Eq. 7-4 to

$$Z_{sc}(z') = Z_0 \tanh \gamma z' \tag{7-13}$$

As in the open-circuited case, direct-current excitation gives an expression which is easier to evaluate. (See Prob. 7-1.)

Should the line be lossless, but the frequency not equal to zero, Eq. 7-13 becomes

$$Z_{sc}(z') = jZ_0 \tan \beta z' \tag{7-14}$$

Observations comparable to those made concerning Eq. 7-12 may be made here; one should refer to Fig. 6-9.

d. Lossless Line with Purely Reactive Termination

The most direct way of getting a compact expression for input impedance to a lossless line terminated with a pure reactance is by dividing the expression for $V(z')$, Eq. 6-29, by that for current, Eq. 6-30:

$$Z_X(z') = -jZ_0 \cot\left(\beta z' - \frac{\theta_K}{2}\right) \tag{7-15}$$

where

$$\theta_K = 2\tan^{-1}\left(\frac{R_0}{X_R}\right) \tag{6-28}$$

Here R_0 is the characteristic impedance, which is a pure resistance $\sqrt{l/c}$, and X_R is the terminating reactance.

Figures 6-10 and 6-11 dealt with a particular example of this type from the standpoint of voltage and current functions; the magnitude and angle of $Z_X(z')$ for the same line and termination are sketched in Fig. 7-2.

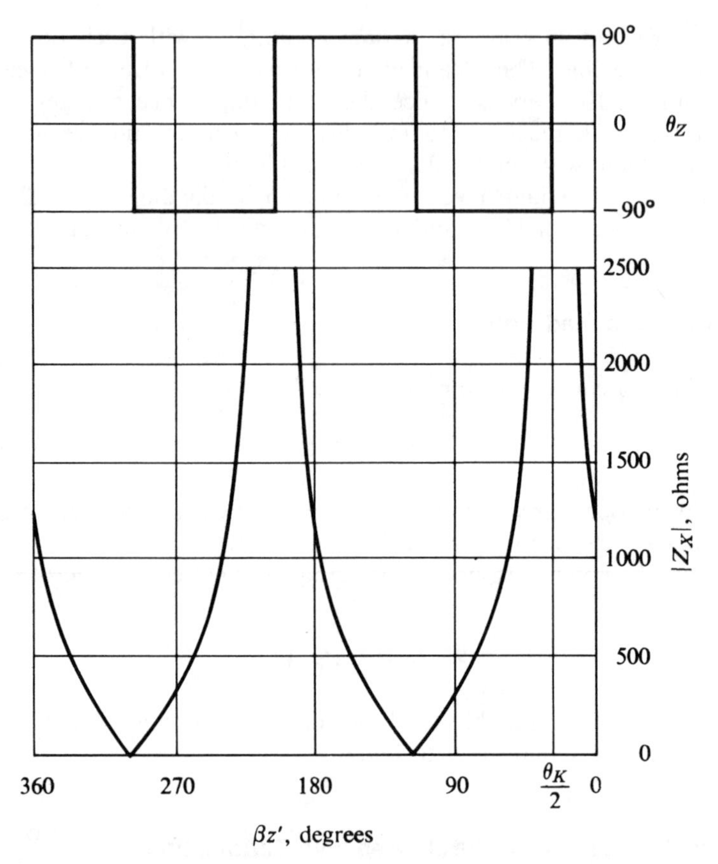

Figure 7-2. Input impedance on lossless line with purely reactive termination: $Z_0 = 600\underline{/0°}$ Ω, $Z_R = 1200\underline{/90.0°}$ Ω; ($K = 1.0\underline{/53.2°}$; see Figs, 6-10 and 6-11).

7-3 LINE PARAMETERS AS FUNCTIONS OF OPEN-CIRCUIT AND SHORT-CIRCUIT IMPEDANCES

The relationship between input impedance and the line parameters may be used in a reciprocal fashion as a means of determining the latter experimentally. Specifically, Eqs. 7-11 and 7-13 may be combined so as to yield the characteristic impedance and a function of the propagation function in terms of two readily measured impedances, $Z_{oc}(z')$ and $Z_{sc}(z')$:

$$Z_0 = \sqrt{Z_{oc}(z_1')Z_{sc}(z_1')} \tag{7-16}$$

$$\tanh \gamma z_1' = \sqrt{\frac{Z_{sc}(z_1')}{Z_{oc}(z_1')}} \tag{7-17}$$

The characteristic impedance is given directly by Eq. 7-16, but one has to find the inverse hyperbolic tangent of the right-hand side of Eq. 7-17, which in general is complex, to evaluate the propagation function.

a. Evaluation of Propagation Function from Hyperbolic Tangent

The inverse hyperbolic tangent of a complex quantity is also complex; it may be converted to the natural logarithm of a complex argument by means of the following identity,[3] and the real and imaginary parts of $\gamma z_1'$ then found:

$$\tanh^{-1}(A + jB) = 0.5 \ln\left(\frac{1 + A + jB}{1 - A - jB}\right) \tag{7-18}$$

Let

$$A + jB = \sqrt{\frac{Z_{sc}(z_1')}{Z_{oc}(z_1')}} \tag{7-19}$$

Then

$$\gamma z_1' = \tanh^{-1}(A + jB)$$

$$= 0.5 \ln\left(\frac{1 + A + jB}{1 - A - jB}\right)$$

Let

$$u\epsilon^{j\theta} = \frac{1 + A + jB}{1 - A - jB} \tag{7-20}$$

$$\gamma z_1' = 0.5 \ln(u\epsilon^{j\theta})$$

$$= 0.5[\ln(u) + j(\theta + 2m\pi)] \tag{7-21}$$

$$\alpha z_1' = 0.5 \ln(u) \tag{7-22}$$

$$\beta z_1' = 0.5(\theta + 2m\pi) \tag{7-23}$$

Here m is any integer; the imaginary part of the logarithm of a complex argument is multiple-valued. An approximate value of the electrical length of the line ($\beta z_1'$) must be known from other information in order to select the one physically meaningful result from among the infinite multitude indicated by Eq. 7-23.

If the physical length of the line (z_1') is known, the results of Eqs. 7-22 and 7-23 may be reduced to α and β.

Illustrative problem—A 280-m section of polyethylene-insulated cable has the following impedances when measured at 1.00 MHz:

$$Z_{oc} = 74.8\angle 29.2^\circ\ \Omega$$
$$Z_{sc} = 34.2\angle -34.4^\circ\ \Omega$$

The velocity of propagation is known to be about 2.0×10^8 m/s. Find the characteristic impedance and the value of the propagation function. From Eqs. 7-16 and 7-17,

$$Z_0 = \sqrt{(74.8\angle 29.2^\circ)(34.2\angle -34.4^\circ)}$$
$$= \sqrt{2.56 \times 10^3\angle -5.2^\circ}$$
$$= 50.6\angle -2.6^\circ\ \Omega$$
$$\tanh \gamma z' = \sqrt{\frac{34.2\angle -34.4^\circ}{74.8\angle 29.2^\circ}}$$
$$= \sqrt{0.457\angle -63.6^\circ}$$
$$= 0.676\angle -31.8^\circ$$
$$= 0.575 - j0.356$$

From Eqs. 7-18, 7-20, and 7-21,

$$\gamma z' = \frac{1}{2}\ln\left(\frac{1.575 - j0.356}{0.425 + j0.356}\right)$$
$$= \frac{1}{2}\ln\left(\frac{1.615\angle -12.7^\circ}{0.554\angle 40.0^\circ}\right)$$
$$= \frac{1}{2}\ln(2.913\angle -52.7^\circ)$$
$$= \frac{1}{2}\left[\ln(2.913) + j\left(\frac{-52.7}{53.7} + 2m\pi\right)\right]$$
$$\alpha z' = \frac{1}{2}\ln(2.913)$$
$$= 0.535 \text{ N (nepers)}$$
$$\alpha = \frac{0.535}{280}$$
$$= 1.91 \times 10^{-3} \text{ N/m}$$
$$\beta z' = -26.3^\circ + m180^\circ$$

All the following are *possible* values, on the basis of the impedance measurements, for $\beta z'$: 153.7°, 333.7°, 513.7°, 693.7°,. . . . An approximate value for β may be found from the frequency and the velocity of propagation. From Eq. 4-28,

$$\beta = \frac{\omega}{v} \tag{7-24}$$

$$\approx \frac{(1.00 \times 10^6)(2\pi)}{2.0 \times 10^8}$$

$$\approx 3.1 \times 10^{-2} \text{ rad/m}$$

$$\beta z' \approx (3.1 \times 10^{-2})(280)$$

$$\approx 8.68 \text{ rad or } 497°$$

(Division of 497° by 360° gives an approximate electrical length of 1.4 wavelengths.) The third value previously listed as "possible" for $\beta z'$, 513.7°, is closer to 497° than any of the others. Therefore

$$\beta z' = 513.7° \text{ or } 8.97 \text{ rad}$$

$$\beta = \frac{8.97}{280}$$

$$= 3.20 \times 10^{-2} \text{ rad/m}$$

$$\gamma = 1.91 \times 10^{-3} + j3.20 \times 10^{-2}$$

$$= 3.21 \times 10^{-2} \angle 86.6° \text{ m}^{-1}$$

b. Line Resistance, Inductance, Capacitance and Conductance

The basic line parameters may be found from the characteristic impedance and the propagation function by the following relations derived from Eqs. 4-16 and 4-31:

$$r + j\omega l = \gamma Z_0 \tag{7-25}$$

$$g + j\omega c = \frac{\gamma}{Z_0} \tag{7-26}$$

The illustrative problem given in Sec. 7-3a may be extended to include evaluation of r, l, g, and c. From Eqs. 7-25 and 7-26,

$$r + j\omega l = (3.21 \times 10^{-2} \angle 86.6^\circ)(50.6 \angle -2.6^\circ)$$

$$= 1.62 \angle 84.0^\circ$$

$$= 0.169 + j1.61\ \Omega/\text{m}$$

$$r = 0.169\ \Omega/\text{m}$$

$$l = \frac{1.61}{(1.00 \times 10^6)(2\pi)}$$

$$= 2.56 \times 10^{-7}\ \text{H/m}$$

$$g + j\omega c = \frac{3.21 \times 10^{-2} \angle 86.6^\circ}{50.6 \angle -2.6^\circ}$$

$$= 6.34 \times 10^{-4} \angle 89.2^\circ$$

$$= 8.9 \times 10^{-6} + j6.34 \times 10^{-4}\ \text{S/m}$$

$$g = 8.9 \times 10^{-6}\ \text{S/m}$$

$$c = \frac{6.34 \times 10^{-4}}{(1.00 \times 10^6)(2\pi)}$$

$$= 1.01 \times 10^{-10}\ \text{F/m}$$

7-4 LOSSLESS LINE: IMPEDANCE FUNCTION AND STANDING-WAVE RATIO

The lossless line is characterized by a repetition of the standing-wave patterns in voltage and current at half-wavelength intervals, regardless of what the terminating impedance may be. This was illustrated in Fig. 6-6 for a purely resistive termination. The principle of half-wavelength repetition also applies to the input impedance along such a line.

Standing-wave-pattern measurements provide a means, when the characteristic impedance is known, of finding the impedance of the termination. This experimental procedure is useful chiefly at radio and microwave frequencies.

a. Impedance Variation along a Lossless Line

Equation 7-5 may be restated as

$$Z(z') = Z_0 \frac{1 + |K|\exp[j(\theta_K - 2\beta z')]}{1 - |K|\exp[j(\theta_K - 2\beta z')]} \tag{7-27}$$

Let

$$\phi = \theta_K - 2\beta z' \tag{7-28}$$

Substitution of Eq. 7-28 in Eq. 7-27 yields

$$Z(z') = Z_0 \frac{1 + |K|\epsilon^{j\phi}}{1 - |K|\epsilon^{j\phi}} \tag{7-29}$$

To derive expressions for the resistive and reactive components of $Z(z')$, Eq. 7-29 should be separated into its real and imaginary parts. First $\epsilon^{j\phi}$ is replaced by $\cos\phi + j\sin\phi$:

$$Z(z') = Z_0 \frac{1 + |K|\cos\phi + j|K|\sin\phi}{1 - |K|\cos\phi - j|K|\sin\phi} \tag{7-30}$$

Equation 7-30 may be rationalized by multiplying numerator and denominator by the conjugate of the denominator. This result reduces to

$$Z(z') = Z_0 \frac{1 - |K|^2 + j2|K|\sin\phi}{1 - 2|K|\cos\phi + |K|^2} \tag{7-31}$$

Let

$$Z(z') = R(z') + jX(z') \tag{7-32}$$

It should be remembered that for a lossless line, the characteristic impedance is a pure resistance, equal to $\sqrt{l/c}$:

$$R(z') = Z_0 \frac{1 - |K|^2}{1 - 2|K|\cos\phi + |K|^2} \tag{7-33}$$

$$X(z') = Z_0 \frac{2|K|\sin\phi}{1 - 2|K|\cos\phi + |K|^2} \tag{7-34}$$

On the other hand, expressions for the polar components of $Z(z')$ may be desired.

Let
$$Z(z') = |Z(z')| \exp[j\theta_z(z')] \tag{7-35}$$

An equation for $\theta_Z(z')$ may be found from the numerator of Eq. 7-31:

$$\theta_Z(z') = \tan^{-1}\left(\frac{2|K|\sin\phi}{1-|K|^2}\right) \tag{7-36}$$

A concise form for the magnitude of $Z(z')$ may be found by multiplying the numerator and denominator of Eq. 7-30 by their respective conjugates:

$$|Z(z')| = Z_0\sqrt{\frac{1+2|K|\cos\phi+|K|^2}{1-2|K|\cos\phi+|K|^2}} \tag{7-37}$$

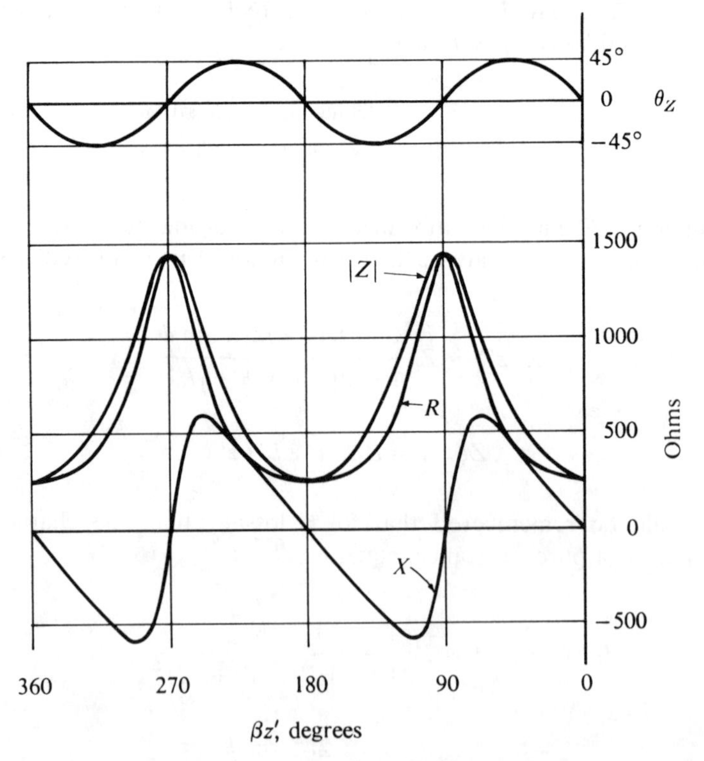

Figure 7-3. Input impedance on lossless line with resistive termination: $Z_0 = 600\angle 0^\circ\ \Omega$, $Z_R = 250\angle 0^\circ\ \Omega$; ($K = 0.412\angle 180.0^\circ$; SWR = 2.40; see Figs. 6-5 and 6-6).

From the definition of angle ϕ, Eq. 7-28, it is apparent that an increase in $\beta z'$ of π radians will reduce ϕ by 2π and leave the trigonometric functions of ϕ unchanged:

$$Z\left(z' + \frac{\pi}{\beta}\right) = Z(z') \tag{7-38}$$

This confirms the property of repetition at half-wavelength intervals which was noted at the beginning of this section.

Plots of $R(z')$, $X(z')$, $\theta_Z(z')$, and $|Z(z')|$ are given in Fig. 7-3 for a lossless line with a purely resistive termination. Voltage and current functions for this line were shown in Fig. 6-6.

b. Standing-Wave Ratio and Impedance Maxima and Minima

The maxima and minima of the magnitude of $Z(z')$ for a lossless line prove to be simple functions of the standing-wave ratio, which was introduced in Sec. 6-3. The maximum possible value of $1 + |K|\epsilon^{j\phi}$ as a function of ϕ is $1 + |K|$, and the minimum possible value is $1 - |K|$. These occur at $\epsilon^{j\phi} = 1$ and $\epsilon^{j\phi} = -1$, respectively, or $\phi = 2n\pi$ and $\phi = (2n - 1)\pi$, where n is any integer. At $\phi = 2n\pi$, $1 - |K|\epsilon^{j\phi}$ is a minimum $(1 - |K|)$, and at $\phi = (2n - 1)\pi$, it is a maximum $(1 + |K|)$. Hence, from Eq. 7-29,

$$|Z(z')|_{\max} = Z_0\frac{1 + |K|}{1 - |K|} \qquad \phi = 2n\pi$$

$$\theta_{Z\max} = 0 \tag{7-39}$$

$$|Z(z')|_{\min} = Z_0\frac{1 - |K|}{1 + |K|} \qquad \phi = (2n - 1)\pi$$

$$\theta_{Z\min} = 0 \tag{7-40}$$

Comparison of these results with Eqs. 6-33 through 6-36 shows that a voltage maximum, current minimum, and impedance maximum occur at the same location on the line, and likewise that a voltage minimum, current maximum, and impedance minimum occur at the same location.

The expression for standing-wave ratio in terms of the magnitude of reflection coefficient, derived in Sec. 6-3, is appropriate as a simplifying substitute:

$$\text{SWR} = \frac{1 + |K|}{1 - |K|} \tag{6-37}$$

$$Z(z')_{\max} = (Z_0)(\text{SWR}) \tag{7-41}$$

$$Z(z')_{\min} = \frac{Z_0}{\text{SWR}} \tag{7-42}$$

As was indicated by Eqs. 7-39 and 7-40, $Z(z')_{\max}$ and $Z(z')_{\min}$ are both purely resistive and they are found alternately along the line at intervals of a quarter wavelength.

c. Quarter-Wavelength Section of Line as Transformer

The results of the preceding section suggest that a quarter-wavelength section of line might be used for impedance transformation. An example is shown in Fig. 7-4, in which two lines of differing characteristic impedances are linked in tandem. It will be assumed for the moment that $Z_{01} > Z_{02}$.

If Z_{01} corresponds to $Z(z')_{\max}$ of the quarter-wavelength section, and Z_{02} to $Z(z')_{\min}$, the value of Z_{0T} indicated in Fig. 7-4 may be found by multiplying Eq. 7-41 by Eq. 7-42:

$$Z_{0T} = \sqrt{Z(z')_{\max} Z(z')_{\min}}$$

$$= \sqrt{Z_{01} Z_{02}}$$

The corresponding result applies, of course, if $Z_{01} < Z_{02}$.

The frequency for which one quarter-wavelength equals the length of the physical section of Z_{0T} line may be designated as the *design frequency*. The input impedance of such a unit is frequency-sensitive because the electrical length is proportional to the frequency actually in use; this effect is the subject of Prob. 7-7. The usefulness of this assembly is limited to systems which will operate at a single carrier frequency and a narrow bandwidth.

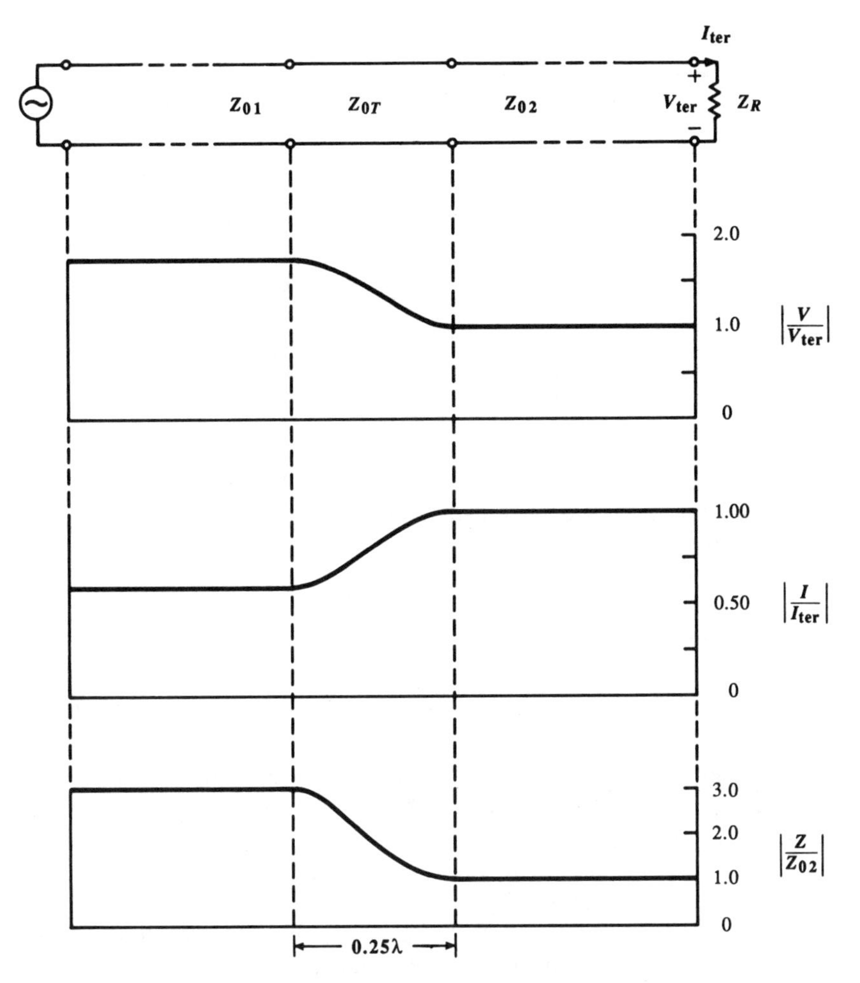

Figure 7-4. Circuit diagram and standing-wave patterns for two transmission lines tandem-linked by a quarter-wavelength impedance-matching section. (Quarter-wavelength transformer.) $Z_{01} = 3.0Z_{02}$; $Z_R = Z_{02}$.

d. Measurement of Terminal Impedances by Means of Standing-Wave Pattern

An experimentally-measured standing-wave pattern of voltage, such as in Fig. 6-6 or 7-5, provides a means of finding the radio-frequency impedance of the termination. The standing-wave ratio may be computed from scaled maximum and minimum values, and Eq. 6-37 may be solved

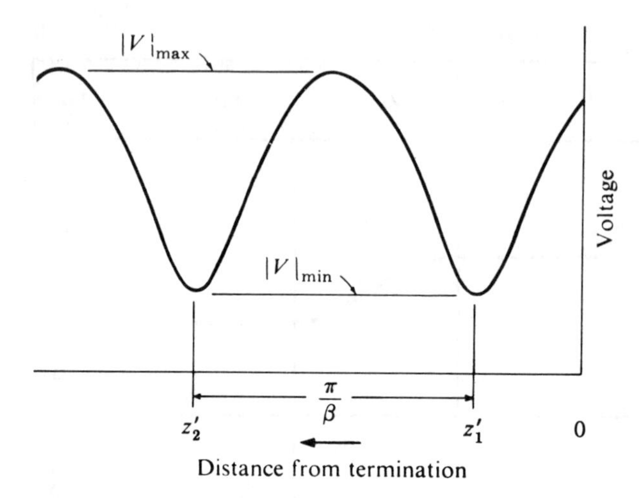

Figure 7-5. Voltage-standing-wave pattern for lossless line with general termination; (SWR = 3.70).

for $|K|$ in terms of SWR:

$$|K| = \frac{(\text{SWR}) - 1}{(\text{SWR}) + 1} \tag{7-43}$$

The possible values for ϕ at voltage maxima and minima have been noted in Eqs. 7-39 and 7-40; hence if the characteristic impedance and the electrical distance $\beta z'$ from the termination to a voltage maximum or minimum were known, θ_K and the terminating impedance ($z' = 0$) could be found from Eqs. 7-27 and 7-28.

As may be seen by inspection of Figs. 6-6 or 7-5, the resultant voltage changes more rapidly with displacement along the line in the vicinity of minima than in the vicinity of maxima; hence the locations of minima may be fixed more accurately from experimental data than can the locations of maxima. Accordingly a procedure for finding the terminating impedance will be detailed on the basis of using the locations of minima for distance calculations.

The distance between two adjacent voltage minima, $z'_2 - z'_1$ as shown in Fig. 7-5, is half a wavelength; hence the electrical distance from the termination to the nearer of these minima is given by the following:

$$\beta z'_1 = \frac{2\pi z'_1}{2(z'_2 - z'_1)} \quad \text{(radians)}$$

$$= \frac{360 z'_1}{2(z'_2 - z'_1)} \quad \text{(degrees)} \tag{7-44}$$

At a voltage minimum, ϕ is an odd multiple of π, as noted in Eq. 7-40. The electrical distance $\beta z_1'$ is necessarily positive because of the coordinate plan used, and for convenience θ_K should be stated in terms of the range $-\pi$ to $+\pi$. In accordance with the definition of ϕ in Eq. 7-28, the following equation for ϕ_1 will satisfy these requirements:

$$\phi_1 = -\pi - 2m\pi$$

This is equivalent to

$$\theta_K = 2\beta z_1' - \pi - 2m\pi \tag{7-45}$$

If $\beta z_1'$ is the electrical distance from the termination to the *nearest* voltage minimum, $m = 0$; otherwise m is equal to the number of voltage minima between $\beta z' = 0$ and $\beta z_1'$. Since $|K|$ and θ_K are now known, Z_R may be found in terms of Z_0 by setting z' equal to zero in Eq. 7-27; thus:

$$Z_R = Z_0 \frac{1 + |K| \exp(j\theta_K)}{1 - |K| \exp(j\theta_K)} \tag{7-46}$$

Example—As an illustration of the foregoing, suppose that the standing-wave ratio on a lossless transmission line is 3.70, and that the distances z_1' and z_2' are 2.10 and 7.70 m, respectively. Assume that the characteristic impedance of the line is 600 Ω:

$$|K| = \frac{3.70 - 1}{3.70 + 1}$$

$$= 0.573$$

$$\beta z_1' = \frac{2.10\pi}{7.70 - 2.10}$$

$$= 1.18 \text{ rad}$$

$$\theta_K = (2)(1.18) - 3.14$$

$$= -0.78 \text{ rad, or } -44.7^\circ$$

$$Z_R = 600 \frac{1 + 0.573\underline{/-44.7^\circ}}{1 - 0.573\underline{/-44.7^\circ}}$$

$$= 1224\underline{/-50.2^\circ}\ \Omega$$

7-5 NOMOGRAPHIC CHART FOR IMPEDANCE CALCULATIONS

Graphical construction is a technique which in many instances may be used to circumvent lengthy calculations. Several charts have been devised to expedite the determination of the impedance along a transmission line.[1, 2, 4–14]

Perhaps the most generally useful chart for this purpose is that developed by Smith.[9–12] Basically this chart portrays the variation of the normalized impedance or admittance with the angle of a generalized reflection coefficient. It is directly applicable to analysis of a lossless line but, with some auxiliary calculations, problems involving a lossy line may be solved too.

a. Impedance and Generalized Reflection Coefficient

The input impedance for the lossless line will be restated here:

$$Z(z') = Z_0 \frac{1 + |K| \exp[j(\theta_K - 2\beta z')]}{1 - |K| \exp[j(\theta_K - 2\beta z')]} \tag{7-27}$$

As z' approaches zero, $Z(z')$ approaches the terminating impedance Z_R (Eq. 7-46). If the latter equation were solved for $|K|\exp(j\theta_K)$, Eq. 6-12, which is repeated below, would result:

$$|K|\exp(j\theta_K) = \frac{Z_R - Z_0}{Z_R + Z_0} \tag{6-12}$$

As in Sec. 7-4a, let

$$\phi = \theta_K - 2\beta z' \tag{7-28}$$

Substitution of a particular distance z'_1 into Eq. 7-27, followed by substitution of Eq. 7-28, yields

$$Z(z'_1) = Z_0 \frac{1 + |K| \exp(j\phi_1)}{1 - |K| \exp(j\phi_1)} \tag{7-47}$$

Solving this equation for $|K| \exp(j\phi_1)$ will yield

$$|K| \exp(j\phi_1) = \frac{Z(z'_1) - Z_0}{Z(z'_1) + Z_0} \tag{7-48}$$

As suggested in Fig. 7-6, the impedance as viewed from any point z'_2 on the line to the left of z'_1 is the same whether the line extends to $z' = 0$

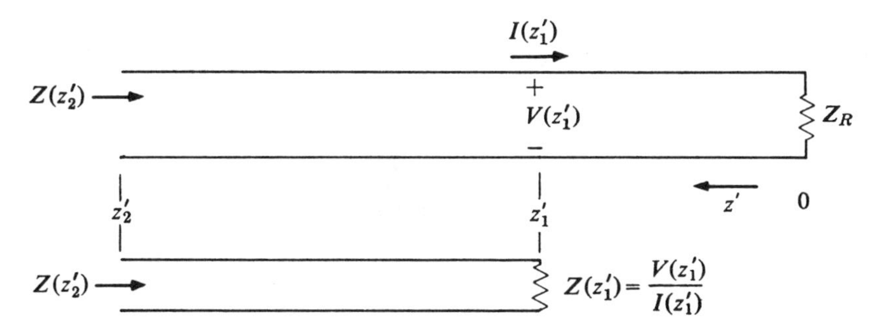

Figure 7-6. Two transmission lines of equivalent performance when viewed at locations $z' > z_1'$.

and is terminated with Z_R, or whether it is cut off at z_1' and terminated with a lumped impedance equal to $Z(z_1')$. The magnitudes of the reflection coefficients shown in Eqs. 6-12 and 7-48 for the respective terminations are the same, but their angles differ by $2\beta z_1'$. It would seem that use of a polar form of chart, with $|K|$ as the radial distance coordinate and $2\beta z'$ as an angular coordinate of direction or bearing, could capitalize on this convenient relationship. Loci of particular values of resistance and reactance could be plotted on the chart, and the locus of the complex impedance function $Z(z')$ for a lossless line with a specific termination would be a circle concentric with the origin. This is the essence of the Smith impedance-admittance chart. (The historical evolution of the Smith chart is detailed in ref. 11.)

b. Derivation of Loci of Transmission-Line Impedance Chart

To derive the equations of loci of constant value of the resistive and reactive components of input impedance, Eq. 7-47 is separated into its real and imaginary parts, as was done with Eq. 7-29. The subscript 1 may be deleted; then the results found in Eqs. 7-33 and 7-34 are immediately applicable. The impedance components may be stated in normalized form by dividing the respective equations by Z_0. Let

$$\frac{Z(z')}{Z_0} = \mathscr{R} + j\mathscr{X} \tag{7-49}$$

or

$$\mathscr{R} = \frac{R(z')}{Z_0}$$

$$\mathscr{X} = \frac{X(z')}{Z_0} \tag{7-50}$$

Here $\mathscr{R}$ and $\mathscr{X}$ are the normalized resistance and reactance, respectively; both are functions of ϕ and hence of z'. Substituting Eq. 7-50 into Eqs. 7-33 and 7-34 yields

$$\mathscr{R} = \frac{1 - |K|^2}{1 - 2|K|\cos\phi + |K|^2} \tag{7-51}$$

$$\mathscr{X} = \frac{2|K|\sin\phi}{1 - 2|K|\cos\phi + |K|^2} \tag{7-52}$$

Loci of constant values of $\mathscr{R}$ and $\mathscr{X}$ as specified in Eqs. 7-51 and 7-52 prove to be circles, as shown below.

(1) $\mathscr{R}$-*Loci Parameters.* The radii and the location of the centers of the $\mathscr{R}$ loci may be determined as follows. Let

$$u = |K|\cos\phi \tag{7-53}$$

$$v = |K|\sin\phi \tag{7-54}$$

$$|K|^2 = u^2 + v^2$$

The coordinates u, v, K, and ϕ are illustrated in Fig. 7-7.

Substitution of Eqs. 7-53 and 7-54 in Eq. 7-51, followed by multiplication of both sides by the denominator of the right-hand side, yields

$$\mathscr{R}(1 + u^2 + v^2 - 2u) = 1 - u^2 - v^2$$

$$(\mathscr{R} + 1)u^2 - 2u\mathscr{R} + (\mathscr{R} + 1)v^2 = 1 - \mathscr{R}$$

To complete the square in u, the quantity $\mathscr{R}^2/(\mathscr{R} + 1)$ is added to both sides, and the terms grouped as follows:

$$(\mathscr{R} + 1)\left[u^2 - \frac{2u\mathscr{R}}{\mathscr{R} + 1} + \frac{\mathscr{R}^2}{(\mathscr{R} + 1)^2}\right] + (\mathscr{R} + 1)v^2 = 1 - \mathscr{R} + \frac{\mathscr{R}^2}{\mathscr{R} + 1}$$

$$= \frac{1}{\mathscr{R} + 1}$$

This equation may be divided by $\mathscr{R} + 1$, and the quadratic involving u

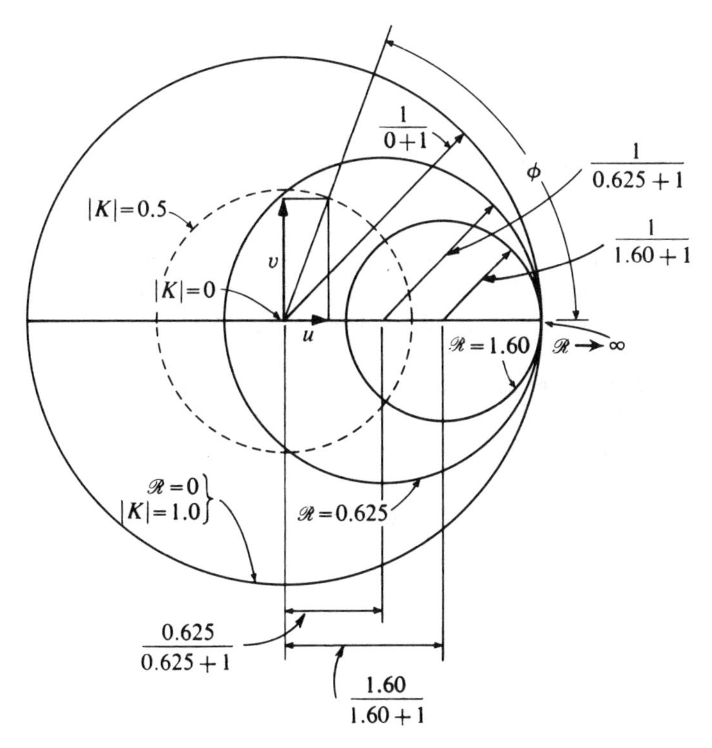

Figure 7-7. Loci of generalized reflection coefficient and normalized resistance on Smith impedance chart.

factored:

$$\left(u - \frac{\mathscr{R}}{\mathscr{R} + 1}\right)^2 + v^2 = \frac{1}{(\mathscr{R} + 1)^2} \tag{7-55}$$

This is the equation for a circle. The radius of the locus circle for a particular value of normalized resistance $\mathscr{R}_1$ is $1/(\mathscr{R}_1 + 1)$. The displacement of its center in the v direction is zero, and hence the locus center is on the axis along which ϕ is zero. It is displaced from the origin in the u direction by the distance $\mathscr{R}_1/(\mathscr{R}_1 + 1)$.

Two other distances of interest are the *intercepts*, or distances from the chart center along the axis of ϕ equal to zero (or 180°) to the intersections with the locus circle. The right-hand intercept of every $\mathscr{R}$-locus circle, found by adding the radius to the displacement of the locus center, proves to be unity. Thus all $\mathscr{R}$-locus circles are tangent to one another at the point $|K| = 1$, $\phi = 0$. The left-hand intercept, found by subtracting the radius from the displacement of center, is at $K_{\text{int}} = (\mathscr{R} - 1)/(\mathscr{R} + 1)$. The circle for $\mathscr{R}$ equal to unity passes through the center of the chart ($|K| = 0$). A significant reciprocity relationship is that the magnitude of the left-hand intercept for any $\mathscr{R}_1$ is equal to the corresponding intercept magnitude for its inverse, $1/\mathscr{R}_1$. Those intercepts are located in opposite directions from the chart center, however. For $\mathscr{R}$ equal to zero, the center of the locus circle coincides with the center of the chart, and the locus circle has unit radius. At the other extreme, the locus for $\mathscr{R}$ approaching infinity shrinks to a point at $|K| = 1$, $\phi = 0$.

(2) $\mathscr{X}$-*Loci Parameters.* The $\mathscr{X}$ loci may be examined in a corresponding manner. Equations 7-53 and 7-54 are substituted in Eq. 7-52 and the denominator multiplied out:

$$\mathscr{X}(1 + u^2 + v^2 - 2u) = 2v$$

$$\mathscr{X}(u^2 - 2u + 1) + \mathscr{X}\left(v^2 - \frac{2v}{\mathscr{X}}\right) = 0$$

The quantity $1/\mathscr{X}$ is added to both sides to complete the square in v:

$$\mathscr{X}(u^2 - 2u + 1) + \mathscr{X}\left(v^2 + \frac{2v}{\mathscr{X}} + \frac{1}{\mathscr{X}^2}\right) = \frac{1}{\mathscr{X}}$$

This is divided by $\mathscr{X}$ and the exact squares factored:

$$(u - 1)^2 + \left(v - \frac{1}{\mathscr{X}}\right)^2 = \frac{1}{\mathscr{X}^2} \tag{7-56}$$

Equation 7-56 describes a circle. The $\mathscr{X}$ loci have radii of $1/\mathscr{X}$; their centers are displaced by unit distance to the right of the chart center, and each is displaced vertically by a distance equal to the radius of the particular circle. Thus all the $\mathscr{X}$ loci pass through the point $|K| = 1$, $\phi = 0$. For the case of $\mathscr{X}$ equal to zero, the displacement of the center and the radius both approach infinity; hence the locus becomes the axis of $\phi = 0$ (or 180°). As $\mathscr{X}$ approaches infinity, the locus shrinks to a point at $|K| = 1$, $\phi = 0$. This is illustrated in Fig. 7-8.

The angle ϕ increases in the counterclockwise direction; from Eq. 7-28 it appears that movement along the transmission line from the terminated end toward the generator, which is the increasing direction of z' (see Fig. 6-1), corresponds to a clockwise motion on the chart.

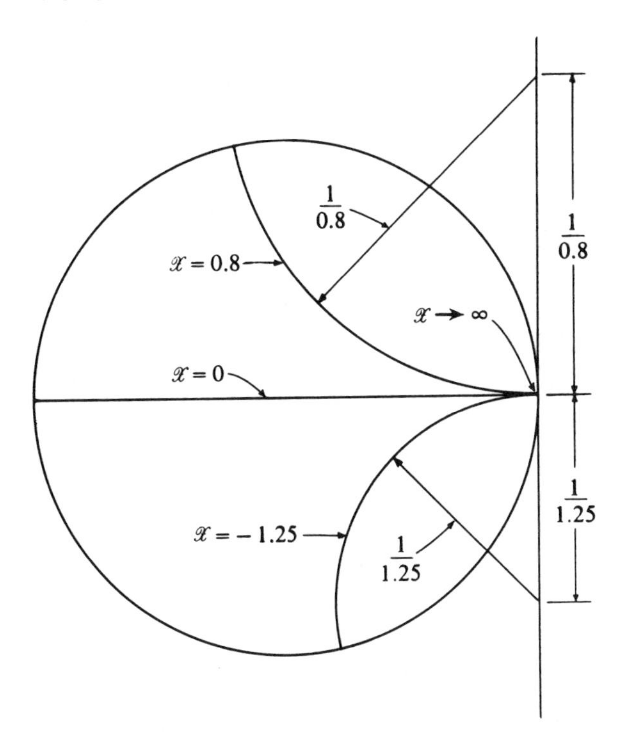

Figure 7-8. Loci of normalized reactance on Smith impedance chart.

The $\mathscr{X}$ loci in the upper half of the chart are inductive, while those in the lower half are capacitive. This may be verified by reference to Eq. 7-9, the input impedance to a short-circuited lossless line. This impedance is a pure reactance, increasing through the inductive range from zero to infinity as $\beta z'$ increases from zero to $\pi/2$ radians. The zero-impedance locus point on the chart is at the extreme left, at $|K| = 1$, $\phi = 180°$. Clockwise movement by $2\beta z'$ carries one along the upper half of the perimeter during the first quarter-wavelength change in $\beta z'$.

c. Impedance-Admittance Reciprocity Property of Chart

Admittance is a more convenient function to work with than impedance when the the problem involves shunt rather than series combinations of network elements. An impedance chart scaled for normalized values may, as shown below, be used as a normalized admittance chart through the following substitutions: (1) the $\mathscr{R}$ circles become loci of normalized conductance $\mathscr{G}$, (2) the $\mathscr{X}$ circles become loci of normalized susceptance $\mathscr{B}$, and (3) the upper half of the chart represents capacitive admittances rather than inductive impedances, and lower half represents inductive admittances rather than capacitive impedances.

Admittance and impedance are reciprocal quantities, and the relation of this property to the chart may be demonstrated as follows. Dividing Eq. 7-47 by Z_0 yields

$$\frac{Z(z_1')}{Z_0} = \frac{1 + |K|\exp(j\phi_1)}{1 - |K|\exp(j\phi_1)} \tag{7-57}$$

Let

$$Y(z_1') = \frac{1}{Z(z_1')} \tag{7-58}$$

$$Y_0 = \frac{1}{Z_0} \tag{7-59}$$

Substituting Eqs. 7-58 and 7-59 into Eq. 7-57 and inverting the result gives

$$\frac{Y(z_1')}{Y_0} = \frac{1 - |K|\exp(j\phi_1)}{1 + |K|\exp(j\phi_1)} \tag{7-60}$$

Let

$$\phi_1' = \phi_1 + \pi \tag{7-61}$$

$$\exp(j\phi_1') = -\exp(j\phi_1) \tag{7-62}$$

Substitution of Eq. 7-62 into Eq. 7-60 gives the following expression, which parallels Eq. 7-57 for normalized impedance:

$$\frac{Y(z_1')}{Y_0} = \frac{1 + |K|\exp(j\phi_1')}{1 - |K|\exp(j\phi_1')} \tag{7-63}$$

The normalized admittance $Y(z_1')/Y_0$ may be stated in terms of normalized conductance $\mathscr{G}$ and normalized susceptance $\mathscr{B}$ as follows:

$$\frac{Y(z_1')}{Y_0} = \mathscr{G} + j\mathscr{B} \tag{7-64}$$

(It should be noted that capacitive susceptance is considered positive, and inductive susceptance negative, under the definition set forth in Eq. 7-64. Usage on this point is divided; to avoid confusion it is well, when stating a value for susceptance separately, to designate it as "inductive" or "capacitive" rather than relying on an algebraic sign to convey this information.)

The derivation carried out in Eqs. 7-31 through 7-34, and Eqs. 7-49 through 7-56, may be paralleled in terms of admittance, beginning with Eqs. 7-63 and 7-64. An item-for-item correspondence may thereby be

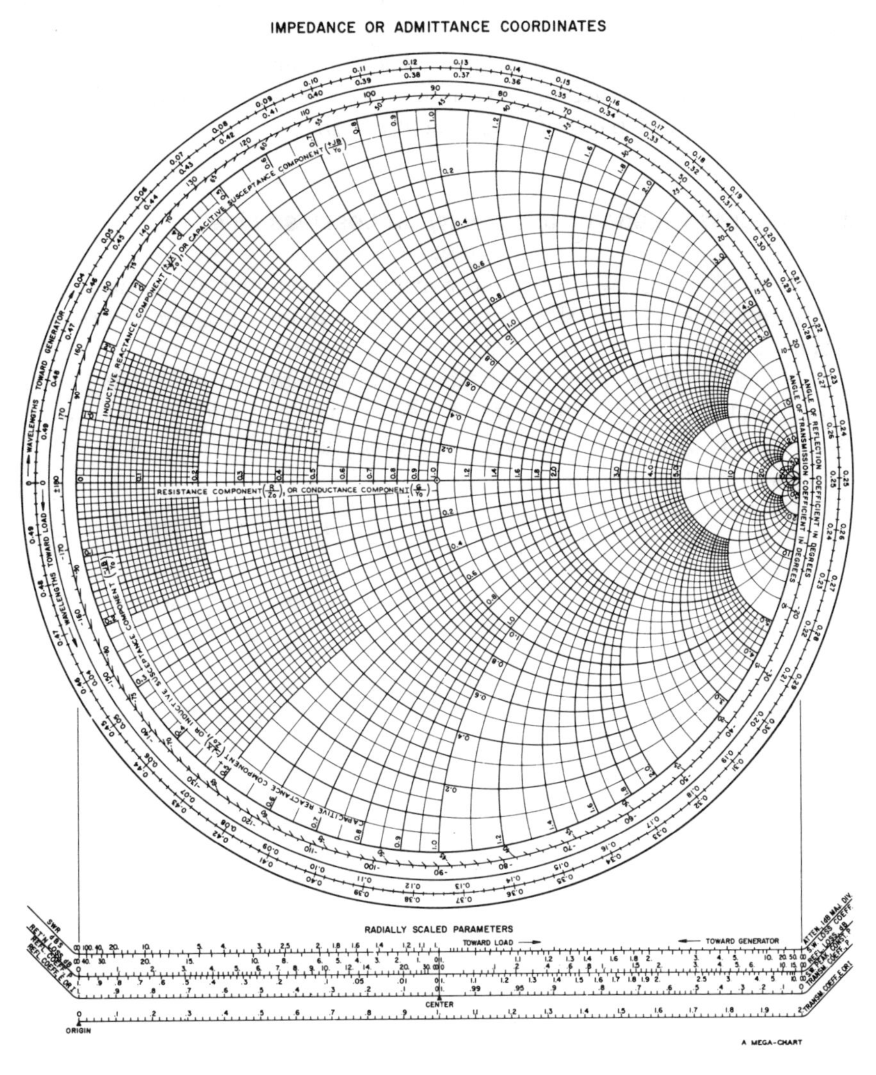

Figure 7-9. Commercial form of Smith impedance chart. Reproduced by courtesy of Analog Instruments Co., New Providence, NJ.

established between the normalized impedance interpretation and the normalized admittance interpretation.

As indicated in Eq. 7-61, the angle ϕ'_1 used in Eq. 7-63 differs by π radians from the angle ϕ_1 of the generalized reflection coefficient. Since the magnitude $|K|$ is the same in Eqs. 7-57 and 7-63, the locus points of the normalized impedance and the normalized admittance at any location z'_1 are at the same radius and are separated by a semicircle; in other words, they lie on a line passing through the center of the chart, at equal distances from the center.

A commercially published form of this chart has been reproduced in Fig. 7-9. For convenience the circumference is scaled not only in degrees of ϕ but in decimal parts of a wavelength in $\beta z'$.

d. Quarter-Wavelength Reciprocity on Lossless Line

An alternation between voltage maxima and minima at quarter-wavelength intervals on a lossless line, and an accompanying alternation between current minima and maxima, were illustrated in Fig. 6-6. The corresponding alternation of maxima and minima in impedance was brought out in Sec. 7-4b. In terms of the impedance or admittance chart, a displacement of a quarter wavelength yields a change in ϕ or ϕ' of π radians, or a semicircle.

Let

$$\beta z'_2 = \beta z'_1 + \frac{\pi}{2} \tag{7-65}$$

Then, from Eq. 7-28,

$$\phi_2 = \phi_1 - \pi \tag{7-66}$$

From Eqs. 7-27 and 7-28,

$$\frac{Z(z'_2)}{Z_0} = \frac{1 + |K| \exp(j\phi_2)}{1 - |K| \exp(j\phi_2)}$$

From Eq. 7-66,

$$\exp(j\phi_2) = -\exp(j\phi_1)$$

Hence

$$\frac{Z(z'_2)}{Z_0} = \frac{1 - |K| \exp(j\phi_1)}{1 + |K| \exp(j\phi_1)} \tag{7-67}$$

This is the reciprocal of the normalized impedance at point z'_1, Eq. 7-57.

Hence
$$\frac{Z[z' + (\pi/2\beta)]}{Z_0} = \frac{Z_0}{Z(z')} \tag{7-68}$$

As noted in Sec. 7-5c, the locus points for reciprocal complex quantities on the chart are on a line passing through the center of the chart, at equal distances from the center.

e. Impedance Locus of Transmission Line with Loss

When account is taken of line losses, the impedance locus changes from a circle whose radius has a constant magnitude $|K|$ to a spiral which converges toward the center of the chart with increasing z' or diminishing ϕ.

Beginning with Eq. 7-2 the following equations, paralleling Eqs. 7-27, 7-47, and 7-48, may be written

$$Z(z') = Z_0 \frac{1 + |K| \exp[-2\alpha z' + j(\theta_K - 2\beta z')]}{1 - |K| \exp[-2\alpha z' + j(\theta_K - 2\beta z')]} \tag{7-69}$$

$$Z(z'_1) = Z_0 \frac{1 + |K| \exp(-2\alpha z'_1 + j\phi_1)}{1 - |K| \exp(-2\alpha z'_1 + j\phi_1)} \tag{7.70}$$

$$|K| \exp(-2\alpha z'_1 + j\phi_1) = \frac{Z(z'_1) - Z_0}{Z(z'_1) + Z_0} \tag{7-71}$$

Thus the effective reflection coefficient at location z'_1 (the ratio of the phasor reflected voltage at location z'_1 to the phasor incident voltage at the same location) has a magnitude of $|K| \exp(-2\alpha z'_1)$. This quantity is equal to the normalized radial coordinate of the impedance locus on the chart at angle ϕ_1. Physically the result just noted is to be expected because the magnitude of $V_{inc}(z'_1)$ is $\exp(\alpha z'_1)$ times $|V_{inc}(0)|$, whereas $|V_{ref}(z'_1)|$ is $\exp(-\alpha z'_1)$ times $|V_{ref}(0)|$.

The characteristic impedance of a lossy line is usually slightly capacitive, as noted after Eq. 4-56. Equation 6-12 for the reflection coefficient, and Eqs. 7-69, 7-60, and 7-71, and chart solutions based on them, are all valid for a complex Z_0. Conversion of load impedances or input impedances from ohmic values to normalized form, or vice versa, may be carried out most easily in polar form if Z_0 is complex. An example will be given in Sec. 7-6d.

7-6 SOME APPLICATIONS OF THE SMITH CHART

Procedures for use of a chart are perhaps made most meaningful if they are conveyed in the course of working some illustrative problems. Four such problems which will be examined here are (1) calculation of terminating impedance from standing-wave measurements, (2) input admittance to a line which has a resistive shunt between the input and the termination (see Fig. 6-13), (3) single-stub matching on a radio-frequency line, and (4) input impedance to a lossy, open-circuited line. Problems 7-8 through 7-16 amplify various aspects of these examples.

a. Terminating Impedance Found from Standing-Wave Pattern

The illustrative problem at the close of Sec. 7-4d, which was worked by means of Eqs. 7-43 through 7-46, will be reworked here by means of the chart. As shown in Eq. 7-41, the normalized impedance at a voltage maximum is resistive and is equal to the standing-wave ratio. Thus the chart may be entered at $\mathscr{R} = 3.70$, $\mathscr{X} = 0$, as shown in Fig. 7-10. At one quarter-wavelength displacement from this location is a voltage minimum,

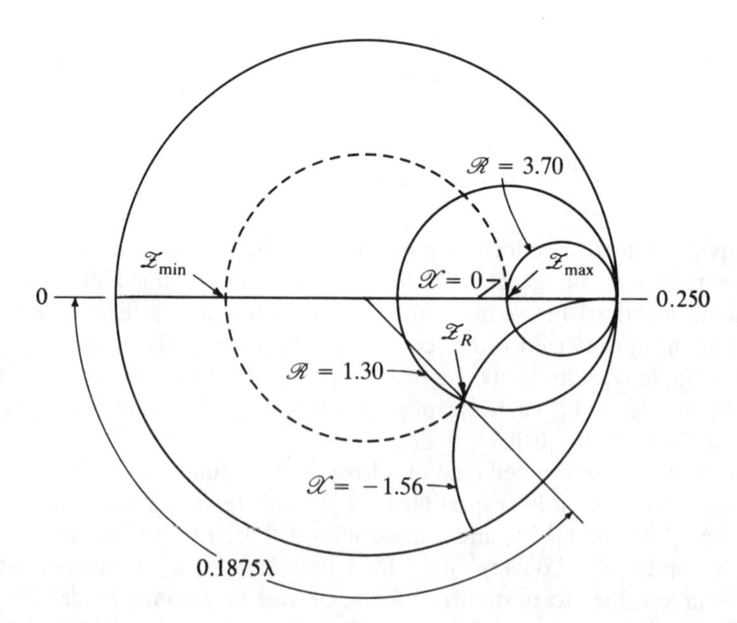

Figure 7-10. Graphical solution for terminating impedance from standing wave pattern (SWR = 3.70).

which would correspond to point z_1' on the line. (It may be noted that at the latter location $\mathscr{R} = 0.270$.) To locate the terminal-impedance locus point, one proceeds counterclockwise by an arc corresponding to $2.10/[(2)(7.70 - 2.10)]$, or 0.1875, wavelength. The normalized impedance of the termination may be scaled as $1.30 - j1.56$. The characteristic impedance was given as 600 Ω; hence the terminal impedance is $780 - j936$, or $1220\underline{/-50.2^\circ}$, Ω. The latter result compares favorably with that found by numerical calculation in Sec. 7-4d.

b. Lossless Line with Shunt Discontinuity

Problem 6-8, which is related to Fig. 6-13, was to be worked on the basis of voltage and current functions; it will be reworked here on an admittance basis with the Smith chart.

The two lumped impedances may be converted into normalized admittances as follows:

$$\mathscr{Y}_R = \frac{Z_0}{Z_R} = \frac{600}{250} = 2.40 + j0$$

$$\mathscr{Y}_{\text{sh}} = \frac{Z_0}{R_B} = \frac{600}{400} = 1.50 + j0$$

The chart is entered at $2.40 + j0$, as shown in Fig. 7-11. A radial line drawn through this point intersects the electrical distance scale at the periphery of the chart at the value of 0.250 wavelength. The distance between Z_R and R_B, $\pi/3\beta$, is equivalent to $\frac{1}{6}$, or 0.1667, wavelength, so one moves clockwise to a scale value of $0.250 + 0.1667$ or 0.4167, wavelength. A radial line drawn from the chart center to this peripheral point proves to intersect a circle drawn through the $\mathscr{Y}_R$ point at the admittance loci combination of $0.525 - j0.450$. This is the normalized admittance $\mathscr{Y}_{SR}$ as one looks to the right from the right-hand side of R_B in Fig. 6-13. To

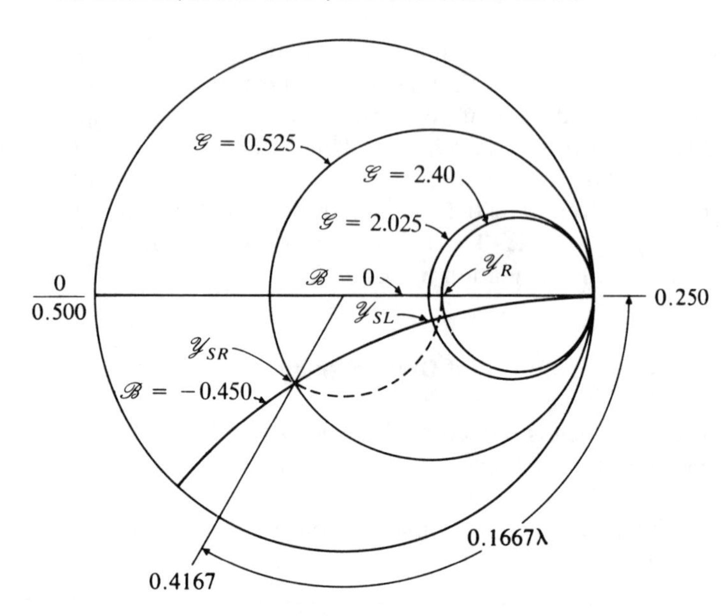

Figure 7-11. Graphical solution for input impedance to line with shunt resistor (see Fig. 6-13).

get the admittance as viewed from the left-hand side of R_B, $\mathscr{Y}_{SL}$, one adds the normalized admittance of R_B:

$$\begin{aligned}\mathscr{Y}_{SL} &= 0.525 - j0.450 + 1.500 \\ &= 2.025 - j0.450\end{aligned}$$

If input admittance viewed from some point on the line extending to the left of R_B is desired, the chart is entered at the point $2.025 - j0.450$, as shown in Fig. 7-11, and a circular arc swung from there.

c. Single-Stub Admittance Matching

A common application for radio-frequency transmission lines is as a feeder connection between a transmitter and an antenna. Such a line may extend many wavelengths, and best performance in terms of power transfer is obtained if the transmitter and antenna impedances are matched to the line. Under this condition no standing waves will exist on the line. Commonly, however, the input impedance to the antenna itself is not equal to the characteristic impedance of the line, and a matching circuit is necessary. An impedance-matching technique which will eliminate standing waves in all but the immediate vicinity of the load, and which is

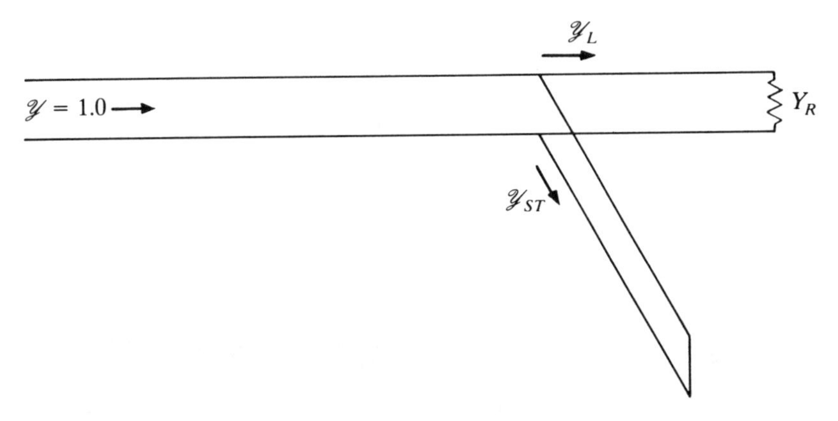

Figure 7-12. Transmission line with short-circuited stub connected in shunt.

feasible for wavelengths in the order of 5 m or less, is the connecting of a short-circuited stub in electrical parallel with (but spatially at right angles to) the given line. This is illustrated in Fig. 7-12. Determination of the correct location for the stub and its proper length provides an interesting example for utilization of the Smith chart.

A short-circuited stub will have as its input admittance a pure susceptance. If it is connected at a point on the line at which the input admittance viewed toward the load has a conductance component equal to the characteristic admittance of the line, and the susceptance component of that admittance is offset by that of the stub, the resultant admittance viewed from the generator side of the junction will be equal to the characteristic admittance.

If the normalized admittance of the load is known, one enters the chart at the locus point, as indicated in Fig. 7-13. In the example shown, $\mathscr{Y}_R = 3.00 - j1.00$, and the corresponding reading on the peripheral scale for line location is 0.2675. A circular arc is swung clockwise to locate intersections with the $\mathscr{G} = 1.0$ circle. Two such intersections occur; one of the normalized admittances, $\mathscr{Y}_{L1}$, is $1.0 - j1.30$, and the other, $\mathscr{Y}_{L2}$, $1.00 + j1.30$. The corresponding line-location scale readings are 0.3290 and 0.1710. The first-stated admittance is inductive. Hence a capacitive stub would be needed if that location is chosen, whereas the reverse is true if the second location is used. The two cases will be examined separately.

(1) *Capacitive-Stub Solution.* In this instance the distance from the end of the line to the stub location is 0.3290 − 0.2675, or 0.0615, wavelength. The stub should present a normalized admittance, $\mathscr{Y}_{\text{st}1}$, of $0 + j1.30$; to

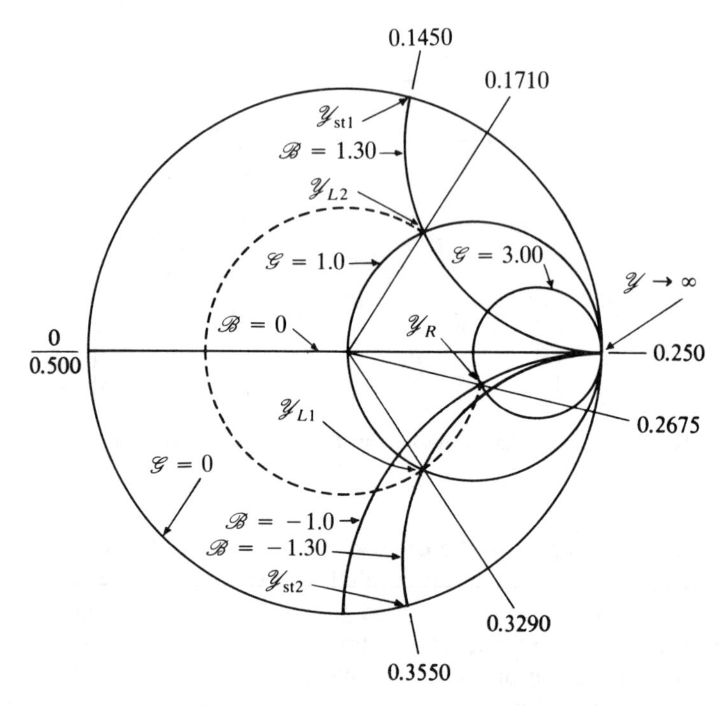

Figure 7-13. Single-stub matching of transmission-line admittance; solution by chart.

find the correct length one enters the chart at the locus point of infinite admittance, which is at $|K| = 1$ and at 0.250 on the line-location scale. Clockwise progression to the point $0 + j1.30$ covers a distance of more than a quarter wavelength; the line-location scale reading is 0.1450. The stub length is 0.1450 + 0.500 − 0.250, or 0.3950, wavelength.

(2) *Inductive-Stub Solution.* Since the stub location is more than a quarter wavelength from the load, the distance is found as follows: 0.1710 + 0.5000 − 0.2675, or 0.4035 wavelength. The stub length is found by starting at the point of infinite admittance and proceeding clockwise to the locus point of $0 - j1.30$. The corresponding line-location scale reading is 0.3350, and the stub length is 0.3550 − 0.2500, or 0.1050, wavelength.

d. Open-Circuited Lossy Line

The open-circuit impedance of a lossy line was discussed in Sec. 7-2b, and calculated values were plotted in Fig. 7-1 for the specific case in which the ratio α/β is 0.1 N/rad, and the characteristic impedance, $600\underline{/0^\circ}\ \Omega$. Suppose that these results are to be checked with the Smith chart for the

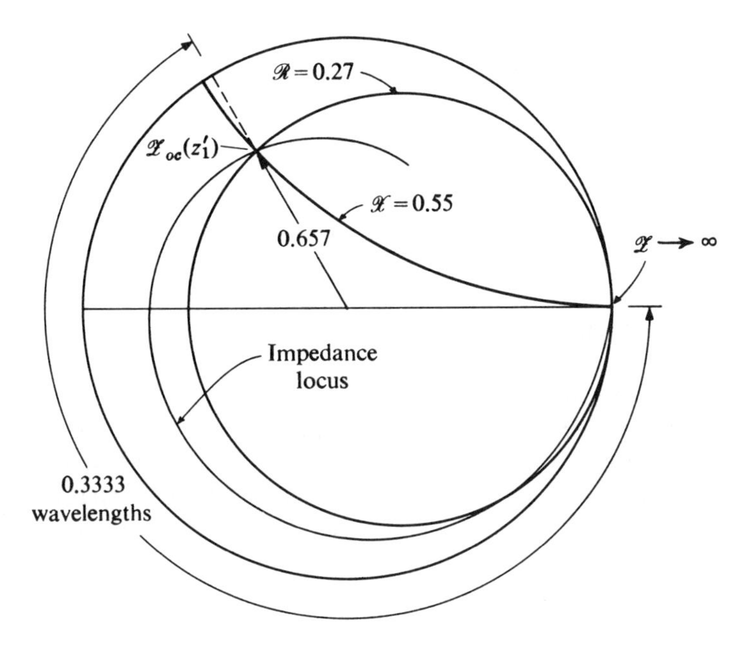

Figure 7-14. Smith chart analysis of open-circuit impedance of lossy line.

particular location $\beta z_1' = 120°$ ($2\pi/3$ rad, or 0.3333 wavelength). The impedance of the open-circuit termination is infinite; this corresponds to the point at the extreme right of the chart, as shown in Fig. 7-14, with $|K| = 1.0$. [The spiral locus for the normalized open-circuit impedance $\mathcal{Z}_{oc}(z')$ has been drawn in Fig. 7-14 as a continuous function, although this step is not necessary.] The value of 0.0833 on the peripheral scale (0.3333 − 0.2500) corresponds to the location z_1'; hence a radial line is drawn there, and then the radial distance to the locus is computed and scaled:

$$|K|\exp(-2\alpha z_1') = 1.0\exp[-(2)(0.1)(2\pi/3)]$$
$$= 0.657$$

From the $\mathcal{R}$ and $\mathcal{X}$ circles on the chart

$$\mathcal{Z}_{oc}(z_1') = 0.27 + j0.55 = 0.612\underline{/63.9°}$$
$$Z_{oc}(z_1') = (0.612\underline{/63.9°})(600\underline{/0°})$$
$$= 368\underline{/63.9°}\ \Omega$$

If the characteristic impedance were complex, say $700\underline{/-5.0°}\ \Omega$, but the α/β ratio the same as before, the chart solution for $\mathcal{Z}_{oc}(z_1')$ would be the same as that just given, because the normalized impedance of an open circuit is infinite regardless of the value of Z_0. However, the ohmic value $Z_{oc}(z_1')$ would differ; thus

$$Z_{oc}(z_1') = (0.612\underline{/63.9^\circ})(700.\underline{/-5.0^\circ})$$

$$= 428.\underline{/58.9^\circ}\ \Omega$$

7-7 CONCLUSIONS

The complex ratio of the voltage phasor to the current phasor at any location on a transmission line with a static termination is defined as the *input impedance* at that location. (See Eqs. 7-2 and 7-4.) Many aspects arising out of reflections on a transmission line may be analyzed more expediently by means of input impedance or admittance than directly in terms of standing waves of voltage and current.

The open-circuit and short-circuit impedances of a line together furnish enough experimental data to determine the characteristic impedance and the attenuation of the line. (See Eqs. 7-16 through 7-22.) If, in addition, the electrical length of the line is known approximately and the physical length is known, values may be found for the phase function and the line parameters. (See Eqs. 7-23, 7-25, and 7-26.)

On a lossless line the impedance function recurs every half-wavelength. The normalized impedance at a voltage maximum is equal to the standing-wave ratio (SWR), and the normalized impedance at a voltage minimum is equal to the reciprocal of that ratio.

The Smith chart is a nomographic means for evaluating the normalized impedance or admittance on a transmission line as a function of location. It is based on the concept of a generalized reflection coefficient in that radial distance on the chart is proportional to $|K|$, and angular distance is proportional to electrical distance on the line, with one revolution corresponding to a half-wavelength. On a lossless line the ratio of the magnitude of the reflected voltage phasor to that of the incident voltage phasor is everywhere the same, hence the locus of the line impedance on the chart is a circle of radius $|K|$.

PROBLEMS

7-1. Find expressions in terms of r and g for $Z_{oc}(z')$ and $Z_{sc}(z')$ for a lossy line with direct-current excitation. Sketch these impedances as functions of z', noting asymptotes and any maxima or minima.

7-2. Simplify Eq. 7-11 for Z_{oc} for the particular locations $\beta z' = 180°$ and $\beta z' = 270°$. Construct a phasor diagram showing the incident, reflected, and resultant voltages and currents at those locations and at $\beta z'$ equal zero.

7-3. Verify the following expression for input impedance, corresponding to Eq. 7-4:

$$Z(z') = Z_{oc}(z') \frac{Z_R + Z_{sc}(z')}{Z_R + Z_{oc}(z')}$$

7-4. Given the following values for the short-circuit impedance and the open-circuit impedance of a section of transmission line of length z'_s:

$$Z_{sc}(z'_s) = 538\epsilon^{j21.2°}\ \Omega$$

$$Z_{oc}(z'_s) = 464\epsilon^{-j39.2°}\ \Omega$$

Find the characteristic impedance, $\alpha z'_s$, and $\beta z'_s$. It is known that $(z'_s/\lambda) \approx 0.65$.

7-5. The following value has been derived from open-circuit and short-circuit impedances at 0.500 MHz on a 440-m section of cable: $\tanh \gamma z'_1 = 1.00\angle{-75.0°}$. From measurements with a cathode ray oscilloscope and a pulse generator, the round-trip time for a wave to travel to the far end of the cable and back has been found to be 3.4 μs. Find the attenuation function and the phase velocity at the test frequency. Sketch the standing-wave patterns of $|V_{oc}|$ and $|V_{sc}|$.

7-6. Assume that the line shown in Fig. 7-15 is lossless, and that the velocity of propagation in 3×10^8 m/s. $R_G = 500\ \Omega$, $Z_0 = 500\ \Omega$, $z'_1 = 75.0$ m, $z'_2 = 37.5$ m,

$$v_g(t) = 10.0\cos(2\pi \times 10^6 t) + 4.0\cos(4\pi \times 10^6 t)\ \text{V}$$

Find $i(z'_1, t)$, $v(z'_1, t)$, $i(z'_2, t)$, $v(z'_2, t)$, and $v(0, t)$. *Hint*: Apply the phasor method to each frequency component separately.

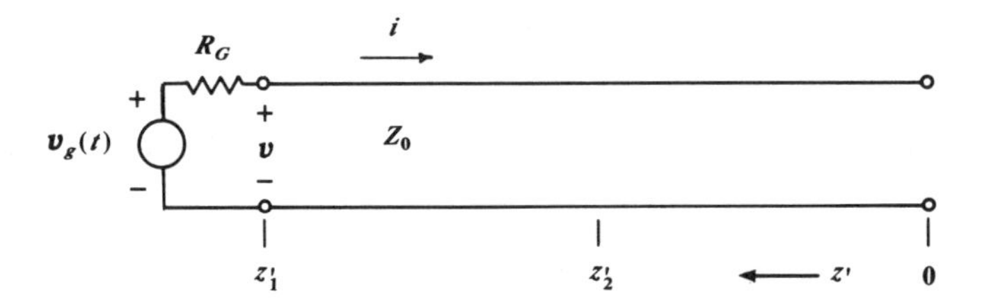

Figure 7-15. Resonant section of line (see Prob. 7-6).

7-7. Refer to Fig. 7-4, with the circuit involving a quarter-wavelength transformer.

(a) Sketch the standing wave patterns which would result if the line were energized at (1) twice the design frequency and (2) three times the design frequency. Find the SWR in the left-hand line in each case.

(b) Find the input impedance, in terms of Z_{01}, of the Z_{0T} section if the frequency is 1% greater than the design frequency. Compute the corresponding reflection coefficient and SWR.

7-8. Apply the Smith chart to the circuits in the examples in Secs. 6-5a and b to find the respective SWRs to the left of the discontinuities.

7-9. A lossless open-wire transmission line with $Z_0 = 150\angle 0°\ \Omega$ is terminated with $Z_R = 120 + j180\ \Omega$; $f = 10.0$ MHz. Find the distance to the nearest voltage minimum, and the SWR. If the rms voltage at the minimum point is 2.0 V, what will the rms current there be?

7-10. Show that the Smith-chart locus of the impedance of a lossy line is a spiral of constant slope and that it intersects every $|K|$ circle at the angle whose tangent is α/β nepers per radian. (*Hint*: The slope is equal to a differential displacement in the radial direction, $d|K|$, divided by the corresponding circumferential distance, $|K|\,d\phi$.)

7-11. For the single-stub matching example given in Sec. 7-6c(2), draw a phasor diagram showing voltage and currents at the stub junction. Sketch the standing waves of voltage and current on the line with the stub attached, paying particular attention to continuity or discontinuity of each function at the stub junction.

7-12. Refer to the example in Sec. 7-6c(2). (Line matched with a single inductive stub.) The solution shown is based on the assumption that the characteristic impedances of the line and the stub were equal. Assume that $Z_{0\,\mathrm{Line}} = 500$ Ω, but that through error in fabrication, $Z_{0\,\mathrm{stub}} = 410\ \Omega$; find the SWR that would result between the stub junction and the supply if the stub length indicated (0.1050λ) were used.

7-13. Refer to the example in Sec. 7-6c(2). Assume that the stub consists of an open-circuited section (characteristic impedance equal to that of the line) 0.20 wavelengths in length, to which a short-circuiting clamp was to be attached. The stub is located in accordance with the design, but through oversight the short-circuiting clamp was not connected to the stub. Find the normalized admittance of the stub as actually installed, and the resulting SWR between the stub junction and the supply.

7-14. A lossless transmission line, operating at 40.0 MHz, has a characteristic impedance of 350 Ω, and the standing-wave ratio has been found from measurements to be 4.50. It is proposed to use a short-circuited stub to reduce the standing waves. Find the distance from a voltage minimum at which the stub should be located. How long should the stub be? (Find both inductive-stub and capacitive-stub solutions.) Work this by the Smith chart and also analytically.

7-15. A lossless short-circuited stub with a characteristic impedance of 350 Ω is to have an input reactance of 450 Ω. The design frequency corresponds to a wavelength of 3.5 m. How long should the stub be if it is to be inductive? How long if capacitive? If stubs of those two lengths are operated at 5% above the design frequency, what will their reactances be? Express the changes in reactances as percentages.

7-16. Figure 7-16 illustrates a technique which has been used for supporting a rigid open-wire line at microwave frequencies. Assume that line and stub are

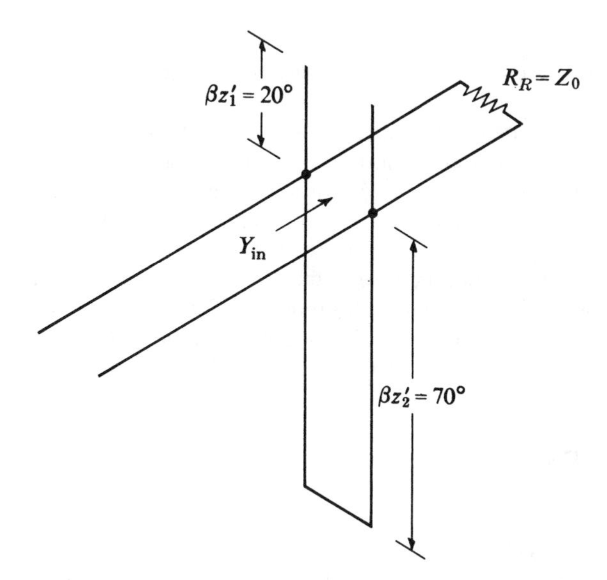

Figure 7-16. Short-circuited stub support for rigid line (see Prob. 7-16).

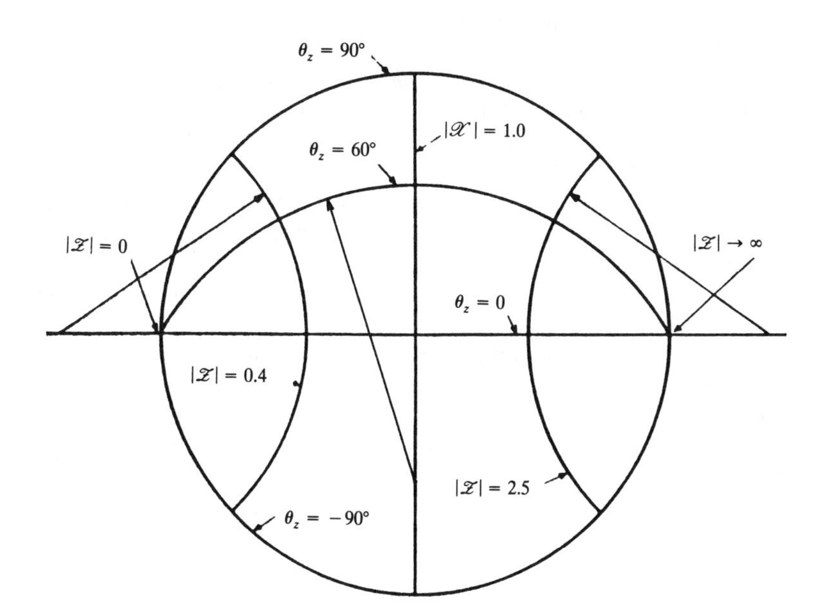

Figure 7-17. Polar (Carter) type of transmission-line impedance chart (see Prob. 7-17).

lossless and have the same characteristic impedance Z_0. Find Y_{in} in terms of Z_0. Draw a phasor diagram showing the voltage and the currents at the junction.

7-17. Figure 7-17 shows an impedance chart[1] similar to the Smith chart, but one in which the loci represent constant values of the polar coordinates of normalized impedance, $|\mathscr{Z}|$ and θ_z. Show (a) that the loci are circles, (b) that the u offsets of the $|\mathscr{Z}|$ circles are $(|\mathscr{Z}|^2 + 1)/(|\mathscr{Z}|^2 - 1)$, (c) that the radii of the $|\mathscr{Z}|$ circles are $2|\mathscr{Z}|/(|\mathscr{Z}|^2 - 1)$, (d) that the v offsets of the θ_z circles are $-\cot \theta_z$, and (e) that the radii of the θ_z circles are $\csc \theta_z$. (*Hint*: For the $|\mathscr{Z}|$ loci, square Eq. 7-37 and substitute Eqs. 7-53 and 7-54. For the θ_z loci, restate Eq. 7-36 in terms of the tangent of θ_z and substitute Eqs. 7-53 and 7-54.)

REFERENCES

1. Carter, P. S., Charts for transmission-line measurements and computations, *RCA Rev.*, 3, 355, 1939.
2. Creamer, W. J., *Communication Networks and Lines*, Harper and Brothers, New York, 1951, 315.
3. Dwight, H. B., *Tables of Integrals and Other Mathematical Data*, 4th ed., MacMillan, New York, 1961, 168.
4. Everitt, W. L. and Anner, G. E., *Communication Engineering*, 3rd ed., McGraw-Hill, New York, 1956, 370.
5. Kennelly, A. E., *Chart Atlas of Complex Hyperbolic and Circular Functions*, Harvard University Press, Cambridge, MA, 1914, 1921, 1924.
6. Kimbark, E. W., *Electrical Transmission of Power and Signals*, Wiley, New York, 1949, 185, 204, 205.
7. Members of the Staff of the Radar School, Massachusetts Institute of Technology, *Principles of Radar*, 3rd ed., rev. by Reintjes, J. F. and Coate, G. T. McGraw-Hill, New York, 1952, 502.
8. Ragan, G. L., *Microwave Transmission Circuits*, Radiation Laboratory Series No. 9, McGraw-Hill, New York, 1948.
9. Smith, P. H., An improved transmission line calculator, *Electronics*, 17, no. 1, 130, 318, 1944.
10. Smith, P. H., Transmission line calculator, *Electronics*, 12, no. 1, 29, 1939.
11. Smith, P. H., *Electronic Applications of the Smith Chart*. Krieger Publishing, Melbourne, FL: Reprint 1983.
12. Southworth, G. C., More about Phil Smith and his diagram, *Microwave J.*, 1, no. 2, 26, 1958.
13. Wheeler, H. A., Reflection charts relating to impedance matching, *IEEE Trans. Microwave Theory Tech.*, MTT-32, no. 9, 1008, Sept. 1984.
14. Woodruff, L. F., *Principles of Electric Power Transmission*, 2nd ed., Wiley, New York, 1938.

CHAPTER 8

The Ladder-Network Delay Line Or Artificial Line

A section of lossless transmission line has, as indicated in Chap. 3, the property of delaying an input signal by a time interval that is proportional to the line length. (A common application of signal-delaying is between the input and the deflection plates of an oscilloscope, so that the incoming signal may trigger the sweep circuit, and then be displayed in full on the screen.) A special lumped-parameter network that is more compact than, say, a reel of cable, yet which may be used to accomplish the same general result, is the *ladder-network delay line*, which is sketched in Fig. 8-1. It is equivalent to a tandem assembly of elementary sections of the *symmetrical, two-terminal-pair* type. (It may be noted that the term "port" is used synonymously with "terminal pair" in network theory.) Such a network is inherently a low-pass filter,* but, at frequencies well below cutoff, voltage and current relationships at the nodes, or midway between them, approach closely those at corresponding locations on a distributed parameter line.

Networks of this type, with resistances in series with the inductances, were used in earlier years under the name of *artificial lines* in analog-type

*The transmission properties of the network structure in Fig. 8-1 (and those of other ladder networks) may be reduced by the *image parameter* formulation (Sec. 8-2) to concise algebraic results. Before the development of high-speed electronic computers, this was a distinct advantage in the chore of designing filters. Better characteristics for filtering may be obtained by other approaches which involve more numerical compilation. An excellent summary is presented by H. W. Hale and E. C. Jones, Jr., in the *Standard Handbook for Electrical Engineers*, 12th ed., D. G. Fink and H. W. Beaty, eds., McGraw-Hill, New York, 1987, 2-54.

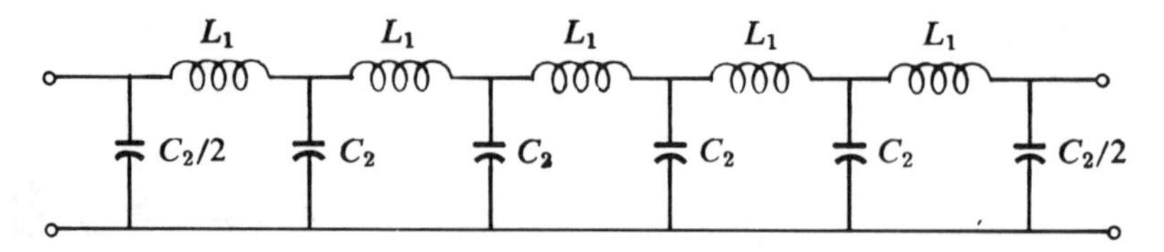

Figure 8-1. Ladder-network delay line.

models of transmission systems.* They have also been used to simulate transmission lines in computer programs for power-system networks.

When a step-function input is impressed on a lossless or low-loss lumped-parameter delay line, each voltage and current (within the network or at the output terminals) is characterized by an oscillatory or *ringing* component whose frequency depends on the simulated length of line *per section*.

A lumped-parameter delay line is usually "highly compressed" in the electromagnetic-wave sense; that is, the time required for energy propagation (at 3×10^8 m/s) between any two locations in the network is extremely short compared to the periods of natural oscillation. In tandem-connected lumped networks, the resolution of voltage and current functions into equivalent traveling-wave components, although a useful technique, should be regarded primarily as a mathematical operation, and it is better to reason in terms of the *delay time per section* rather than in terms of wave or signal velocities.

8-1 SYMMETRICAL NETWORKS: GENERAL PROPERTIES

The basic symmetrical two-terminal-pair section may have either a "T" or pi form, as shown in Fig. 8-2. This analysis will not be concerned with feedback systems, and it is contemplated that a network or transmission line will be connected to terminals 1 and 2, and another network or line to terminals 3 and 4, but that no external connections will be made from

---

**Network calculating boards*, assemblies of adjustable resistors, inductors, and capacitors, together with phase-shifting transformers to simulate the generated voltages in alternators, were used to find the steady-state voltages and currents in electric power systems. Conditions which were modeled commonly included (a) normal operation under selected load conditions, (b) the immediate effects of a short circuit at a chosen location, and (c) step-by-step relative phase changes in the generated voltages as the prime movers adjusted to a new loading condition.

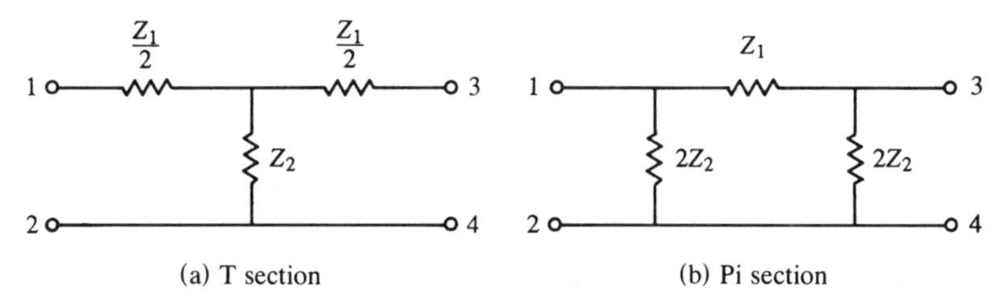

Figure 8-2. Symmetrical two-terminal-pair networks.

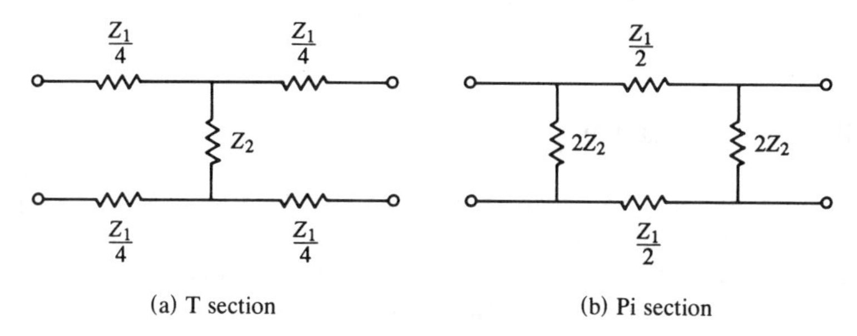

Figure 8-3. Balanced symmetrical two-terminal-pair networks.

terminals 1 or 2 or points to the left of them, to terminals 3 or 4 or points to the right of them. The customary notation is, as indicated, Z_1 for the total series impedance per section and Z_2 for the resultant shunt impedance per section. The section may be drawn in balanced form, as shown in Fig. 8-3, with the series impedance divided equally between the top and bottom conductors. Impedances Z_1 and Z_2 may each be composed of a single resistor, inductor, or capacitor, or any combination of elements of any two types, or of all three types.

If several symmetrical sections are connected in tandem so as to form a ladder network, as suggested in Fig. 8-4, a degree of uniformity is introduced. To an observer concerned with voltage-current relationships at the junction points between sections, this network resembles a succession of discrete lengths along a distributed-parameter line. From this analogy two

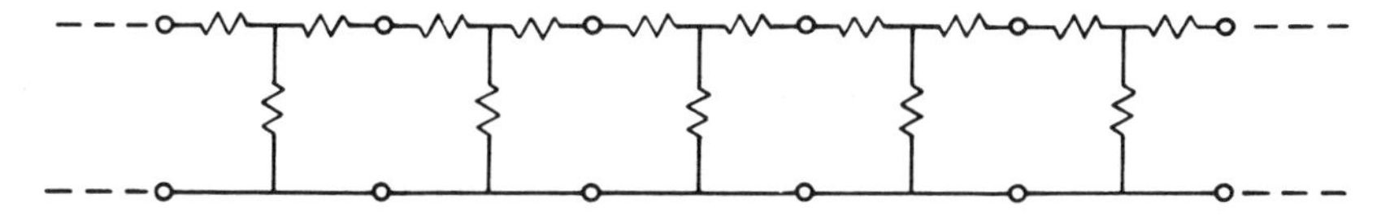

Figure 8-4. Ladder network of symmetrical T sections.

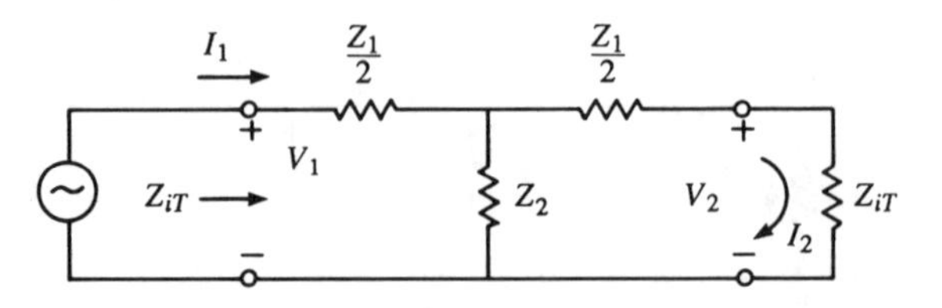

Figure 8-5. Symmetrical T section with iterative-impedance termination.

derived parameters may be formulated, the iterative impedance and the iterative transfer function.

a. Iterative Impedance

It was found in the study of the transmission line that if the line is terminated in an impedance known as the characteristic impedance, Eq. 4-31, the input impedance has that same value, independent of the line length. By analogy, a ladder network might consist of an integral number of symmetrical sections, and an impedance should exist which, if used to terminate the network, would yield the same value for the input impedance. That impedance will be called the *iterative impedance*. If one section is terminated in its iterative impedance, a second, identical section connected to the input terminals of the first section would also be terminated in iterative impedance, and hence would have the same value as its input impedance.

Hence, in accordance with Fig. 8-5, for a T section,

$$Z_{iT} = \frac{Z_1}{2} + \frac{Z_2(Z_1/2 + Z_{iT})}{Z_2 + (Z_1/2) + Z_{iT}} \tag{8-1}$$

This may be solved for Z_{iT} to yield

$$Z_{iT} = \sqrt{Z_1 Z_2 + \frac{Z_1^2}{4}} \tag{8-2}$$

$$Z_{iT} = \sqrt{Z_1 Z_2}\sqrt{1 + \frac{Z_1}{4Z_2}} \tag{8-3}$$

For a pi section the admittance point of view is more expedient:

$$Y_{i\pi} = \frac{Y_2}{2} + \frac{Y_1(Y_2/2 + Y_{i\pi})}{Y_1 + (Y_2/2) + Y_{i\pi}}$$

This leads to

$$Y_{i\pi} = \sqrt{Y_1 Y_2 + \frac{Y_2^2}{4}} \tag{8-4}$$

This may be restated in terms of impedances:

$$Z_{i\pi} = 1 \bigg/ \sqrt{\frac{1}{Z_1 Z_2} + \frac{1}{4Z_2^2}}$$

$$= \sqrt{Z_1 Z_2} \big/ \sqrt{1 + (Z_1/4Z_2)} \tag{8-5}$$

b. Iterative Transfer Function

The propagation function of a transmission line is a complex quantity of which the real part is the attenuation function, and the imaginary part, the phase function, of a traveling wave. A single traveling wave can exist on a line of finite length Δz if the line is terminated with its characteristic impedance; then the sending-end and receiving-end voltage and current phasors would be related to the propagation function as follows:

$$\epsilon^{\gamma \Delta z} = \frac{V_S}{V_R} \tag{8-6}$$

$$= \frac{I_S}{I_R} \tag{8-7}$$

By analogy, the *iterative transfer function* Γ of a single symmetrical section of a ladder network will be defined as follows, where it is understood that the section is terminated with its iterative impedance:

$$\epsilon^{\Gamma} = \frac{V_1}{V_2} \tag{8-8}$$

$$= \frac{I_1}{I_2} \tag{8-9}$$

$$\Gamma = \ln\left(\frac{V_1}{V_2}\right) \tag{8-10}$$

$$= \ln\left(\frac{I_1}{I_2}\right) \tag{8-11}$$

The ratio of the current phasors in the T section may be found by examination of Fig. 8-5:

$$I_1 = I_2 \frac{Z_{iT} + (Z_1/2)}{Z_2} + I_2$$

$$\frac{I_1}{I_2} = 1 + \frac{Z_{iT}}{Z_2} + \frac{Z_1}{2Z_2} \tag{8-12}$$

$$= 1 + \sqrt{\frac{Z_1}{Z_2}}\sqrt{1 + \frac{Z_1}{4Z_2}} + \frac{Z_1}{2Z_2}$$

The same result may be found for the pi section by using voltage phasors and admittances. Substitution of Eq. 8-12 in Eq. 8-11 gives

$$\Gamma = \ln\left(1 + \frac{Z_1}{2Z_2} + \sqrt{\frac{Z_1}{Z_2}}\sqrt{1 + \frac{Z_1}{4Z_2}}\right)$$

$$= 2\ln\left(\sqrt{\frac{Z_1}{4Z_2}} + \sqrt{1 + \frac{Z_1}{4Z_2}}\right) \tag{8-13)*$$

If n identical sections are connected in tandem, and the nth section is terminated in the iterative impedance, the following relationships are applicable:

$$\frac{I_1}{I_{n+1}} = \frac{I_1}{I_2} \cdot \frac{I_2}{I_3} \cdots \frac{I_n}{I_{n+1}}$$

$$= \epsilon^{n\Gamma} \tag{8-15}$$

c. Impedance Functions in Tandem-Connected Networks

For the analysis of a distributed-parameter line which was terminated with an impedance not equal to the characteristic impedance, the concept of two wave sets moving in opposite directions proved convenient. The

*The reader is reminded that the natural logarithm of a complex argument is also complex and has as its real part the natural logarithm of the magnitude of the argument, and as its imaginary part, the angle of the argument, stated in radians:

$$\ln(|A|\underline{/\theta_A}) = \ln|A| + j\theta_A \tag{8-14}$$

same mathematical technique may be used in a tandem-connected network. The analysis in Chaps. 6 and 7 could be repeated item for item if $\gamma z'$ were replaced by $n\Gamma$, and Z_0 with Z_{iT} or $Z_{i\pi}$. Thus the input impedance of an n-section tandem-connected network, corresponding to Eq. 7-3, would be

$$Z(n) = Z_i \frac{Z_R + Z_i \tanh n\Gamma}{Z_i + Z_R \tanh n\Gamma} \tag{8-16}$$

Here Z_i indicates the applicable iterative impedance, Z_{iT} or $Z_{i\pi}$. The equations for open-circuit and short-circuit impedances parallel Eqs. 7-11 and 7-13:

$$Z_{oc}(n) = Z_i \coth n\Gamma \tag{8-17}$$

$$Z_{sc}(n) = Z_i \tanh n\Gamma \tag{8-18}$$

The procedure outlined in Sec. 7-3 for finding Z_0 and $\gamma z_1'$ of a section of line from the measured open-circuit and short-circuit impedances may be utilized here too. Equations 8-17 and 8-18 may be combined to yield the following results:

$$Z_i = \sqrt{Z_{sc}(n) Z_{oc}(n)} \tag{8-19}$$

$$\tanh n\Gamma = \sqrt{\frac{Z_{sc}(n)}{Z_{oc}(n)}} \tag{8-20}$$

Let

$$\tanh n\Gamma = A_i + jB_i \tag{8-21}$$

As in Sec. 7-3a let

$$u_i \exp(j\theta_i) = \frac{1 + A_i + jB_i}{1 - A_i - jB_i} \tag{8-22}$$

The following parallels Eq. 7-21:

$$n\Gamma = 0.5 \ln(u_i) + 0.5j(\theta_i + 2m\pi) \tag{8-23}$$

The comments following Eq. 7-23 regarding the multiple-value nature of the imaginary part of the logarithm are applicable here.

The real and imaginary parts of the iterative transfer function may be designated as the *attenuation per section* ($\bar{\alpha}$, nepers) and the *phase displacement per section* ($\bar{\beta}$, radians), thus

$$\Gamma = \bar{\alpha} + j\bar{\beta} \tag{8-24}$$

8-2 THE LUMPED-PARAMETER DELAY LINE AS A LOW-PASS FILTER

A lossless transmission line has only series inductance and shunt capacitance, and a two-terminal-pair network intended to simulate a short section of such a line would logically have a purely inductive Z_1 and capacitive Z_2. By proper choice of values, including the length of line to be simulated by one section, the correspondence between the properties of the line and the network may be made very good over the band of frequencies between zero and some specified value, yet at higher frequencies the network properties will differ markedly from those of the line. Effectively the network would be a *low-pass filter*, a member of a class known as constant-k filters. The fundamental properties of the low-pass filter of this type will be developed to highlight the differences between a lumped-parameter delay line and a distributed-parameter transmission line.*

a. Definition and Equivalent Values: Derived Parameters

The designator "constant k" applies to any symmetrical two-terminal pair network for which Z_1 and Z_2 are purely reactive and obey the following constraint with respect to frequency:

$$Z_1 Z_2 = k^2 \tag{8-25}$$

Here k is a real number.

For the low-pass filter the following are applicable:

$$Z_1 = j\omega L_1 \tag{8-26}$$

$$Z_2 = \frac{1}{j\omega C_2} \tag{8-27}$$

In terms of a simulated line section of length Δz_{sec}, and distributed inductance and capacitance per unit length of l and c, the parameters L_1 and C_2 would be as follows:

$$L_1 = \Delta z_{\text{sec}}\, l \tag{8-28}$$

$$C_2 = \Delta z_{\text{sec}}\, c \tag{8-29}$$

**Mutual inductance* may exist between the coils of adjacent sections in a practical delay line. The effect of this may be analyzed by means of equivalent circuits and m-derived filter theory.[2, 8, 9, 11]

Equations 8-26 and 8-27 may be substituted into Eqs. 8-3, 8-5, and 8-13:

$$Z_{i\mathrm{T}} = \sqrt{\frac{L_1}{C_2}}\sqrt{1 - \frac{\omega^2 L_1 C_2}{4}} \tag{8-30}$$

$$Z_{i\pi} = \sqrt{\frac{L_1}{C_2}}\frac{1}{\sqrt{1 - \frac{\omega^2 L_1 C_2}{4}}} \tag{8-31}$$

$$\Gamma = 2\ln\left(j\frac{\omega}{2}\sqrt{L_1 C_2} + \sqrt{1 - \frac{\omega^2 L_1 C_2}{4}}\right) \tag{8-32}$$

b. Cutoff Frequency

The expressions just listed have two differing sets of properties depending on whether the discriminant $1 - (\omega^2 L_1 C_2/4)$ is positive or negative, in that (1) the iterative impedances are resistive in the former case but reactive in the latter, and (2) the argument of the logarithm in Eq. 8-32 is complex in the former case but purely imaginary in the latter. Thus two distinct modes of behavior may be expected, depending on the frequency. The radian frequency at which the discriminant vanishes, in other words, the boundary between these two modes, is known as the *cutoff radian frequency*, ω_c:

$$\omega_c = \frac{2}{\sqrt{L_1 C_2}} \tag{8-33}$$

This may be put in terms of a simulated line section by substituting Eqs. 8-28 and 8-29:

$$\omega_c = \frac{2}{\Delta z_{\text{sec}}\sqrt{lc}} \tag{8-34}$$

Thus a simulated line with a stated equivalent length may be designed with a cutoff frequency as high as desired, short of the microwave range. (There the physical components of the network have appreciable electrical length.) Given a specific value of ω_c and the values of l and c of the equivalent line, one may find from Eq. 8-34 the length of line which should be represented by each network section. Division of (a) the total length to be simulated, by (b) the per-section simulated length, yields the number of sections which will be needed in the given case.

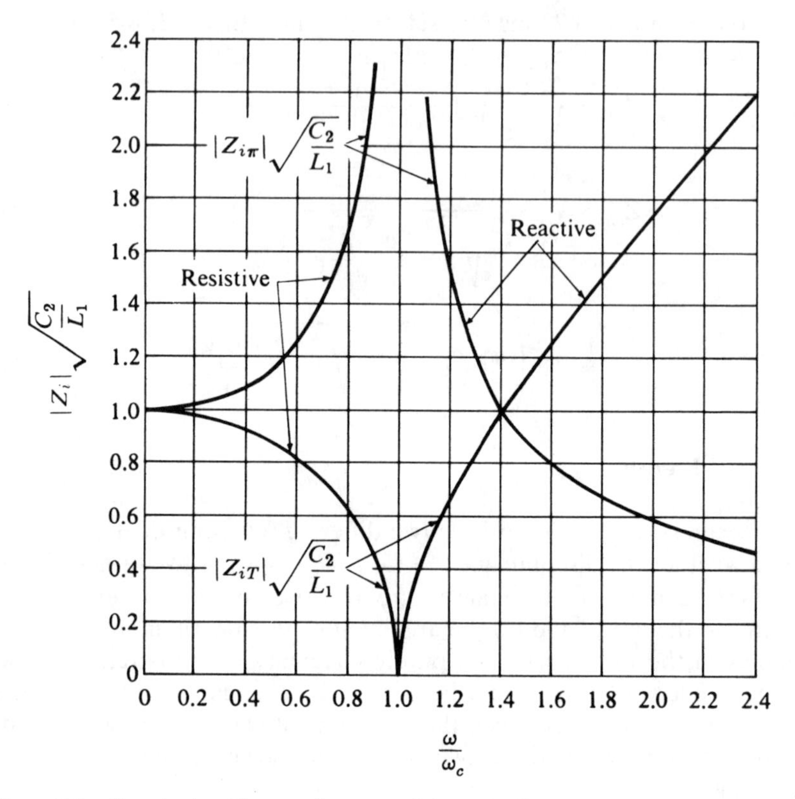

Figure 8-6. Iterative impedances of constant-*k* low-pass filter.

It is convenient to substitute Eq. 8-33 for the L_1C_2 products in Eqs. 8-30 through 8-32. Then

$$Z_{iT} = \sqrt{\frac{L_1}{C_2}}\sqrt{1 - \left(\frac{\omega}{\omega_c}\right)^2} \tag{8-35}$$

$$Z_{i\pi} = \sqrt{\frac{L_1}{C_2}}\frac{1}{\sqrt{1 - \left(\frac{\omega}{\omega_c}\right)^2}} \tag{8-36}$$

$$\Gamma = 2\ln\left[j\frac{\omega}{\omega_c} + \sqrt{1 - \left(\frac{\omega}{\omega_c}\right)^2}\right] \tag{8-37}$$

or

$$\Gamma = 2\ln\left[j\frac{\omega}{\omega_c} + j\sqrt{\left(\frac{\omega}{\omega_c}\right)^2 - 1}\right] \tag{8-38}$$

Frequencies below the cutoff frequency are said to be in the *pass band*, whereas those above it are in the *stop band*. The appropriateness of this terminology is borne out in the discussion of iterative impedance and iterative transfer function as functions of frequency.

c. Iterative Impedance as a Function of Frequency

The iterative impedances for the T and pi sections are plotted versus frequency in Fig. 8-6. Of particular interest is the fact that the iterative impedances are resistive at frequencies below cutoff, but reactive at higher frequencies. (The change is abrupt for a lossless network; if resistance is present, a smooth transition takes place in which the angle of Z_i changes continuously, although not uniformly.)

The *image-parameter* formulation of filter design, which was referred to in the footnote at the beginning of this chapter, assumes that the filter is terminated, at all frequencies, in its iterative impedance. This concept parallels that of termination of a transmission line with its characteristic impedance, and it simplifies the calculation of the transfer function of the filter.*

Below cutoff frequency a set of sections in tandem, if terminated in iterative impedance, will transmit power from the supply to the terminating resistor. This behavior corresponds to that of a lossless transmission line terminated in its characteristic impedance. Average power enters the sending end in both cases, but no power is dissipated in the network or line, because both are composed of pure reactances.

Above cutoff frequency, a system consisting of a set of sections in tandem with an iterative-impedance termination is purely reactive throughout. No average power will be absorbed in the terminating reactance, hence none can be transmitted by the network. Rather the energy movement in the system will be oscillatory, and volt-amperes, wherever measured, will be reactive.

d. Properties of Iterative Transfer Function

The attenuation and phase functions may be designated separately. As before, let

$$\Gamma = \bar{\alpha} + j\bar{\beta} \tag{8-24}$$

---

*Exact calculations of attenuation and input impedance for multisection constant-*k* low-pass filters which are terminated with the zero-frequency value of Z_i yield good agreement with those predicted on the assumption of iterative impedance termination, except in the vicinity of the cutoff frequency.

The iterative transfer function has differing properties in the pass and stop bands.

(1) *Pass Band.* If $(\omega/\omega_c) < 1$, Eq. 8-37 is applicable. The complex argument of the logarithm must be converted to polar form:

$$\left| j\frac{\omega}{\omega_c} + \sqrt{1 - \left(\frac{\omega}{\omega_c}\right)^2} \right| = \left[\left(\frac{\omega}{\omega_c}\right)^2 + 1 - \left(\frac{\omega}{\omega_c}\right)^2 \right]^{1/2} = 1 \quad (8\text{-}39)$$

Hence
$$\Gamma = 2\left\{ 0 + j\tan^{-1}\left[\frac{(\omega/\omega_c)}{\sqrt{1 - (\omega/\omega_c)^2}} \right] \right\} \quad (8\text{-}40)$$

Thus
$$\bar{\alpha} = 0 \quad (8\text{-}41)$$

The expression for $\bar{\beta}$, the imaginary part of Eq. 8-40, may be converted by means of the following identity[1] to a more compact form:

$$\sin^{-1} u = \tan^{-1}\left(\frac{u}{\sqrt{1 - u^2}} \right) \quad (8\text{-}42)$$

$$\bar{\beta} = 2\sin^{-1}\left(\frac{\omega}{\omega_c} \right) \quad (8\text{-}43)$$

A phasor diagram is shown in Fig. 8-7, and the phase function is shown as a function of frequency in Fig. 8-8. For $\omega \ll \omega_c$, a series approximation to Eq. 8-43 is informative:

$$\sin^{-1}x \approx x + \frac{x^3}{6} + \cdots \qquad x^2 < 1$$
$$\bar{\beta} \approx 2\frac{\omega}{\omega_c} + \frac{1}{3}\left(\frac{\omega}{\omega_c}\right)^3 + \cdots \quad (8\text{-}44)$$

The first term represents a straight line and is the low-frequency asymptote of Fig. 8-8.

A related function, of interest as an indicator of the phase distortion produced by a section of the network, is the *delay time*. Ideally it should

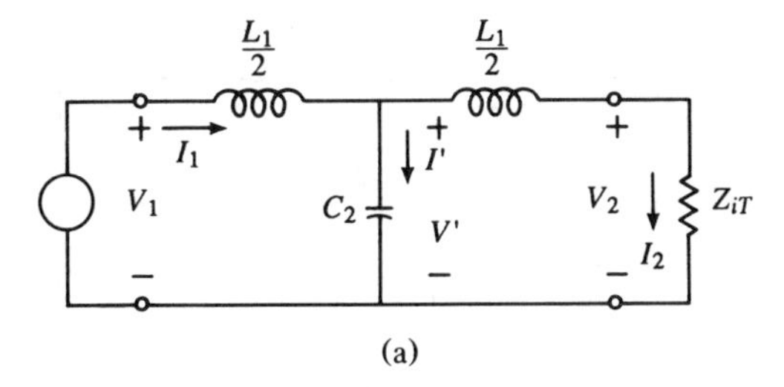

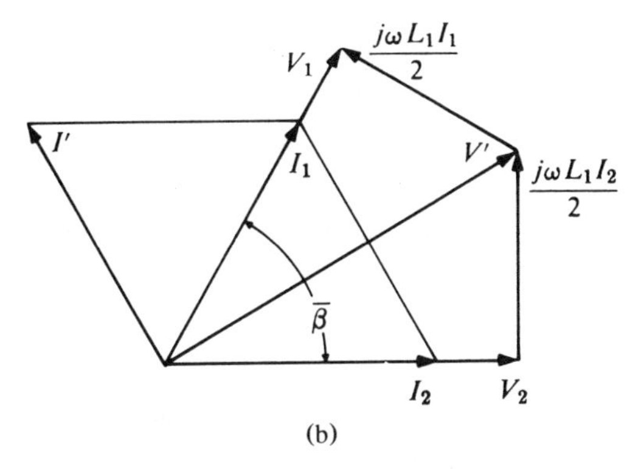

Figure 8-7. Constant-*k* low-pass filter, T section. (a) Circuit diagram. (b) Phasor diagram for pass band.

be independent of frequency. Let

$$v_{\text{in}} = V_0 \cos \omega t \tag{8-45}$$

$$v_{\text{out}} = V_0 \cos(\omega t - \bar{\beta})$$

$$= V_0 \cos \omega \left(t - \frac{\bar{\beta}}{\omega} \right) \tag{8-46}$$

Let the *delay time* t_D be defined as

$$t_D = \frac{\bar{\beta}}{\omega} \tag{8-47}$$

If $\bar{\beta}$ were directly proportional to ω, the delay time would be the same for all frequencies, and a composite wave consisting of components of several

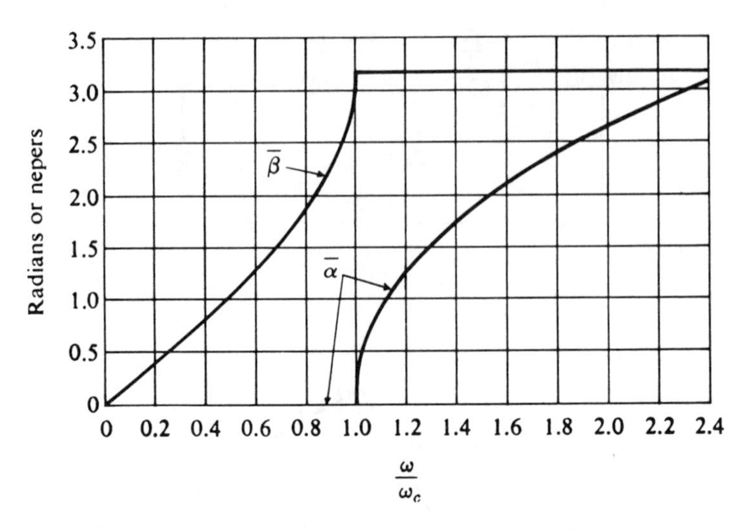

Figure 8-8. Attenuation and phase functions of constant-k low-pass filter section.

frequencies would undergo no phase distortion in passing through the network. This condition is satisfied approximately in the lower part of the pass band to the extent indicated by Eq. 8-44. If Eq. 8-33 is substituted for ω_c in Eq. 8-44, and the first term of that result in Eq. 8-47, the following approximation is obtained:

$$t_D \approx \sqrt{L_1 C_2} \text{ s/section} \qquad \omega \ll \frac{2}{\sqrt{L_1 C_2}} \tag{8-48}$$

(2) *Stop Band.* For $(\omega/\omega_c) > 1$, Eq. 8-38 is the appropriate form rather than Eq. 8-37. In this instance the argument of the logarithm will be an imaginary number. Thus

$$\Gamma = 2\left\{\ln\left[\frac{\omega}{\omega_c} + \sqrt{\left(\frac{\omega}{\omega_c}\right)^2 - 1}\right] + j\frac{\pi}{2}\right\} \tag{8-49}$$

Hence

$$\bar{\beta} = \pi \tag{8-50}$$

The real part of Eq. 8-49 may be converted to a more compact form by means of the identity[1]

$$\cosh^{-1} u = \ln\left(u + \sqrt{u^2 - 1}\right) \tag{8-51}$$

$$\bar{\alpha} = 2\cosh^{-1}\left(\frac{\omega}{\omega_c}\right) \tag{8-52}$$

The attenuation function is also shown in Fig. 8-8 as a function of frequency. Since the iterative impedance in the stop band is purely reactive, "attenuation" consists of a reduction in volt-amperes reactive between input and output, rather than an absorption of power.

The behavior of the constant-k low-pass filter *in the stop band* has no resemblance to that of a distributed-parameter transmission line.

8-3 STEP-FUNCTION RESPONSE OF LADDER-NETWORK DELAY LINE[5, 6]

The transform of the step function includes the complete frequency spectrum from zero to infinity, part of which is necessarily above the cutoff frequency of any given ladder-network delay line. Whereas the phase function of a lossless distributed-parameter line (Eq. 4-29) is directly proportional to frequency throughout the spectrum, and attenuation is zero, Fig. 8-8 indicates that linearity is approached for the phase function of the ladder-network delay line only between zero and a rather small fraction of the cutoff frequency, and that attenuation is nonzero above the cutoff frequency. Marked differences in the respective step-function responses may therefore be anticipated.

For a study of the key features of the transient response of a lumped-parameter delay line, a lossless line with an infinite number of sections will be postulated. It may have either a "T" or π input configuration, as indicated in Fig. 8-9. In either case a step-function voltage source is connected directly to the line,* and the line is initially uncharged.

---

*The formal solution for a more complicated situation, that in which the source includes a series lumped resistance equal to $\sqrt{L_1/C_2}$, is derived in Carslaw and Jaeger.[4]

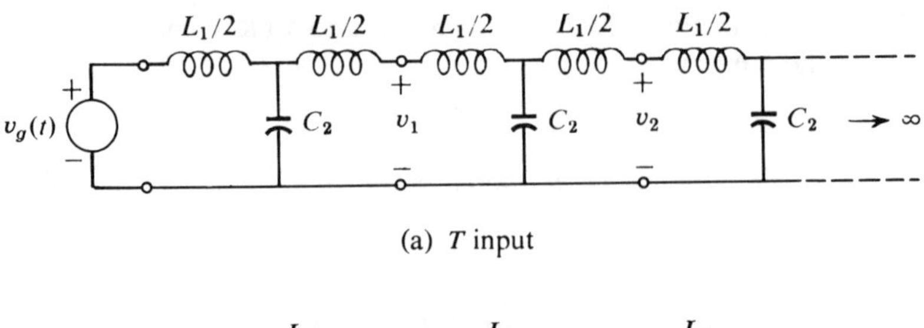

(a) T input

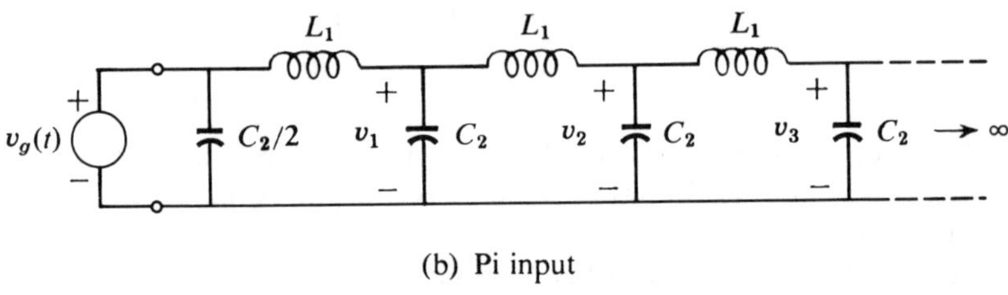

(b) Pi input

Figure 8-9. Infinite ladder-network delay lines.

a. Transform Solution for Voltages

The input impedance of the portion of the line to the right of any section junction is equal to the iterative impedance; hence in either network the voltages at consecutive junctions will be related by the operational iterative transfer function. This may be obtained by substituting the following functions in Eq. 8-13:

$$Z_1(s) = L_1 s \tag{8-53}$$

$$Z_2(s) = \frac{1}{C_2 s} \tag{8-54}$$

$$\epsilon^{\Gamma} = \frac{L_1 C_2}{4}\left(\sqrt{s^2 + \frac{4}{L_1 C_2}} + s\right)^2 \tag{8-55}$$

$$\epsilon^{-\Gamma} = \frac{1}{\epsilon^{\Gamma}} \tag{8-56}$$

If Eq. 8-55 is substituted in Eq. 8-56, and numerator and denominator are both multiplied by $(\sqrt{s^2 + 4/L_1C_2} - s)^2$, we have

$$\epsilon^{-\Gamma} = \frac{L_1C_2}{4}\left(\sqrt{s^2 + \frac{4}{L_1C_2}} - s\right)^2 \tag{8-57}$$

The following relates the voltage transform at the right-hand terminals of the nth section to that at its left-hand terminals:

$$V_n(s) = \epsilon^{-\Gamma}V_{n-1}(s) \tag{8-58}$$

The relationship between the former voltage transform and that at the line input is

$$V_n(s) = \epsilon^{-n\Gamma}V_g(s) \tag{8-59}$$

Let

$$\mathrm{V_g(s)} = \frac{\mathrm{V_0}}{\mathrm{s}} \tag{8-60}$$

Substitution of Eqs. 8-57 and 8-60 in Eq. 8-59 yields

$$V_n(s) = \frac{V_0}{s}\left(\frac{L_1C_2}{4}\right)^n\left(\sqrt{s^2 + \frac{4}{L_1C_2}} - s\right)^{2n} \tag{8-61}$$

The following tabulated transform corresponds to $sV_n(s)$[1]:

$$\mathscr{L}^{-1}\left[\left(\sqrt{s^2 + a^2} - s\right)^m\right] = \frac{ma^mJ_m(at)}{t}U(t) \tag{8-62}$$

The inverse transform of $V_n(s)$ is related to the inverse transform of $sV_n(s)$ in the following manner:

$$\mathscr{L}^{-1}[V_n(s)] = \int_0^t\{\mathscr{L}^{-1}[sV_n(s)]\}\,d\tau \tag{8-63}$$

Hence

$$v_n(t) = 2nV_0\left[\int_0^t J_{2n}\left(\frac{2\tau}{\sqrt{L_1C_2}}\right)\frac{d\tau}{\tau}\right]U(t) \tag{8-64}$$

(Extrema of this integral occur when the function $J_{2n}(2t/\sqrt{L_1C_2})$ is zero. Corresponding values of the Bessel function argument have been tabulated[1] for orders zero through 8.) The integral may be simplified by a change of variables. Let

$$T = \frac{2\tau}{\sqrt{L_1C_2}} \tag{8-65}$$

$$v_n(t) = 2nV_0\left[\int_0^{2t/\sqrt{L_1C_2}} J_{2n}(T)\frac{dT}{T}\right]U(t) \tag{8-66}$$

The integral expression in Eq. 8-66 may be replaced by a series of $n + 1$ terms[1]:

$$v_n(t) = V_0\left[1 - \frac{\sqrt{L_1C_2}}{t}\sum_{k=1}^{n}(2k-1)J_{2k-1}\left(\frac{2t}{\sqrt{L_1C_2}}\right)\right]U(t) \tag{8-67}$$

The latter form is convenient for small values of $2t/\sqrt{L_1C_2}$, those within the range of tables.

b. Asymptotic Approximation for Voltage

In order to study the voltage functions for large values of time, in particular for t approaching infinity, the asymptotic approximation for $J_m(u)$ is useful[1]:

$$J_m(u) \approx \sqrt{\frac{2}{\pi u}}\left[\cos\left(u - \frac{\pi}{4} - \frac{m\pi}{2}\right) - \frac{4m^2 - 1}{8u}\sin\left(u - \frac{\pi}{4} - \frac{m\pi}{2}\right) + \cdots\right] \tag{8-68}$$

The first term within the brackets will yield a satisfactory approximation if

$$\frac{4m^2 - 1}{8u} \ll 1$$

In terms of the problem at hand, this inequality corresponds to

$$\frac{\left[4(2k-1)^2 - 1\right]\sqrt{L_1C_2}}{16t} \ll 1$$

or, roughly

$$\frac{t}{\sqrt{L_1 C_2}} \gg k^2$$

The parameter m in Eq. 8-68 may be replaced by $2k$:

$$\cos\left(u - \frac{\pi}{4} - k\pi\right) = (-1)^k \cos\left(u - \frac{\pi}{4}\right) \tag{8-69}$$

The following relationship may be verified by trial:

$$\sum_{k=1}^{n} (2k - 1)(-1)^k = n(-1)^n \tag{8-70}$$

Substitution of Eqs. 8-68, 8-69, and 8-70 in Eq. 8-67 yields the following asymptotic approximation:

$$v_n(t) \approx V_0\left[1 - \frac{(L_1C_2)^{3/4}}{\sqrt{\pi}\, t^{3/2}} n(-1)^n \cos\left(\frac{2t}{\sqrt{L_1C_2}} - \frac{\pi}{4}\right)\right] t \Big/ \sqrt{L_1C_2} \gg n^2 \tag{8-71}$$

The limiting radian frequency indicated in Eq. 8-71 is equal to the cutoff radian frequency, ω_c (Eq. 8-33). In accordance with the comment following Eq. 8-64, and by applying Eq. 8-68 to find the intervals in u between consecutive zeros of $J_m(u)$, the "instantaneous frequency" of the oscillation during the limiting process may be investigated. The cosine and sine components of Eq. 8-68 may be combined to yield

$$J_m(u) \approx \sqrt{\frac{2}{\pi u}} \sqrt{1 + \left(\frac{4m^2 - 1}{8u}\right)^2} \cdot \cos\left[u - \frac{\pi}{4} - \frac{m\pi}{2} + \tan^{-1}\left(\frac{4m^2 - 1}{8u}\right)\right] \tag{8-72}$$

Let
$$\tan^{-1}\left(\frac{4m^2 - 1}{8u}\right) \approx \frac{4m^2 - 1}{8u} \tag{8-73}$$

Two consecutive roots of $J_m(u)$ in this range may be designated by u_{r1} and u_{r2}. The corresponding arguments of the cosine function in Eq. 8-72 would differ by π radius. Let

$$u_{r1} - u_{r2} = \pi + \Delta u \tag{8-74}$$

Substitution yields

$$\Delta u \approx -\frac{4m^2 - 1}{8}\left(\frac{1}{u_{r2}} - \frac{1}{u_{r1}}\right)$$

$$\approx -\frac{4m^2 - 1}{8}\left(\frac{1}{u_{r1} + \pi} - \frac{1}{u_{r1}}\right)$$

$$\Delta u \approx \frac{4m^2 - 1}{8} \cdot \frac{\pi}{(u_{r1} + \pi)u_{r1}} \tag{8-75}$$

Thus the individual oscillations all have longer periods than that for ω_c, but at increasing values of u they progressively approach the latter. This is consistent with computed differences between tabulated values[1] for the roots of $J_m(u)$; they yield, for $m > 0$, intervals that are initially greater than π, by amounts which increase with m, but these decrease with increasing u.

c. Voltage Waveform Characteristics

Figure 8-10 shows some computed results which may be compared with wavefronts on a lossless distributed-parameter line, Fig. 3-2. In the distributed-parameter line a time $z\sqrt{lc}$ elapses before any voltage appears at location z, after which the voltage impressed by the source is reproduced at delayed time $t - z\sqrt{lc}$. In the lumped-parameter delay line: (1) Voltage, exceedingly small perhaps, but definitely nonzero, appears in every section of the network as soon as t becomes positive. (2) The voltage functions do not rise instantaneously, but have slopes which change at finite rates. (3) The time at which $v_n(t)$ reaches $0.5V_0$ increases approximately linearly with n. (4) After the rise in voltage at a given junction, a ringing oscillation ensues. The asymptotic approximation, Eq. 8-71, indicates a radian frequency equal to $2/\sqrt{L_1C_2}$, the same as the cutoff radian frequency ω_c (Eq. 8-33) for steady-state operation. This oscillation is, in a lossless delay line, not exponentially damped, but decreases in amplitude as $t^{-3/2}$.

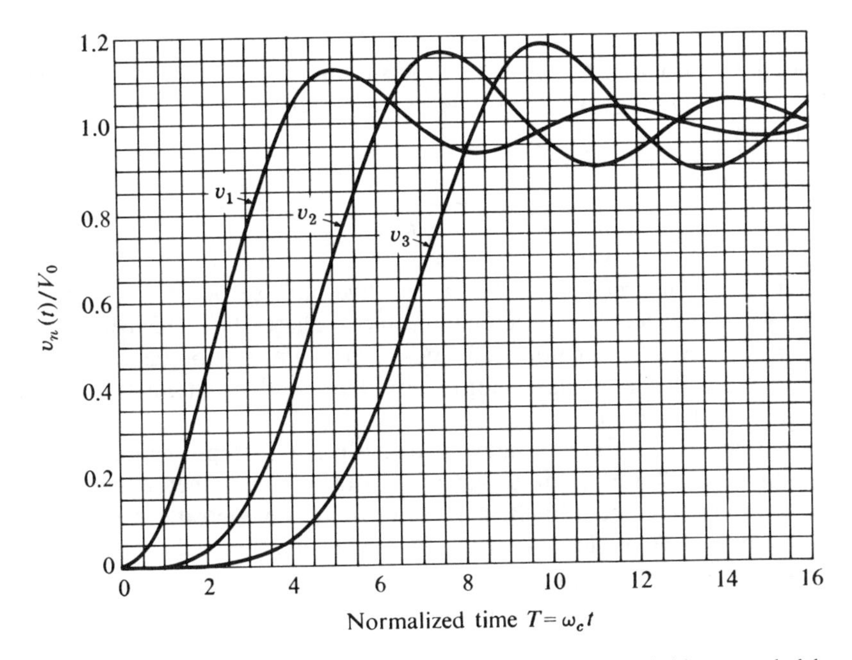

Figure 8-10. Responses to step-function voltage on infinite lossless ladder-network delay line.

Practical delay lines have resistive losses, but realistic theoretical analysis of that effect would be difficult, since skin effect and proximity effect in the coils will make the resistance and inductance per section frequency-dependent.[9] Solutions which include frequency-independent resistance and conductance have been derived in the literature.[3, 4, 12]

8-4 CONCLUSIONS

The behavior of symmetrical two-terminal-pair networks may be described in a manner resembling that for a distributed-parameter transmission line. The iterative impedance and iterative transfer function are derived parameters analogous to the characteristic impedance and the propagation function.

A tandem assembly of symmetrical two-terminal-pair sections in which the series elements are inductive and the shunt elements are capacitive is known as a ladder-network delay line. It has current and voltage relationships at the junctions between sections which correspond closely to those on a transmission line, provided the frequency is sufficiently low. Such a network is also a low-pass filter (specifically, a constant-k filter if its elements are purely reactive), and is characterized by a radian cutoff frequency ω_c:

$$\omega_c = \frac{2}{\sqrt{L_1 C_2}} \qquad (8\text{-}33)$$

Below the cutoff frequency a purely reactive filter section has a resistive iterative impedance, zero attenuation, and a phase shift which increases with frequency. Above cutoff frequency such a filter section has a reactive iterative impedance, a phase shift of π radians, and an attenuation which increases with frequency.

The response of a ladder-network delay line to a step-function voltage has inherent differences from that of a distributed-parameter line. Voltages on the delay line initially increase at finite, varying rates rather than instantaneously. On a lossless or low-loss delay line they will overshoot the final-value asymptote and oscillate about it at a frequency which, at large values of time, approaches the cutoff frequency.

PROBLEMS

8-1. Sketch a phasor diagram similar to Fig. 8-7(b) for a pi-section constant-k low-pass filter in the pass band. Accompany it with an adequately labeled circuit diagram.

8-2. A ladder-network delay line is to have a nominal (low-frequency asymptote) delay time of 1.5 μs and $Z_{i\pi}(0) = 50\ \Omega$. If the cutoff frequency is to be 25 MHz, what values of inductance and capacitance should be used, and how many sections in tandem will be needed?

8-3. The multisection delay line shown in Fig. 8-1 is to be terminated with a resistance equal to $Z_{i\pi}(0)$. In terms of equivalent traveling-wave theory, (a) what is the value of the reflection coefficient at $\omega = 0.5\omega_c$ and (b) what is the input impedance at the other end of the line at that frequency in terms of $Z_{i\pi}(0)$?

8-4. A symmetrical π section, Fig. 8-2(b) is to represent exactly, at a single frequency, the terminal relationships of a transmission line of length z_r. Verify that the following values are correct:

$$Z_1 = Z_0 \sinh \gamma z_r$$

$$Z_2 = 0.5 Z_0 \coth(\gamma z_r/2)$$

(*Hint:* Calculate the elements of the corresponding *ABCD* matrix, and compare those results with Eq. 6-72.)

8-5. Repeat the preceding problem, using the following values for the impedances in Fig. 8-2(a) of a symmetrical "T" section:

$$Z_1 = 2Z_0 \tanh(\gamma z_r/2)$$

$$Z_2 = Z_0 \operatorname{csch} \gamma z_r$$

8-6. Suppose that the delay line which was designed in Prob. 8-2 is energized by a step-function voltage $V_0U(t)$. By scaling from Fig. 8-10, find the times at which voltages v_1, v_2, and v_3 would reach $0.5V_0$.

REFERENCES

1. Abramowitz, M. and Stegun, I. A., *Handbook of Mathematical Functions*, Dover, New York, 1965, 73, 411, 484, 1025.
2. Blackburn, J. F., *Components Handbook*, MIT Radiation Laboratory Series, vol. 17, McGraw-Hill, New York, 1949, 191.
3. Bohn, E. V., *The Transform Analysis of Linear Systems*, Addison-Wesley, Reading, MA, 1963.
4. Carslaw, H. S. and Jaeger, J. C., *Operational Methods in Applied Mathematics*, Oxford University Press, London, 1948.
5. Carson, J. R. and Zobel, O. J., Transient oscillations in electric wave filters, *Bell Syst. Tech. J.*, 2, 1, 1923.
6. Carson, J. R., *Electric Circuit Theory and Operational Calculus*, McGraw-Hill, New York, 1926, 117.
7. Grivet, P., *The Physics of Transmission Lines at High and Very High Frequencies*, vol. 1, *Primary and Secondary Parameters, Traveling Waves, Pulses*. Masson et Cie, Paris, 1969 (in French). English translation Academic, New York, 1970, 389.
8. Lewis, I. A. D. and Wells, F. H. *Millimicrosecond Pulse Techniques*, Pergamon, New York, 1959, 59.
9. Medhurst, R. G., HF resistance and self capacitance of single layer solenoids, *Wireless Engineer*, 24, 35, 80, February and March, 1947.
10. Millman, J. and Taub, H. *Pulse, Digital and Switching Waveforms*, McGraw-Hill, New York, 1965, 800.
11. Turner, A. H., Artificial lines for video distribution and delay, *RCA Review*, X, no. 4, 477, 1949.
12. Weber, E., *Linear Transient Analysis*, vol. II, Wiley, New York, 1956.

CHAPTER 9

Electromagnetic Fields and Maxwell's Equations

Classical macroscopic electromagnetic theory does not take account of quantized, atomic-scale effects, but postulates a set of mathematically described *vector fields* that are "smooth" and continuous in space. This assumption is generally valid for transmission line and waveguide applications at all frequencies from zero through the microwave range.* Interrelationships among the fields may then be formulated within an incremental-sized region, the region may be shrunk to a geometric point by limiting processes, and the results expressed as a set of differential equations known as *Maxwell's equations*.

The principal fields will be designated as (1) magnetic flux density† ($\boldsymbol{B}$, teslas); (2) magnetic field intensity ($\boldsymbol{H}$, amperes per meter); (3) electric flux density ($\boldsymbol{D}$, coulombs per square meter); and (4) electric field intensity ($\boldsymbol{E}$, volts per meter). They must be related to one another and to the circuit quantities of current and voltage in such a manner as to account for observed mechanical forces and other effects within any "electrically activated" system.

---

*In at least one important realm in transmission-line technology, namely, the behavior of metals at low temperature, physical electric phenomena arise which are not consistent with the basic assumption just cited. Two of these are: (1) *superconductivity*, or the disappearance of any detectable resistivity, and (2) *anomalous skin effect*, which will cause the effective resistance to alternating current to be higher than that predicted by macroscopic theory. These effects are discussed briefly in Appendix E.

†Those vectors which have physical connotations will be designated by boldface italic capital letters; the boldface italic lowercase letter $\boldsymbol{a}$, with any subscript, will indicate a *unit vector*, one which has unit magnitude (dimensionless) and simply defines direction. A unit vector may be multiplied by a dimensioned scalar constant or function to represent a physical quantity.

To obtain explicit solutions for the fields in particular geometrical situations, such as the space between two infinitely long concentric cylinders (coaxial cable) or within an infinitely long rectangular tube (rectangular waveguide), coordinate systems are introduced.

9-1 AN OVERVIEW OF CLASSICAL ELECTROMAGNETICS[3, 7, 8, 10]

It is helpful to review some of the basic concepts of the electromagnetic fields, with emphasis on those aspects which bear directly on wave propagation. In so doing it will be assumed that the reader has studied (a) the operations and functions of vector analysis (key definitions and formulas are summarized in Appendix A) and (b) the basic concepts of static and quasistatic electric and magnetic field theory. The latter will be recapitulated here in order to provide a logical base for applications of Maxwell's equations and the accompanying boundary conditions.

The following elementary phenomena indicate the *space pervasiveness* of electromagnetic effects:

1. The presence of current in two circuits in proximity to each other is accompanied by mechanical forces on each conductor, forces which change if either current is changed.
2. A changing of the current in either of two such circuits is accompanied by an induced voltage in the other.
3. Capacitors consisting of metallic spheres or other conducting bodies suspended in vacuum or in an insulating medium may be charged and later discharged. During these processes wire-borne current flows onto one sphere and off the other.
4. The presence of electric charges on two bodies is accompanied by a mechanical force on each, forces which change if either charge is changed.

Mechanical-force effects, items 1 and 4, have been mentioned primarily because they help in assigning directions to the fields. Item 2 states an application of Faraday's law, and item 3 describes a situation in which the displacement current, postulated by Maxwell, complements a discontinuous conduction current to yield, as a resultant, a composite current which is continuous.

a. Directional Properties of the Electric and Magnetic Fields

The direction to be assigned to a vector field is an important part of the definition of the function and is chosen in the light of the physical situation to which it relates. It is rather obvious for the electric field, but less so for the magnetic field.

Curves may be faired in a vector field such that they are continuously tangent to the local direction of the field; such lines are called *field-direction lines*, and for many physical fields, have easily identifiable properties.

(1) *Electric Field.* The direction of a static electric field at a given location is defined as that of the force which would be exerted on a small, positively charged test sphere. The direction lines of a static electric field extend from charges of one polarity to charges of the opposite polarity.

(2) *Magnetic Field.* A small, circular test coil may be visualized for investigation of the vector-related aspects of the magnetic-field effects noted in items 1 and 2. If current is flowing in the test coil while the coil is in an external field, the differential forces acting on the various parts of the coil will yield, in general, a resultant translational force and a torque. A coil orientation can be found, at each location, for which the torque vanishes.

Let the generator supplying current to the test coil be replaced by a ballistic galvanometer,[5, 6] and the external field suddenly strengthened or weakened. In accordance with Faraday's law, a current surge will pass through the galvanometer. The galvanometer response, as a function of the orientation of the test coil at the given location, will be greatest when the orientation is the same as that for torque equilibrium in the preceding test. The normal to the plane of the test coil when it is thus oriented is designated as the alignment of the field.

Direction lines of the static magnetic field (if one excludes the phenomenon of permanent magnetism) form closed loops around current-carrying conductors or parts of such conductors. Conventionally the positive direction of the magnetic field is defined as clockwise around a current which is flowing away from the observer.

b. Quantitative Properties

For quantitative study of vector fields, the line integral and surface integral, which are reviewed in Sec. A-1b, are useful in that they may be identified with circuit-type quantities.

(1) *Static Fields in a Homogeneous Medium.* For the present, let the electric-field magnitude at a given location be defined as proportional to the magnitude of the force which would be exerted on a small, positively charged test sphere, divided by the magnitude of the test charge. The magnetic-field-direction experiment with a current-carrying test coil may be extended in the following manner to measure the field strength under static conditions: If one rotates the plane of the test coil say 90° from the equilibrium orientation, the resulting torque will be proportional to the magnetic field.*

The following results are obtained if the medium surrounding the charged bodies or the coil is homogeneous:

1. The surface integral of the magnetic field over an enclosing surface is zero.
2. The surface integral of the electric field over an enclosing surface is proportional to the charge enclosed.
3. The line integral of the magnetic field around a closed path is proportional to the current encircled.
4. The line integral of the electric field around a closed path is zero, and the line integral between any two points is independent of the path of integration chosen.

Results (2) and (3) are the bases for the definition of the magnitudes of the $\boldsymbol{D}$ and $\boldsymbol{H}$ fields; in the SI, or International System of Units, those proportionalities are equalities:

$$\oiint \boldsymbol{D} \cdot d\boldsymbol{S} = \sum q \tag{9-1}$$

Here $\boldsymbol{D}$ has the units coulombs per square meter, area is measured in square meters, and q has the units coulombs.

$$\oint \boldsymbol{H} \cdot d\boldsymbol{L} = i \tag{9-2}$$

Here $\boldsymbol{H}$ has the units amperes per meter, incremental length of path is in meters, and i is in amperes. This is Ampère's law for static magnetic fields.

---

*The philosophy behind these modes of definition is that of developing elementary theory in a self-contained manner. Laboratory methods for precise measurement of fields utilize results of more extensive analysis.

(2) Properties of Physical Media and Their Effects. As was mentioned in Sec. 1-3a, various physical media have differing properties with respect to electric and magnetic fields; these may be expressed in terms of three parameters, the permittivity ϵ (farads per meter), the conductivity σ (siemens per meter), and the permeability μ (henrys per meter). In the discussion that follows, the parameters ϵ, σ, and μ will be assumed to be constant at any given location (not functions of the strengths of the $\boldsymbol{E}$, $\boldsymbol{D}$, $\boldsymbol{H}$, or $\boldsymbol{B}$ fields) and frequency-independent.* They will further be assumed to have the same values for all orientations of the fields, and hence each may be described with a scalar.†

An increase in permittivity in the region of an electric field, leaving the geometry and the charge distribution unchanged, will not affect the $\boldsymbol{D}$ field but will decrease the force exerted on the test charge by the ratio of the permittivities. Specifically, let the $\boldsymbol{E}$ field be defined as follows:

$$\boldsymbol{E} = \boldsymbol{D}/\epsilon \tag{9-3}$$

The units of $\boldsymbol{E}$ are volts per meter; those of ϵ and $\boldsymbol{D}$ have been noted previously.

The force on the test charge is directly proportional to the $\boldsymbol{E}$ field, regardless of ϵ.

In composite media the line integral of a static $\boldsymbol{E}$ field proves to be independent of path, whereas that is not true in general of $\boldsymbol{D}$. The line integral of $\boldsymbol{E}$ in a static field is defined as the *voltage*:

$$V_{ab} = -\int_b^a \boldsymbol{E} \cdot d\boldsymbol{L}$$

$$= \int_a^b \boldsymbol{E} \cdot d\boldsymbol{L} \tag{9-4}$$

---

*The permittivity and conductivity of a solid dielectric commonly vary to some extent with frequency.[2, 4] (See the first footnote in Chap. 5 for comments and references.) If this is to be taken into account for fields varying other than sinusoidally with time, Maxwell's equations should be formulated on the basis of harmonic time dependence, with superposition by means of the Laplace transform, a process similar to that followed in Chap. 5 for the skin-effect-lossy transmission line.

†It is necessary in some physical media that one or more of these quantities be *tensors*, that is to say, that a given proportionality described in Maxwell's equations is not the same in all directions at a given point, and in general the two fields related by the given parameter are not collinear. Such a material is said to be *anisotropic* with respect to that property. *Ferrites*[1, 9] constitute a class of anisotropic materials important in microwave techniques, but their analysis is beyond the scope of the present text. Anisotropy is also a cardinal aspect of such phenomena as magnetostriction and the piezoelectric behavior of crystals.

Likewise the forces arising from magnetic fields are altered by the presence of ferromagnetic materials, and it is expedient to define the $\boldsymbol{B}$ field as follows:

$$\boldsymbol{B} = \mu \boldsymbol{H} \tag{9-5}$$

Here $\boldsymbol{B}$ has the units teslas, and those of μ and $\boldsymbol{H}$ have been noted previously.

In composite media the enclosed surface integral of $\boldsymbol{B}$ always vanishes, whereas that is not true in general for $\boldsymbol{H}$:

$$\oiint \boldsymbol{B} \cdot d\boldsymbol{S} = 0 \tag{9-6}$$

(3) *Current: Conduction and Displacement.* Conduction current consists of the movement of charge. When attention is directed to its distribution within a conducting mass, it is properly regarded as a vector field $\boldsymbol{J}$, with units of amperes per square meter. It is related to the static electric field in a conducting medium as follows:

$$\boldsymbol{J} = \sigma \boldsymbol{E} \tag{9-7}$$

As noted in the discussion of surface integrals in Sec. A-1b, $\boldsymbol{J}$ and i are related in the following manner:

$$\iint \boldsymbol{J} \cdot d\boldsymbol{S} = i \tag{A-7}$$

In terms of the charging of a capacitor, the positive charge accumulated on one conducting surface by reason of current flowing onto that plate is

$$q = \int i\, dt \tag{9-8}$$

The integral of $\boldsymbol{D}$ over a surface which surrounds that plate, except for the cross section of the wire which brings in the conduction current, would increase at the same rate as q increases

$$\iint \frac{\partial \boldsymbol{D}}{\partial t} \cdot d\boldsymbol{S} = i$$

The time rate of change of $\boldsymbol{D}$ thus has the nature of a current density, and proves to have corresponding properties in terms of magnetic-field effects. It is the displacement-current density, and will be designated $\boldsymbol{J}_D$:

$$\boldsymbol{J}_D = \frac{\partial \boldsymbol{D}}{\partial t} \tag{9-9}$$

Its units, like those of $\boldsymbol{J}$, are amperes per square meter. Under the dynamic conditions of time-varying fields, Eq. 9-2 must be generalized to the following:

$$\oint \boldsymbol{H} \cdot d\boldsymbol{L} = \iint \left(\frac{\partial \boldsymbol{D}}{\partial t} + \boldsymbol{J} \right) \cdot d\boldsymbol{S}$$

$$= \iint \left(\epsilon \frac{\partial \boldsymbol{E}}{\partial t} + \sigma \boldsymbol{E} \right) \cdot d\boldsymbol{S} \tag{9-10}$$

This is the general form of Ampère's law.

(4) *Electromagnetic Induction.* The changing of current in a coil changes the magnetic field associated with it, and the latter is regarded as the mechanism whereby voltage is induced in a nearby circuit. The voltage induced in a single-turn coil is equal to the time rate of change in magnetic flux passing through the coil, or, in terms of line and surface integrals,

$$\oint \boldsymbol{E} \cdot d\boldsymbol{L} = - \iint \frac{\partial \boldsymbol{B}}{\partial t} \cdot d\boldsymbol{S} \tag{9-11}$$

This is Faraday's law of electromagnetic induction.

Lenz's law serves as a memory aid with respect to the negative sign in this equation; it states that the voltage induced by a change in magnetic field is so directed as to tend to produce a current which will create a component magnetic field in opposition to the impressed change.

c. Absolute and Relative Permittivity and Permeability[5, 6]

The absolute permittivity and permeability of physical media are related to the respective values for free space, ϵ_0 and μ_0 as follows:

$$\epsilon = \kappa_\epsilon \epsilon_0 \tag{9-12}$$

$$\mu = \kappa_\mu \mu_0 \tag{9-13}$$

Here κ_ϵ and κ_μ, which are dimensionless, are the relative permittivity and relative permeability, respectively.

The permeability of free space μ_0 is defined in the SI system as

$$\mu_0 = 4\pi \times 10^{-7} \text{ H/m} \tag{9-14}$$

As is shown in Sec. 10-1d, the velocity of propagation of a plane electromagnetic wave in space is related to the permeability and permittivity in the following manner:

$$\nu = \frac{1}{\sqrt{\mu_0 \epsilon_0}} \tag{9-15}$$

This velocity has been determined experimentally as 2.998×10^8 m/s. The permittivity of space ϵ_0 is evaluated from Eq. 9-15:

$$\begin{aligned} \epsilon_0 &= \frac{1}{(4\pi \times 10^{-7})(2.998 \times 10^8)^2} \\ &= 8.854 \times 10^{-12} \text{ F/m} \\ &\approx \frac{10^{-9}}{36\pi} \text{ F/m} \end{aligned} \tag{9-16}$$

9-2 REDUCTION OF FIELD EQUATIONS TO DIFFERENTIAL FORM

Equations 9-1, 9-6, 9-10, and 9-11, which involve line integrals and surface integrals, may be designated as Maxwell's equations in the integral form. Their application to many problems may be expedited by restating them as differential equations. Two differential functions of a vector field are introduced for this purpose, the divergence and the curl. (These are defined in Sec. A-1c, and their forms in rectangular and cylindrical coordinates are listed in Sec. A-2.)

a. Continuity of Electric Flux

$$\oiint \boldsymbol{D} \cdot d\boldsymbol{S} = \sum q \tag{9-1}$$

Electric flux is postulated to emanate from positive charges and to terminate on negative charges. The algebraic sum of such charges within the designated surface of integration in Eq. 9-1 was indicated by Σq. The net outwardly directed flux, $\oint \boldsymbol{D} \cdot d\boldsymbol{S}$, is equal to the algebraic sum of the enclosed charge. For macroscopic applications, charge may be viewed in a mathematical sense which differs from the physical form.

(1) *Electric Charge and Mathematical Functions.* Experimentally, electric charge has been found to be concentrated in electrons and protons, "particles" of submicroscopic, yet nonzero, size. If one accepts the restriction that classical electromagnetic theory will be used only to describe macroscopic phenomena, actual charge distributions may be replaced by forms which are easier to work with mathematically.

The following are idealized charge distributions:

1. A finite charge q concentrated in a geometric point.
2. A finite amount of charge per unit length ρ_L concentrated in a geometric line.
3. A finite charge per unit area ρ_S concentrated in a geometric surface.
4. A continuous distribution of charge density ρ throughout a volume.

Charge density under item 4 constitutes a scalar field.

In the traditional language of vector analysis, built primarily on hydrodynamics, the positive charges are *sources* of electric flux, the negative charges, *sinks*. In some dynamic situations, such as the propagation of energy which has been radiated from an antenna, or certain modes of propagation in wave guides, electric-field-direction lines form closed loops, and there are no sources or sinks. Air and space will be assumed to be charge-free in the field analyses considered in this text, but charges on conducting surfaces will have important roles in satisfying boundary conditions.

If all the charge within a given enclosing surface is in the continuously distributed form, the following substitution may be made:

$$\sum q = \iiint \rho \, dv \tag{9-17}$$

The right-hand side is a volume integral, as described in Sec. A-1b. Equation 9-17 is substituted in Eq. 9-1:

$$\oint\!\!\!\oint \boldsymbol{D} \cdot d\boldsymbol{S} = \iiint \rho \, dv \tag{9-18}$$

(2) *Differential Equation of Continuity.* The following is known as the *divergence theorem*; see Sec. A-1 for references and for the definition of the divergence function, $\nabla \cdot \boldsymbol{D}$:

$$\oint\!\!\oint \boldsymbol{D} \cdot d\boldsymbol{S} = \iiint (\nabla \cdot \boldsymbol{D})\, dv \qquad \text{(A-15)}$$

It is understood that the volume indicated on the right-hand side is that enclosed by the surface indicated on the left-hand side. [As mentioned in Sec. A-1b(3), $d\boldsymbol{S}$ is directed outward.] Equation A-15 may be substituted into Eq. 9-18; the volumes over which the integrals are taken are identical and may include all space or any part thereof. Hence

$$\nabla \cdot \boldsymbol{D} = \rho \qquad \text{(9-19)}$$

In regions devoid of charge, the divergence of $\boldsymbol{D}$ vanishes.

(3) *Normal-Component Constraints on Electric Fields at Boundary Between Dissimilar Media.* A positive surface charge layer (coulombs per square meter, in zero thickness) will serve as the source of the normal component of a $\boldsymbol{D}$ field (also coulombs per square meter), and a negative surface charge layer will terminate one. The presence of a surface charge layer may be anticipated at the boundary between a conducting and a nonconducting medium.

At the boundary between two nonconducting media with different permittivities, if no surface charge layer is present, the component of $\boldsymbol{D}$ normal to the surface must be continuous.

The foregoing constraints have been summarized as equations in Sec. 9-3 at the end of this chapter.

b. Continuity of Magnetic Flux

A corresponding constraint upon the magnetic field was noted in Sec. 9-1b(2):

$$\oint\!\!\oint \boldsymbol{B} \cdot d\boldsymbol{S} = 0 \qquad \text{(9-6)}$$

This is a mathematical statement of the physical observation that sources and sinks of magnetic flux are unknown. The surface integral of $\boldsymbol{B}$ in Eq. 9-6 may be replaced by the volume integral of its divergence. Because the equation is valid for any surface, including an infinitesimally small one surrounding any given point in space, the divergence of $\boldsymbol{B}$ must be identically zero:

$$\nabla \cdot \boldsymbol{B} = 0 \tag{9-20}$$

In order to satisfy this requirement at the boundary between two media with different permeabilities, the normal component of the $\boldsymbol{B}$ field must be continuous.

c. Ampère's Law

The following relationship concerning the magnetic field which accompanies current, whether of the conduction or displacement type, was stated in Sec. 9-1b(3):

$$\oint \boldsymbol{H} \cdot d\boldsymbol{L} = \iint \left(\epsilon \frac{\partial \boldsymbol{E}}{\partial t} + \sigma \boldsymbol{E} \right) \cdot d\boldsymbol{S} \tag{9-10}$$

The surface of integration may be any finite surface which is bounded by the closed curve used as a path for the line integral of $\boldsymbol{H}$.

The following is known as *Stokes' theorem*; see Sec. A-1 for references and for the definition of the curl function, $\nabla \times \boldsymbol{H}$:

$$\oint \boldsymbol{H} \cdot d\boldsymbol{L} = \iint (\nabla \times \boldsymbol{H}) \cdot d\boldsymbol{S} \tag{A-16}$$

It is understood that the surface indicated on the right-hand side is bounded by the closed curve indicated on the left-hand side. Equation A-16 may be substituted in Eq. 9-10:

$$\iint (\nabla \times \mathbf{H}) \cdot d\boldsymbol{S} = \iint \left(\epsilon \frac{\partial \boldsymbol{E}}{dt} + \sigma \boldsymbol{E} \right) \cdot d\boldsymbol{S}$$

The integrations are over the same area, which may extend over any part of space.

$$\nabla \times \boldsymbol{H} = \epsilon \frac{\partial \boldsymbol{E}}{\partial t} + \sigma \boldsymbol{E} \tag{9-21}$$

In order for the curl of $\boldsymbol{H}$ to be finite, the line integral around a closed path must approach zero as the enclosed area approaches zero. Thus if the path is a narrow rectangle whose longer sides are parallel to and on opposite sides of the boundary surface between two media of different permeabilities (and conductivities which are zero or finite), this constrains the components of $\boldsymbol{H}$ in the two media which are tangential to the boundary surface to be equal.

d. Faraday's Law

The principle of electromagnetic induction was given as follows in Sec. 9-1b(4):

$$\oint \boldsymbol{E} \cdot d\boldsymbol{L} = -\iint \frac{\partial \boldsymbol{B}}{\partial t} \cdot d\boldsymbol{S} \tag{9-11}$$

Stokes' theorem, Eq. A-16, may, if written in terms of $\boldsymbol{E}$, be substituted for the left-hand side, giving the following:

$$\iint (\nabla \times \boldsymbol{E}) \cdot d\boldsymbol{S} = -\iint \frac{\partial \boldsymbol{B}}{\partial t} \cdot d\boldsymbol{S}$$

As with Ampère's law, the surface of integration is the same on both sides of the equality sign and may be any surface in space:

$$\nabla \times \boldsymbol{E} = -\frac{\partial \boldsymbol{B}}{\partial t} \tag{9-22}$$

Substitution of Eq. 9-5 yields

$$\nabla \times \boldsymbol{E} = -\mu \frac{\partial \boldsymbol{H}}{\partial t} \tag{9-23}$$

If two media have different conductivities or permittivities, or both, the components of $\boldsymbol{E}$ which are tangential to the boundary surface must be equal there.

9-3 CONCLUSIONS

Macroscopic electromagnetic-wave phenomena may be described by means of (1) four vector fields, $\boldsymbol{B}$, $\boldsymbol{H}$, $\boldsymbol{D}$, and $\boldsymbol{E}$; (2) the parameters of permittivity ϵ, permeability μ, and conductivity σ (constant scalars for the media of interest in this text); and (3) electric-charge volume density ρ (a scalar field). Two of the vector fields are related to the other two by direct proportion:

$$\boldsymbol{D} = \epsilon \boldsymbol{E} \tag{9-3}$$

$$\boldsymbol{B} = \mu \boldsymbol{H} \tag{9-5}$$

Maxwell's differential equations constrain and interrelate the fields, independently of any specific coordinate system, by means of two rate-of-change-with-distance functions of a vector field, the divergence and the curl. These equations are

$$\nabla \cdot \boldsymbol{D} = \rho \tag{9-19}$$

$$\nabla \cdot \boldsymbol{B} = 0 \tag{9-20}$$

$$\nabla \times \boldsymbol{H} = \sigma \boldsymbol{E} + \epsilon \frac{\partial \boldsymbol{E}}{\partial t} \tag{9-21}$$

$$\nabla \times \boldsymbol{E} = -\mu \frac{\partial \boldsymbol{H}}{\partial t} \tag{9-23}$$

Two auxiliary vector fields which are convenient in some applications are the conduction-current density $\boldsymbol{J}$ and the displacement-current density $\boldsymbol{J}_D$:

$$\boldsymbol{J} = \sigma \boldsymbol{E} \tag{9-7}$$

$$\boldsymbol{J}_D = \frac{\partial \boldsymbol{D}}{\partial t} \tag{9-9}$$

Actual electrical-charge distributions are approximated by finite, continuous volume-charge-density distributions, or by point charges, line charges, or surface charges.

At the boundary surface between two media for which one or more of the corresponding parameters μ, ϵ, and σ differ, Maxwell's equations reduce to the following constraints on the normal and tangential components of the various fields. (The positive sense of the normal components will be taken as from medium 1 into medium 2.)

From Eqs. 9-20 and 9-23:

$$B_{1\text{nor}} = B_{2\text{nor}} \tag{9-24}$$

$$E_{1\tan} = E_{2\tan} \tag{9-25}$$

The following from Eq. 9-21 is applicable if σ_1 and σ_2 are zero or finite:

$$H_{1\tan} = H_{2\tan} \tag{9-26}$$

(The limiting case in which the conductivity of one medium is infinite will be examined in Sec. 10-2d.)

At the boundary between a conducting medium ($\sigma_1 \neq 0$) and a nonconducting medium ($\sigma_2 = 0$), the requirement for continuity of $\boldsymbol{D}$ reduces to the following for a static field:

$$D_{2\text{nor}} = \rho_S \tag{9-27}$$

For time-varying fields, if medium 1 is a metal, $D_{1\text{nor}}$ is negligible compared to $D_{2\text{nor}}$, and Eq. 9-27 is applicable.

If σ_1 and σ_2 are both zero, and no surface-charge layer is present, this follows from Eq. 9-19:

$$D_{1\text{nor}} = D_{2\text{nor}} \tag{9-28}$$

Coordinate systems are introduced as aids in the mathematical description of the particular fields which exist when specific boundary surfaces are present. The various operations of vector algebra and vector calculus may be stated in terms of components of the appropriate vectors in any of several coordinate systems. Those for the rectangular and circular cylindrical systems are listed in Sec. A-2.

PROBLEMS

9-1. Given the following two vectors, directed outwardly from the origin:

$$\boldsymbol{K} = 7\boldsymbol{a}_x + 10\boldsymbol{a}_y - 5\boldsymbol{a}_z$$

$$\boldsymbol{L} = 8\boldsymbol{a}_x + 6\boldsymbol{a}_y - 9\boldsymbol{a}_z$$

find the angle between $\boldsymbol{K}$ and $\boldsymbol{L}$, (1) by using the scalar product, and (2) by using the vector product.

9-2. The following formula is purported to represent the static magnetic field produced by a direct-current segment $I\Delta z\boldsymbol{a}_z$ (of infinitesimal cross section) in a uniform medium of infinite extent and conductivity σ; $\mu = \mu_0$; $\epsilon = \epsilon_0$.*

$$\Delta\boldsymbol{H} = \frac{I\Delta z r}{4\pi(r^2 + z^2)^{3/2}}\boldsymbol{a}_\phi$$

*The formula given in Prob. 9-2 is the *Biot–Savart law* stated in cylindrical coordinates. In nonconducting media, it is applicable as a differential form for the derivation of the resultant field produced by a closed-path direct current.

(a) Find (1) the current-density field $\boldsymbol{J}$ and (2) the electric field $\boldsymbol{E}$.

(b) Verify, by substitution in Eqs. A-36 and A-37 as appropriate, that the following are satisfied: (1) $\boldsymbol{B}$ is continuous (Eq. 9-20), (2) $\boldsymbol{E}$ is conservative (Eq. 9-23), and (3) the region is charge-free (Eq. 9-19).

9-3. Direct current is flowing in a nonhomogeneous circular conductor of radius r_a. $\boldsymbol{E} = E_0\boldsymbol{a}_z$. The conductivity is a function of radius: $\sigma = \sigma_0[1 + (r^2/8r_a^2)]$. Write an expression for the conduction-current density, and evaluate a surface integral to find the current in the conductor.

9-4. Given two closely spaced, infinitely long conductors carrying equal currents (not varying with time) in opposite directions. The following formula is purported to represent the magnetic field at a radial distance which is large compared to the separation distance h:

$$\boldsymbol{H} = \frac{Ih}{2\pi r^2}\left(\cos\phi\,\boldsymbol{a}_r + \sin\phi\,\boldsymbol{a}_\phi\right) \qquad r \gg h$$

(a) Evaluate $\int \boldsymbol{H} \cdot \boldsymbol{dL}$ for the following path, where $r_2 > r_1$: (1) from r_1 to r_2, along the radial line $\phi = 0$; (2) from $\phi = 0$ to $\phi = \pi/2$, at the constant radius r_2; (3) from r_2 to r_1, along the radial line $\phi = \pi/2$; and (4) from $\phi = \pi/2$ to $\phi = 0$, at the constant radius r_1. For each segment, state explicitly the expression for $\boldsymbol{dL}$, and the corresponding contribution to the integral.

(b) Evaluate the divergence of $\boldsymbol{B}$ by differentiation.

(c) Evaluate the curl of $\boldsymbol{H}$ by differentiation.

9-5. The following potential function is applicable within the region $r < r_a$, $-\infty < z < \infty$. $V = V_0[1 - (r/r_a)^2]$, $\epsilon = \epsilon_0$.

(a) Find $\boldsymbol{E}$ and the volume charge density within that region, and sketch $|\boldsymbol{E}|$ versus r and the volume charge density versus r.

(b) Evaluate (separately) the following integrals for the cylindrical surface bounded by $r = 0.5r_a$, $z = 0$ and $z = 5r_a$: (1) $\oint \boldsymbol{D} \cdot \boldsymbol{dS}$ and (2) $\iiint \rho\, dv$.

9-6. Given the following information on the static field in coaxial cylindrical system:

$$\boldsymbol{E} = 0 \qquad r < r_a$$

$$V = 25.0 \ln(r_b/r)\ \text{V} \qquad r_a < r < r_b, \quad \epsilon = 4\epsilon_0$$

$$V = 100.0 \ln(r_b/r)\ \text{V} \qquad r_b < r, \quad \epsilon = \epsilon_0$$

(a) Find V for $r < r_a$.

(b) Find $\boldsymbol{E}$ for $r > r_a$.

(c) Find the volume charge density, if any, in each region.

(d) Find the surface charge density, if any, at $r = r_a$ and $r = r_b$.

(e) Sketch the foregoing as functions of r for $0 \leq r \leq 1.5r_b$.

9-7. A coaxial line has a solid inner conductor of radius r_a and an outer conductor with inner and outer radii r_b and r_c. Direct current I is flowing in the $\boldsymbol{a}_z$ direction in the inner conductor and in the $-\boldsymbol{a}_z$ direction in the outer. Evaluate the net currents enclosed by circles of radii r_1, r_2, and r_3, where $r_1 < r_a$, $r_a < r_2 < r_b$, and $r_b < r_3 < r_c$. Find the magnetic field in the respective regions, using Ampère's law. Take the curl of the function found for $\boldsymbol{H}$ in each region and verify that it is equal to the corresponding current density.

9-8. Given a vector field $\boldsymbol{A}_1$:

$$\boldsymbol{A}_1 = \frac{A_{10}}{3} r^2 \sin\phi \, \boldsymbol{a}_r + A_{10} r^2 \cos\phi \, \boldsymbol{a}_\phi \quad \text{(cylindrical coordinates)}$$

(a) Find, by differentiation, divergence $\boldsymbol{A}_1$ and curl $\boldsymbol{A}_1$.
(b) If A_{10} had the units of amperes per cubic meter, could $\boldsymbol{A}_1$ represent a static (non-time-varying) magnetic $\boldsymbol{H}$ field in a medium of uniform permeability? (Test with Maxwell's equations.)
(c) If A_{10} had the units of volts per cubic meter, could $\boldsymbol{A}_1$ represent a static electric $\boldsymbol{E}$ field in a medium of uniform permittivity? (Test with Maxwell's equations.)

9-9. A given coaxial line is carrying direct current, and is short-circuited at $z = 0$ with a sheet of infinite conductivity. The electric field in the line conductors (both of finite conductivity) are as follows:

$$\boldsymbol{E} = -E_0 \boldsymbol{a}_z \qquad 0 \le r \le r_a, \; z > 0$$

$$\boldsymbol{E} = E_0 \boldsymbol{a}_z \qquad r_b \le r \le r_c, \; z > 0$$

The volume charge density between the conductors is zero and the permittivity everywhere is ϵ_0. The following is purported to describe the field between the conductors:

$$\boldsymbol{E} = \frac{2E_0}{\ln(r_b/r_a)} \left[\left(\frac{z}{r} \right) \boldsymbol{a}_r + \ln\left(\frac{r}{\sqrt{r_a r_b}} \right) \boldsymbol{a}_z \right] \qquad r_a < r < r_b, \; z > 0$$

Determine whether the latter expression satisfies Maxwell's equations and the boundary conditions.

9-10. Demonstrate the following identities by direct substitution in the rectangular and the circular cylindrical coordinate systems.
(a) $\nabla \times (\nabla V) = 0$
(b) $\nabla \cdot (\nabla \times A) = 0$

REFERENCES

1. Button, K. J., Microwave ferrite devices: The first ten years, *IEEE Trans. Microwave Theory and Techniques*, MTT-32, no. 9, 1088, 1984.
2. Daniel, V. V., *Dielectric Relaxation*, Academic Press, New York, 1967.

3. Dibner, B., *Ten Founding Fathers of the Electrical Science*, Burndy Library, Norwalk, CT, 1954.
4. Freeman, R. L., *Reference Manual for Telecommunications Engineering*, Wiley, New York, 1985, 571.
5. Harris, F. K., *Electrical Measurements*, Wiley, New York, 1952.
6. Kinnard, I. F., *Applied Electrical Measurements*, Wiley, New York, 1956.
7. Moon, P. and Spencer, D. E., *Field Theory for Engineers*, D. Van Nostrand, Princeton, NJ, 1961.
8. Moon, P. and Spencer, D. E., *Foundations of Electrodynamics*, D. Van Nostrand, Princeton, NJ, 1960.
9. Waldron R. A., *Ferrites, An Introduction for Microwave Engineers*, Van Nostrand, London, 1961.
10. Whittaker, E., *A History of the Theories of Aether and Electricity—The Classical Theories*, Philosophical Library, New York, 1951.

CHAPTER 10

Plane Electromagnetic Waves

The simplest illustration of electromagnetic-wave propagation is the hypothetical one of sinusoidal plane waves of infinite extent in a homogenous, charge-free medium. A considerable portion of the wave front of radiation from an antenna closely approaches this hypothetical configuration at long distances from the source, and the solution is applicable in localized regions in other problems.

An elementary situation involving a nonhomogeneous region for the waves is that of two uniform media of semi-infinite extent, with a planar boundary between them. A traveling-wave set of plane fields incident on such a boundary surface will give rise to a reflected field set and a transmitted field set.

Boundary surfaces of high conductivity figure importantly in many applications; a useful limiting case is that in which one medium is assumed to have infinite conductivity.

10-1 SINUSOIDAL TRAVELING-WAVE FIELDS IN INFINITE MEDIUM

Plane waves are defined as waves that are uniform in direction and magnitude in planes of a stated orientation. If rectangular coordinates are adopted one may begin the analysis by assuming that the $\boldsymbol{E}$ field is (1) everywhere parallel to the X axis, and (2) a function of time and of displacement in the $\boldsymbol{a}_z$ direction, but (3) not a function of displacement in the $\boldsymbol{a}_x$ or $\boldsymbol{a}_y$ directions:

$$\boldsymbol{E} = E_x(z, t)\boldsymbol{a}_x \tag{10-1}$$

It may be seen by inspection that this function satisfies the requirement of zero divergence of the $\boldsymbol{D}$ field (Eqs. 9-19 and 9-3, with zero charge-density postulated) because E_x, the only nonzero component of $\boldsymbol{E}$, is not a function of x (Eq. A-25 for the divergence function).

a. Derivation of Wave Equation: Analogy with Transmission Line

By substitution of the assumed $\boldsymbol{E}$ into Maxwell's curl equations successively, $\boldsymbol{H}$ may be eliminated from the set and a scalar differential equation in E_x obtained. The curl of Eq. 10-1 is found by means of Eq. A-26:

$$\nabla \times \boldsymbol{E} = \frac{\partial E_x(z,t)}{\partial z}\boldsymbol{a}_y \tag{10-2}$$

Substitution of this into Maxwell's equation in curl $\boldsymbol{E}$, Eq. 9-23, gives

$$-\frac{\partial \boldsymbol{H}}{\partial t} = \frac{1}{\mu}\frac{\partial E_x(z,t)}{\partial z}\boldsymbol{a}_y \tag{10-3}$$

Thus the accompanying $\boldsymbol{H}$ field has only one component, $H_y(z,t)$. In other words, the $\boldsymbol{E}$ and $\boldsymbol{H}$ fields are in space quadrature with each other.

The curl of Eq. 10-3 is

$$\nabla \times \left(\frac{\partial \boldsymbol{H}}{\partial t}\right) = -\frac{\partial}{\partial z}\left(\frac{\partial H_y}{\partial t}\right)\boldsymbol{a}_x$$

$$= \frac{1}{\mu}\frac{\partial^2 E_x(z,t)}{\partial z^2}\boldsymbol{a}_x$$

Continuity of the functions and their derivatives will be assumed; hence the order of differentiation of $\boldsymbol{H}$ may be interchanged:

$$\frac{\partial}{\partial t}(\nabla \times \boldsymbol{H}) = \frac{1}{\mu}\frac{\partial^2 E_x(z,t)}{\partial z^2}\boldsymbol{a}_x \tag{10-4}$$

The time derivative of Maxwell's equation in curl $\boldsymbol{H}$, Eq. 9-21, may be substituted for the left-hand side of Eq. 10-4 to eliminate the $\boldsymbol{H}$ function:

$$\frac{\partial}{\partial t}(\nabla \times \boldsymbol{H}) = \epsilon\frac{\partial^2 \boldsymbol{E}}{\partial t^2} + \sigma\frac{\partial \boldsymbol{E}}{\partial t}$$

The following partial differential equation in E_x results:

$$\frac{\partial^2 E_x(z,t)}{\partial z^2} = \epsilon\mu\frac{\partial^2 E_x(z,t)}{\partial t^2} + \sigma\mu\frac{\partial E_x(z,t)}{\partial t} \tag{10-5}$$

If one starts with an assumed function $H_y(z,t)$, $\boldsymbol{E}$ may be eliminated from the set and the same equation as 10-5 obtained in terms of $H_y(z,t)$ instead of $E_x(z,t)$.

Comparison of Eq. 10-5 with Eq. 4-5 for the general transmission line discloses a term-for-term correspondence between the two as listed below, except that r in Eq. 4-5 has no counterpart in Eq. 10-5:

$$\begin{aligned} v(z,t) &\sim E_x(z,t) \\ g &\sim \sigma \\ l &\sim \mu \\ c &\sim \epsilon \end{aligned} \qquad (10\text{-}6)$$

Here the symbol $\sim$ indicates "analogous to."

Two physical systems that have a mathematical parallelism such as that just indicated between (1) the relationships among the dependent variables, the independent variables and the parameters of the first system, and (2) the corresponding relationships in the second system, are known as *analogs*. This property exists despite different physical interpretations and different dimensions for the members of a given pair of analogous quantities from the two systems.* One pair of analogous quantities indicated in set (10-6) consists of (1) the voltage in volts between two transmission-line conductors, and (2) the electric-field intensity in volts per meter of an electromagnetic disturbance of (mathematically) infinite transverse extent. The parallelism extends to the solutions of the differential equations.

b. Sinusoidal-Traveling-Wave General Solution

The same technique of assuming sinusoidal-traveling-wave solutions and resolving into complex exponential form that was followed in Chap. 4 may be used here. Only the key equations will be given. Let

$$\boldsymbol{E}_1(z,t) = E_{x1M}\epsilon^{-az}\cos(\omega t - \beta z)\boldsymbol{a}_x \qquad (10\text{-}7)$$

or

$$\boldsymbol{E}_1(z,t) = \mathrm{Re}\left[E_{x1M}\epsilon^{-(\alpha+j\beta)z}\epsilon^{j\omega t}\right]\boldsymbol{a}_x \qquad (10\text{-}8)$$

$$\boldsymbol{E}_2(z,t) = \mathrm{Re}\left[E_{x2M}\epsilon^{(\alpha+j\beta)z}\epsilon^{j\omega t}\right]\boldsymbol{a}_x \qquad (10\text{-}9)$$

---

*Applications of analogs to wave propagation and diffusion phenomena in electromagnetic, mechanical, and thermal systems are given by R. K. Moore in *Traveling Wave Engineering* (McGraw-Hill, New York, 1960) and *Wave and Diffusion Analogies* (McGraw-Hill, New York, 1964).

Substitution of either Eq. 10-8 or 10-9 into Eq. 10-5 yields the following relationship between the propagation function and the parameters μ, ϵ, σ, and ω:

$$\alpha + j\beta = \sqrt{j\omega\mu(\sigma + j\omega\epsilon)} \tag{10-10}$$

The magnetic-field-intensity function $\boldsymbol{H}_1(z,t)$ corresponding to $\boldsymbol{E}_1(z,t)$ may be found by differentiating Eq. 10-8 with respect to z, substituting in Eq. 10-3, and then integrating with respect to time:

$$\boldsymbol{H}_1(z,t) = \mathrm{Re}\left[\frac{(\alpha + j\beta)}{j\omega\mu} E_{x1M}\epsilon^{-(\alpha+j\beta)z}\epsilon^{j\omega t}\right]\boldsymbol{a}_y \tag{10-11}$$

From Eq. 10-10,

$$\frac{(\alpha + j\beta)}{j\omega\mu} = \sqrt{\frac{(\sigma + j\omega\epsilon)}{j\omega\mu}} \tag{10-12}$$

The reciprocal of this quantity has the units of ohms, and is given the name *intrinsic impedance* (η). It is analogous to the characteristic impedance of a transmission line. For a conducting medium, η will be complex:

$$|\eta| \exp(j\theta_\eta) = \sqrt{\frac{j\omega\mu}{\sigma + j\omega\epsilon}} \tag{10-13}$$

Equation 10-11 may be rewritten as follows:

$$\boldsymbol{H}_1(z,t) = \mathrm{Re}\left\{\frac{E_{x1M}}{|\eta|} \exp\left[-(\alpha + j\beta)z + j(\omega t - \theta_\eta)\right]\right\}\boldsymbol{a}_y \tag{10-14}$$

$$\boldsymbol{H}_1(z,t) = \frac{E_{x1M}}{|\eta|}\epsilon^{-az}\cos(\omega t - \theta_\eta - \beta z)\boldsymbol{a}_y \tag{10-15}$$

Similarly

$$\boldsymbol{H}_2(z,t) = \mathrm{Re}\left\{\frac{-E_{x2M}}{|\eta|} \exp\left[(\alpha + j\beta)z + j(\omega t - \theta_\eta)\right]\right\}\boldsymbol{a}_y \tag{10-16}$$

These results should be compared with the corresponding expressions for current on a lossy transmission line, Eqs. 4-32, 4-33, and 4-34.

c. TEM-Mode Designation: Polarization

For both traveling-wave solutions which were examined here, the $\boldsymbol{E}$ and $\boldsymbol{H}$ fields are everywhere perpendicular to the direction of motion, $\boldsymbol{a}_z$ or $-\boldsymbol{a}_z$. A wave set which has these properties is said to be in the *transverse-electromagnetic* (TEM) *mode*. This designation will prove applicable in Chap. 11, where the fields of specific transmission lines are analyzed.

The functions $\boldsymbol{E}_1$ and $\boldsymbol{E}_2$ in Eqs. 10-7 and 10-9 are special solutions within the class of sinusoidal waves which could propagate in the assigned directions in that each electric field is everywhere parallel to a single reference direction, $\boldsymbol{a}_x$. As such they are known as *linearly polarized* waves. Consider the following special solutions, either (or both) of which could exist simultaneously with $\boldsymbol{E}_1$:

$$\boldsymbol{E}_3(z,t) = E_{y3M}\epsilon^{-az}\cos(\omega t - \beta z)\boldsymbol{a}_y \tag{10-17}$$

$$\boldsymbol{E}_4(z,t) = E_{y4M}\epsilon^{-az}\sin(\omega t - \beta z)\boldsymbol{a}_y \tag{10-18}$$

Addition of Eqs. 10-7 and 10-17 yields

$$\boldsymbol{E}_1(z,t) + \boldsymbol{E}_3(z,t) = \epsilon^{-az}\cos(\omega t - \beta z)(E_{x1M}\boldsymbol{a}_x + E_{y3M}\boldsymbol{a}_y) \tag{10-19}$$

This too is a linearly polarized wave, even though the direction of polarization is not parallel to one of the coordinate axes. On the other hand, addition of $\boldsymbol{E}_1$ and $\boldsymbol{E}_4$ yields an effect known as *elliptical polarization*:

$$\boldsymbol{E}_1(z,t) + \boldsymbol{E}_4(z,t)$$
$$= \epsilon^{-az}\left[E_{x1M}\cos(\omega t - \beta z)\boldsymbol{a}_x + E_{y4M}\sin(\omega t - \beta z)\boldsymbol{a}_y\right] \tag{10-20}$$

At any given value of z the resultant field never goes to zero but varies in magnitude between the extremes of $E_{x1M}\epsilon^{-az}$ and $E_{y4M}\epsilon^{-az}$. The direction of the resultant field changes continuously in the xy plane. A special case is that in which the magnitudes E_{x1M} and E_{y4M} are equal; then the resultant is constant in magnitude and rotates uniformly. This is called *circular polarization*. Some antennas are designed to produce circularly polarized fields, and elliptical polarization may be inadvertently produced in practical communication systems.

d. Approximations and Limiting Forms of Propagation Parameters

Whether a given conductivity is "high" or "low" in wave-propagation problems may be a relative matter with respect to frequency; specifically it may be considered "high" if σ is much greater than $\omega\epsilon$, or "low" if $\omega\epsilon$ is much greater than σ. Both situations may be examined with power-series approximations.

(1) *Medium of Low Conductivity.* If conductivity is low, Eqs. 10-10 and 10-13 may be rearranged and expanded in series as follows:

$$\alpha + j\beta = j\omega\sqrt{\mu\epsilon}\sqrt{1 - \frac{j\sigma}{\omega\epsilon}}$$

$$\approx j\omega\sqrt{\mu\epsilon}\left(1 - \frac{j\sigma}{2\omega\epsilon}\right)$$

$$\approx \frac{\sigma}{2}\sqrt{\frac{\mu}{\epsilon}} + j\omega\sqrt{\mu\epsilon} \tag{10-21}$$

$$\eta = \sqrt{\frac{\mu}{\epsilon}}\frac{1}{\sqrt{1 + \sigma/j\omega\epsilon}} \approx \sqrt{\frac{\mu}{\epsilon}}\left(1 + \frac{j\sigma}{2\omega\epsilon}\right)$$

$$\approx \sqrt{\frac{\mu}{\epsilon}}\exp\left[j\tan^{-1}\left(\frac{\sigma}{2\omega\epsilon}\right)\right]$$

$$\approx \sqrt{\frac{\mu}{\epsilon}}\exp\left(\frac{j\sigma}{2\omega\epsilon}\right) \tag{10-22}$$

In the limiting case of zero conductivity, these reduce to

$$\begin{aligned} \alpha &= 0 \\ \beta &= \omega\sqrt{\mu\epsilon} \\ \eta &= \sqrt{\frac{\mu}{\epsilon}} \\ \theta_\eta &= 0 \end{aligned} \tag{10-23}$$

The following relationship between phase function and phase velocity was noted in the discussion of transmission lines:

$$v = \frac{\omega}{\beta} \tag{4-28}$$

Substitution of Eq. 10-23 gives the following for the phase velocity of electromagnetic waves in a nonconducting medium:

$$\nu = \frac{1}{\sqrt{\mu\epsilon}} \tag{10-24}$$

From Eqs. 10-23 and 10-24 the following observations may be made concerning the propagation of $\boldsymbol{E}$ and $\boldsymbol{H}$ waves through a nonconducting medium: (1) both fields travel without attenuation, (2) all frequency components of both fields travel with the same velocity, (3) for each frequency component the $\boldsymbol{E}$ and $\boldsymbol{H}$ fields are in time phase, and (4) the proportion between the magnitudes of the $\boldsymbol{E}$ and $\boldsymbol{H}$ fields is the same for all frequencies.

These findings are similar to those in Chap. 2 for voltage and current on a lossless line; here, as in that instance, the results may be generalized to traveling-wave fields that vary with time in an arbitrary manner.

(2) *Medium of High Conductivity.* For this case, Eqs. 10-10 and 10-13 may be rearranged and expanded in series as follows:

$$\begin{aligned}\alpha + j\beta &= \sqrt{j\omega\mu\sigma}\sqrt{1 + \frac{j\omega\epsilon}{\sigma}} \\ &\approx \sqrt{j\omega\mu\sigma}\left(1 + \frac{j\omega\epsilon}{2\sigma}\right) \\ &\approx \sqrt{\frac{\omega\mu\sigma}{2}}\left(1 - \frac{\omega\epsilon}{2\sigma}\right) + j\sqrt{\frac{\omega\mu\sigma}{2}}\left(1 + \frac{\omega\epsilon}{2\sigma}\right)\end{aligned} \tag{10-25}$$

$$\begin{aligned}|\eta|\exp(j\theta_\eta) &= \sqrt{\frac{j\omega\mu}{\sigma}}\frac{1}{\sqrt{1 + j\omega\epsilon/\sigma}} \\ &\approx \sqrt{\frac{j\omega\mu}{\sigma}}\left(1 - \frac{j\omega\epsilon}{2\sigma}\right) \\ &\approx \sqrt{\frac{\omega\mu}{\sigma}}\,\underline{/\frac{\pi}{4} - \frac{\omega\epsilon}{2\sigma}}\end{aligned} \tag{10-26}$$

$$\nu \to \sqrt{\frac{2\omega}{\mu\sigma}} \qquad (\text{as } \omega\epsilon/\sigma \to 0) \tag{10-27}$$

It should be recognized that these approximations are not valid for frequencies approaching infinity.

Conclusions to be noted from Eqs. 10-25, 10-26, and 10-27 for media for which $\omega\epsilon/\sigma \to 0$ are: (1) attenuation, phase velocity, and magnitude of intrinsic impedance vary markedly with frequency; (2) attenuation in nepers per meter is equal to the phase function in radians per meter; and (3) for each separate frequency component, the $\boldsymbol{H}$ field lags the $\boldsymbol{E}$ field in time phase by 45°.

e. Poynting's Vector

It was noted in the derivation of the wave-equation solutions for the plane wave that $\boldsymbol{E}$ and $\boldsymbol{H}$ are in space quadrature with respect to each other and with respect to the direction of propagation. This suggests that the vector product of those two fields might be used to indicate the direction in which the waves are traveling. Furthermore, the product of $|\boldsymbol{E}|$, volts per meter, and $|\boldsymbol{H}|$, amperes per meter, has the units of watts per square meter, those of power density on a surface.

The following vector-field identity makes possible a linking of the vector product just suggested with the rate of change of stored energy and the rate of conversion of electromagnetic energy into heat:

$$\nabla \cdot (\boldsymbol{E} \times \boldsymbol{H}) = \boldsymbol{H} \cdot (\nabla \times \boldsymbol{E}) - \boldsymbol{E} \cdot (\nabla \times \boldsymbol{H}) \qquad (10\text{-}28)$$

Maxwell's curl equations 9-32 and 9-34, may be substituted into the right-hand side:

$$\nabla \cdot (\boldsymbol{E} \times \boldsymbol{H}) = -\mu \boldsymbol{H} \cdot \frac{\partial \boldsymbol{H}}{\partial t} - \boldsymbol{E} \cdot \left(\epsilon \frac{\partial \boldsymbol{E}}{\partial t} + \sigma \boldsymbol{E} \right) \qquad (10\text{-}29)$$

$$\mu \boldsymbol{H} \cdot \frac{\partial \boldsymbol{H}}{\partial t} = \frac{\partial}{\partial t} \left(\frac{\mu |\boldsymbol{H}|^2}{2} \right) \qquad (10\text{-}30)$$

$$\epsilon \boldsymbol{E} \cdot \frac{\partial \boldsymbol{E}}{\partial t} = \frac{\partial}{\partial t} \left(\frac{\epsilon |\boldsymbol{E}|^2}{2} \right) \qquad (10\text{-}31)$$

$$\sigma \boldsymbol{E} \cdot \boldsymbol{E} = \sigma |\boldsymbol{E}|^2 \qquad (10\text{-}32)$$

Let

$$w_\epsilon = \frac{\epsilon |\boldsymbol{E}|^2}{2} \qquad (10\text{-}33)$$

$$w_\mu = \frac{\mu |\boldsymbol{H}|^2}{2} \qquad (10\text{-}34)$$

The functions w_ϵ and w_μ are scalar fields with the dimensions of energy per unit volume.

The energy stored in the electric field of a capacitor is equal to the volume integral of w_ϵ over all the space occupied by the field, and the energy stored in the magnetic field of a circuit is equal to the volume integral of w_μ over all the space occupied by that field.[3]

If one allocates the field energies on a distributed basis throughout the regions where the fields are nonzero, w_ϵ and w_μ may appropriately be called the *electric-field-energy density* and the *magnetic-field-energy density*, respectively. Substitution into Eq. 10-29 yields

$$\nabla \cdot (\boldsymbol{E} \times \boldsymbol{H}) = -\frac{\partial w_\epsilon}{\partial t} - \frac{\partial w_\mu}{\partial t} - \sigma |\boldsymbol{E}|^2 \qquad (10\text{-}35)$$

Each term in this equation has the units of power per unit volume. The last term represents the rate of conversion of electromagnetic energy into heat.

Equation 10-35 may be integrated over the volume enclosed by some given surface:

$$\iiint \nabla \cdot (\boldsymbol{E} \times \boldsymbol{H})\, dv = -\iiint \left(\frac{\partial w_\epsilon}{\partial t} + \frac{\partial w_\mu}{\partial t} + \sigma |\boldsymbol{E}|^2 \right) dv \qquad (10\text{-}36)$$

The divergence theorem, Eq. A-15, may be substituted for the left-hand side:

$$-\oiint (\boldsymbol{E} \times \boldsymbol{H}) \cdot d\boldsymbol{S} = +\iiint \left(\frac{\partial w_\epsilon}{\partial t} + \frac{\partial w_\mu}{\partial t} + \sigma |\boldsymbol{E}|^2 \right) dv \qquad (10\text{-}37)$$

Thus the negative of the surface integral of $\boldsymbol{E} \times \boldsymbol{H}$ (with $d\boldsymbol{S}$ directed outwardly) is equal to the time rate of increase of energy within the enclosed volume.

Poynting's vector $\boldsymbol{P}$ is defined as

$$\boldsymbol{P} = \boldsymbol{E} \times \boldsymbol{H} \qquad (10\text{-}38)$$

The vector $\boldsymbol{P}$ is commonly interpreted as representing the energy-flow density at a point, and this is in many instances a reasonable and helpful point of view. This interpretation is not always valid. An oft-cited example is that of a simultaneous static magnetic field and static electric field. Together they could have a nonzero Poynting-vector field even though there is no sustained movement of energy; however, the surface integral of that function over any enclosing surface vanishes. As a matter of fact any field which had zero divergence could be added to $\boldsymbol{P}$ and the enclosed-surface integral would not be changed.

For the traveling-wave set described by Eqs. 10-7 and 10-15, Poynting's vector is

$$\begin{aligned} \boldsymbol{P} &= \frac{E_{x1M}^2}{|\eta|}\epsilon^{-2\alpha z}\cos(\omega t - \beta z)\cos(\omega t - \theta_\eta - \beta z)\boldsymbol{a}_z \\ &= \frac{E_{x1M}^2}{2|\eta|}\epsilon^{-2\alpha z}\left[\cos\theta_\eta + \cos(2\omega t - \theta_\eta - 2\beta z)\right]\boldsymbol{a}_z \end{aligned} \tag{10-39}$$

This may be time-averaged over an integral number of cycles to yield $\boldsymbol{P}_{\text{av}}$:

$$\boldsymbol{P}_{\text{av}} = \frac{E_{x1M}^2}{2|\eta|}\epsilon^{-2\alpha z}\cos\theta_\eta\,\boldsymbol{a}_z \tag{10-40}$$

The surface integral of $\boldsymbol{P}_{\text{av}}$ *entering* a closed surface yields the time-averaged power loss within.

Poynting's vector will be used in Secs. 10-4b, 12-6, and 13-2b(3)(b) for the calculation of losses in conducting surfaces.

10-2 REFLECTION AND REFRACTION AT NORMAL INCIDENCE

If electromagnetic waves are propagating in a region containing two or more media with different intrinsic impedances, reflected waves will arise to satisfy requirements of continuity of the fields at the boundary surfaces. If the waves are plane and uniform, and the direction of propagation is normal to the plane of the boundary surface, the result proves to be analogous to the reflections on a terminated transmission line.

a. Component Traveling-Wave Sets

A diagram of coordinate directions and related information is given in Fig. 10-1. It will be assumed that the boundary surface $z = 0$ is of infinite extent and that media 1 and 2 extend from zero to infinity in the $-\boldsymbol{a}_z$ and $+\boldsymbol{a}_z$ directions, respectively. It will be assumed further that an electromagnetic traveling-wave field called the incident wave set, $\boldsymbol{E}_I(z,t)$ and $\boldsymbol{H}_I(z,t)$, is moving through medium 1 to the boundary surface. Medium 1 will be assumed lossless, and appropriate phasor expressions may be obtained from Eqs. 10-8 and 10-14 after setting α and θ_η equal to zero:

$$E_{x\text{inc}}(z) = E_{xIM}\exp(-j\beta_1 z) \tag{10-41}$$

$$H_{y\text{inc}}(z) = \frac{E_{xIM}}{\eta_1}\exp(-j\beta_1 z) \tag{10-42}$$

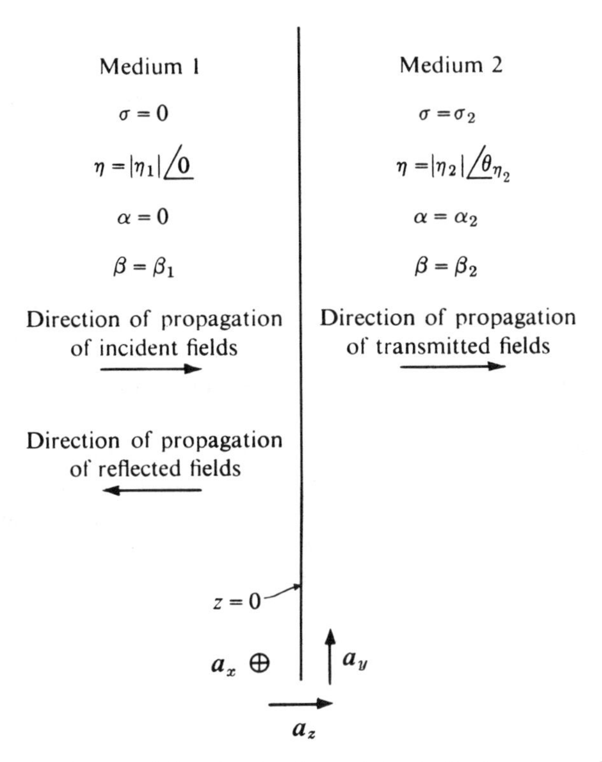

Figure 10-1. Coordinate-and-direction plan for analysis of reflection at normal incidence.*

The principal relationships may be illustrated readily if medium 2 is assumed to have losses as indicated in Fig. 10-1. Equations for the transmitted radiation proceeding from the reflecting surface through medium 2 would be

$$E_{x\text{tr}}(z) = E_{xTM} \exp[j\theta_T - (\alpha_2 + j\beta_2) z] \tag{10-43}$$

$$H_{y\text{tr}}(z) = \frac{E_{xTM}}{\eta_2} \exp[j\theta_T - (\alpha_2 + j\beta_2) z] \tag{10-44}$$

Here θ_T is a phase-shift angle.

---

*In Fig. 10-1 the unit vector $\boldsymbol{a}_x$ is normal to the orthographic projection plane, and is directed away from the observer. The directional code symbol shown (an encircled cross) may be interpreted pictorially as the tail feathers of an arrow. Correspondingly an encircled dot, which is used in Fig. 10-3, means "toward the observer," and suggests the point of an arrow.

The reflected waves travel in medium 1, and hence are unattenuated, but they move in the $-\boldsymbol{a}_z$ direction. The phase-shift angle associated with the reflection process will be taken as θ_K, the same as in Sec. 6-1 for the terminated transmission line:

$$E_{x\mathrm{ref}}(z) = E_{xRM} \exp[j(\theta_K + \beta_1 z)] \tag{10-45}$$

$$H_{y\mathrm{ref}}(z) = \frac{-E_{xRM}}{\eta_1} \exp[j(\theta_K + \beta_1 z)] \tag{10-46}$$

b. Boundary Conditions and Reflection Coefficient

In this problem, $\boldsymbol{E}$ and $\boldsymbol{H}$ are both purely tangential to the boundary surface. The general boundary conditions summarized in Sec. 9-3 require that the tangential components of $\boldsymbol{E}$ on the two sides of the surface be equal, and the same is true of the $\boldsymbol{H}$ field if the conductivities are zero or finite. Hence one may write the following equations, paralleling Eqs. 6-7 and 6-9 for the transmission line:

$$E_{x\mathrm{tr}}(0) = E_{x\mathrm{inc}}(0) + E_{x\mathrm{ref}}(0) \tag{10-47}$$

$$\frac{E_{x\mathrm{tr}}(0)}{\eta_2} = \frac{E_{x\mathrm{inc}}(0)}{\eta_1} - \frac{E_{x\mathrm{ref}}(0)}{\eta_1} \tag{10-48}$$

$E_{x\mathrm{tr}}(0)$ may be eliminated from this set and the resulting equation solved for the reflected phasor in terms of the incident phasor and the intrinsic impedances:

$$E_{x\mathrm{ref}}(0) = E_{x\mathrm{inc}}(0)\frac{\eta_2 - \eta_1}{\eta_2 + \eta_1} \tag{10-49}$$

A reflection coefficient may be defined in the same manner as for the terminated transmission line (Eq. 6-12):

$$|K|\exp(j\theta_K) = \frac{E_{x\mathrm{ref}}(0)}{E_{x\mathrm{inc}}(0)} = \frac{E_{xRM}}{E_{xIM}} \exp(j\theta_K) \tag{10-50}$$

Then, from Eq. 10-49

$$K = \frac{\eta_2 - \eta_1}{\eta_2 + \eta_1} \tag{10.51}$$

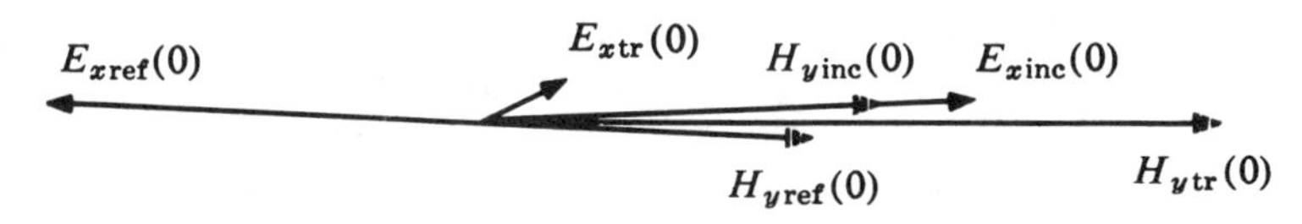

Figure 10-2. Phasor diagram for traveling-wave component fields at reflecting surface for normal incidence, air to seawater (drawn to scale from results of Prob. 10-3).

The relationship between the incident and reflected magnetic fields may be found as follows:

$$H_{y\mathrm{inc}}(z) = \frac{E_{x\mathrm{inc}}(z)}{\eta_1} \tag{10-52}$$

$$H_{y\mathrm{ref}}(z) = -\frac{E_{x\mathrm{ref}}(z)}{\eta_1} \tag{10-53}$$

If z is set equal to zero in Eqs. 10-52 and 10-53 and E_x eliminated by means of Eq. 10-50, the following results:

$$H_{y\mathrm{ref}}(0) = -KH_{y\mathrm{inc}}(0) \tag{10-54}$$

The time-phase pattern among the component $\boldsymbol{E}$ and $\boldsymbol{H}$ fields at the boundary surface may be displayed to advantage by means of a phasor diagram. An example is given in Fig. 10-2, based on the numerical results to Prob. 10-3 in which radio-frequency fields are normally incident upon a poorly conducting medium, seawater. The parameters for that problem were chosen so that the conduction-current density and displacement-current density would be of the same order of magnitude.

c. Reflection Process as a Function of Conductivity

The effect of varying σ, while keeping the other parameters (μ, ϵ, and ω) constant, should be noted. The discussion in this section and the following one will refer to combinations of σ and ω for which macroscopic electromagnetic theory is valid; superconductivity and anomalous skin effect will be commented on in Appendix E.

For the limiting case of σ equal to zero, so that medium 2 is nonconducting, the following simplifications result: (1) the intrinsic impedance η_2 is purely resistive, (2) no attenuation takes place in medium 2, (3) the reflection coefficient is a purely real number, and (4) the phasors corresponding to those shown in Fig. 10-2 are collinear.

If σ is very large compared to $\omega\epsilon$, a situation representative of metals even at microwave frequencies, the following relationships are approached:

$$\eta_2 = \sqrt{\frac{\omega\mu}{\sigma}}\,\epsilon^{j\pi/4} \qquad (10\text{-}55)$$

$$|\eta_2| \ll \eta_1 \qquad (10\text{-}56)$$

The numerator and denominator of Eq. 10-51 may be divided by η_1, and the resulting denominator replaced by the first two terms of the Taylor series:

$$K = \frac{(\eta_2/\eta_1) - 1}{(\eta_2/\eta_1) + 1} \qquad (10\text{-}57)$$

$$K \approx \left(\frac{\eta_2}{\eta_1} - 1\right)\left(1 - \frac{\eta_2}{\eta_1}\right)$$

$$\approx -1 + \frac{2\eta_2}{\eta_1} \qquad (10\text{-}58)$$

From this it may be concluded that the reflected component of $\boldsymbol{E}$ will be essentially equal to the incident component and very nearly in phase opposition with it. The transmitted component of $\boldsymbol{E}$ will be small, but an approximation can be readily obtained:

$$E_{xTM}\exp(j\theta_T) = E_{xIM}(1 + K)$$

Equation 10-57 is substituted for $\boldsymbol{K}$:

$$E_{xTM}\exp(j\theta_T) = E_{xIM}\frac{2\eta_2}{\eta_1 + \eta_2}$$

$$\approx E_{xIM}\frac{2\eta_2}{\eta_1} \qquad (10\text{-}59)$$

As is shown numerically in Prob. 10-4, the electric field transmitted into copper by normally incident waves in air is exceedingly small compared to the incident electric field, but it is definitely nonzero. Whatever the precise magnitude of the ratio, the transmitted electric field at the surface of the metal leads the incident electric field by 45° in time phase.

In accordance with Eq. 10-54 the reflected magnetic field in this instance will be essentially equal to the incident magnetic field in magnitude and very nearly in time phase with it. Thus the transmitted magnetic field, at the surface of the metal, will have approximately twice the magnitude of the incident magnetic field and be almost in time phase with it. The transmitted magnetic field is, of course, related to the transmitted electric field by the intrinsic impedance of medium 2.

The attenuation constant in copper is 15 N/mm at 1 MHz [see Eq. 10-25; also compare with the numerical result of Prob. 10-1(b)]; hence the transmitted fields at radio frequencies are largely confined to a shallow layer immediately below the reflecting surface.

d. Conduction-Current Density in Metals

The current-density function in a metallic reflecting medium should be examined in order to place some commonly used concepts—*surface-current density*, *surface resistance*, and *skin depth*—in good perspective.

Displacement-current density is negligible in comparison with the conduction-current density and hence will be ignored. The latter function is, in accordance with Eq. 9-7,

$$J_{x\text{tr}}(z) = \sigma_2 E_{x\text{tr}}(z) \tag{10-60}$$

Equations 10-43 and 10-59 may be substituted into Eq. 10-60:

$$J_{x\text{tr}}(z) = \sigma_2 \frac{2\eta_2}{\eta_1} E_{xIM} \exp[-(\alpha_2 + j\beta_2)z] \tag{10-61}$$

First-order approximations from Eqs. 10-25 and 10-55 may be substituted for α_2 and η_2:

$$J_{x\text{tr}}(z) = \frac{2\sqrt{\omega\mu_2\sigma_2}}{\eta_1} E_{xIM} \exp\left[-\left(\sqrt{\frac{\omega\mu_2\sigma_2}{2}} + j\beta_2\right)z + \frac{j\pi}{4}\right] \tag{10-62}$$

(1) *Surface-Current Density.* From Eq. 10-62 it appears that an increase in σ_2 will increase the magnitude of conduction-current density at the surface of the metal, but, by increasing the attenuation function, it will also rapidly diminish the magnitude at any given depth in proportion to that at the surface. The limiting condition which is approached is that of a sheet of current of infinite volume density but infinitesimally thin. To obtain a more useful measure of the current field at this limit, consider the integral of conduction-current density with respect to depth. Equation

10-61 will be integrated with respect to z, yielding this result:

$$\int_0^\infty J_{x\text{tr}}(z)\,dz = -\sigma_2 \frac{2\eta_2}{\eta_1} \frac{E_{xIM}}{\alpha_2 + j\beta_2} \exp[-(\alpha_2 + j\beta_2)z]\Big]_0^\infty$$

$$= +\sigma_2 \frac{2\eta_2}{\eta_1} \frac{E_{xIM}}{\alpha_2 + j\beta_2} \tag{10-63}$$

Substitution of Eqs. 10-25 and 10-55, with $\omega\epsilon/\sigma \to 0$, reduces this to

$$\int_0^\infty J_{x\text{tr}}(z)\,dz = \frac{2E_{xIM}}{\eta_1} = 2H_{y\text{inc}}(0)$$

$$= H_{y\text{inc}}(0) + H_{y\text{ref}}(0) \tag{10-64}$$

Thus the resultant found by integrating the conduction-current density with respect to depth has a magnitude which is equal at every instant to the magnitude of the resultant magnetic field just above the surface. It is assigned the name of *surface-current density*: In this instance

$$\boldsymbol{J}_S(t) = \text{Re}\left[\epsilon^{j\omega t} \int_0^\infty J_{x\text{tr}}(z)\,dz\right] \boldsymbol{a}_x \tag{10-65}$$

In terms of the vector current-density field $\boldsymbol{J}_T(z,t)$,

$$\boldsymbol{J}_S(t) = \int_0^\infty \boldsymbol{J}_T(z,t)\,dz \tag{10-66}$$

Surface-current density has the units amperes per meter, and it flows at right angles with respect to the impinging magnetic field. The directions are consistent with the familiar right-hand rule for the magnetic field accompanying a current:

$$\boldsymbol{J}_S(t) \times \boldsymbol{a}_n = \boldsymbol{H}(0,t) \tag{10-67}$$

Here $\boldsymbol{a}_n$ is a unit vector directed outward from the conducting surface.

The boundary relationship when $\boldsymbol{J}_D$ has a component normal to a perfectly conducting surface will be examined in Sec. 10-3b(2).

(2) *Surface Resistance.* The changing of electromagnetic energy into heat when a set of electromagnetic fields is normally incident upon a metallic slab may be expressed in terms of the magnitude of the surface-current density and an equivalent resistance known as the surface resistance. This relationship may be derived from the energy conversion rate of the fields in the metal, stated on a volume-density basis.

The instantaneous rate of changing energy into the heat form in the conducting region is

$$p_{\text{att}}(z,t) = \frac{|\boldsymbol{J}_T(z,t)|^2}{\sigma_2} \quad \text{(watts per cubic meter)} \tag{10-68}$$

Equation 10-61 will be restated as an explicit function of time after substituting Eq. 10-26 for η_2 and replacing E_{xIM}/η_1 with H_{yIM}. This result is then substituted in Eq. 10-68:

$$\boldsymbol{J}_T(z,t) = 2\sqrt{\omega\mu_2\sigma_2}\,H_{yIM}\exp(-\alpha_2 z)\cos\left(\omega t + \frac{\pi}{4} - \beta_2 z\right)\boldsymbol{a}_x \tag{10-69}$$

$$p_{\text{att}}(z,t) = 4\omega\mu_2 H_{yIM}^2 \exp(-2\alpha_2 z)\cos^2\left(\omega t + \frac{\pi}{4} - \beta_2 z\right)$$

$$= 2\omega\mu_2 H_{yIM}^2 \exp(-2\alpha_2 z)\left[1 + \cos 2\left(\omega t + \frac{\pi}{4} - \beta_2 z\right)\right] \tag{10-70}$$

When time-averaged over an integral number of cycles this reduces to

$$P_{\text{att}}(z) = 2\omega\mu_2 H_{yIM}^2 \exp(-2\alpha_2 z) \tag{10-71}$$

The power input to the conducting slab may be found on the basis of watts per square meter of surface by integrating $P_{\text{att}}(z)$ with respect to z:

$$P_{\text{loss}} = -\frac{2\omega\mu_2}{2\alpha_2} H_{yIM}^2 \epsilon^{-2\alpha_2 z}\Big]_0^\infty$$

$$= \sqrt{\frac{2\omega\mu_2}{\sigma_2}}\,H_{yIM}^2 \tag{10-72}$$

As σ_2 approaches infinity, the limit of this expression is zero, a result which seems reasonable. An alternative approach to computing P_{loss} is to evaluate Poynting's vector (see Prob. 10-5).

Let J_{xSM} be the maximum value of the magnitude of $\boldsymbol{J}_S$. From Eqs. 10-64 and 10-65,

$$J_{xSM} = 2H_{yIM} \tag{10-73}$$

Equation 10-72 may be restated as

$$P_{\text{loss}} = \sqrt{\frac{2\omega\mu_2}{\sigma_2}}\left(\frac{J_{xSM}}{2}\right)^2$$

$$= \sqrt{\frac{\omega\mu_2}{2\sigma_2}}\left(\frac{J_{xSM}}{\sqrt{2}}\right)^2 \tag{10-74}$$

The quantity $J_{xSM}/\sqrt{2}$ is the root-mean-square or effective value of the surface-current density, and $\sqrt{\omega\mu_2/2\sigma_2}$ is dimensionally a resistance. It is given the name *surface resistance*, R_S:

$$R_S = \sqrt{\frac{\omega\mu_2}{2\sigma_2}} \tag{10-75}$$

If η_2 as defined in Eq. 10-55 is put into rectangular form,

$$\eta_2 = \sqrt{\frac{\omega\mu_2}{2\sigma_2}} + j\sqrt{\frac{\omega\mu_2}{2\sigma_2}} \tag{10-76}$$

The real part of Eq. 10-76 corresponds to the surface resistance as defined in Eq. 10-75.

(3) *Skin Depth.* Equation 10-75 may be rearranged as

$$R_S = \frac{1}{\sigma_2}\sqrt{\frac{\omega\mu_2\sigma_2}{2}} \tag{10-77}$$

It may be recalled that the DC resistance (R_{dc}) of a slab of metal of length s, width w, thickness δ, and conductivity σ is

$$R_{\text{dc}} = \frac{s}{\delta w \sigma}$$

If the length is equal to the width, so that the slab is a square of thickness δ, with resistance measured from one edge to the opposite one, the resistance is independent of the size of the square. It may be designated R_{dcsq}:

$$R_{\text{dcsq}} = \frac{1}{\delta\sigma} \tag{10-78}$$

Comparison of Eqs. 10-77 and 10-78 suggests that surface-resistance loss may be thought of as that produced in a slab of conductivity σ_2 and thickness $\sqrt{2/(\omega\mu_2\sigma_2)}$, with the surface-current density $J_{xSM}/\sqrt{2}$ distributed uniformly (and in time phase) over the edgewise cross section. The condition of equality between R_S and R_{dcsq} is

$$\delta = \sqrt{\frac{2}{\omega\mu_2\sigma_2}} \tag{10-79}$$

The name *skin depth* is given to δ as just defined. This may be compared with the first-order approximation in Eq. 10-25 for the attenuation function, from which the following may be seen by inspection:

$$\delta = \frac{1}{\alpha} \tag{10-80}$$

The quantity δ is sometimes referred to as the "depth of penetration." This expression is misleading (and so is "skin depth") in that the electric and magnetic fields at a depth of δ have magnitudes of $1/\epsilon$ times the magnitudes which they have at the surface, and hence "penetrate" well beyond that depth; even at 4δ the field strengths are almost 2 percent of what they are at the surface.

10-3 REFLECTION AT OBLIQUE INCIDENCE

Reflection phenomena for oblique angles of incidence is an important subject, even though the mathematical analysis is cumbersome. Radio waves generally strike the earth at orientations other than that of normal incidence, and one may note that the fields in a hollow wave guide are contained within the guide essentially by oblique reflections at the wall surfaces. The treatment here will be limited to a summary of results, with emphasis on those relating to highly conducting reflecting surfaces.

a. Geometry of Oblique-Reflection Problem

The principal geometrical relationships in oblique reflection are shown in Fig. 10-3. Traveling-wave-type incident and reflected fields will be postulated. A transverse-plane traveling-wave-type transmitted field as shown is applicable if the second medium is nonconducting[2, 3]; a more complicated situation ensues if it is of finite conductivity,[3] and a surface-current sheet with no transmitted fields is the result if the medium is of infinite conductivity.

The *plane of incidence* is perpendicular to the reflecting surface and parallel to the direction of propagation, or the *rays*, of the incident waves. Lines *AB* and *DE*, which are parallel to each other, are incident rays, and

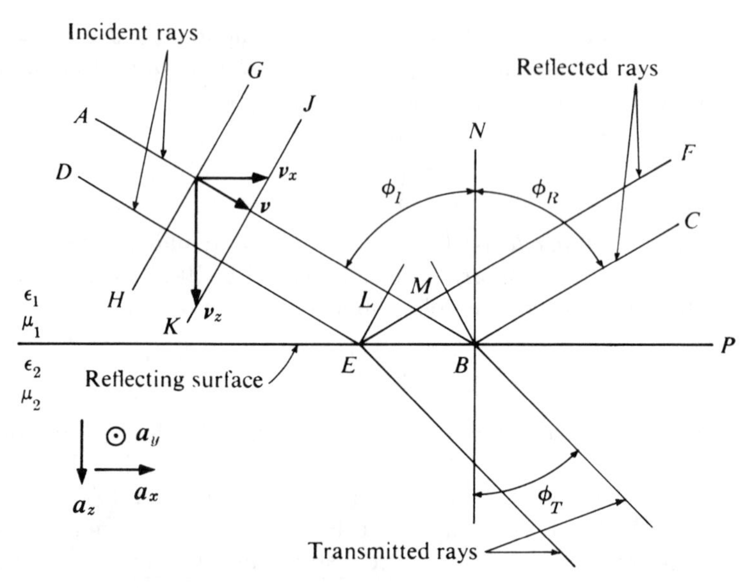

Figure 10-3. Geometry of reflection at oblique incidence.

BC and *EF*, also parallel to each other, are reflected rays. Lines *GH*, *JK* and *LE* represent *equiphase surfaces* of the incident waves; *MB* is an equiphase surface of the reflected waves.

The direction of approach of the incident waves is defined within the plane of incidence by means of the *angle of incidence* ϕ_I, which is measured from a line normal to the reflecting surface to the incident ray. (In some writing it is measured from the plane of reflection; this point should be checked to avoid misunderstandings when consulting other works.)

The necessary relationship between the angles of incidence and reflection may be determined by reference to Fig. 10-3. The equiphase planes of the incident and reflected waves are perpendicular to their respective ray directions, and both sets travel at the same velocity. Hence line segments *LB* and *EM* must be equal and triangles *ELB* and *EBM* must be congruent; thus angles *LBE* and *MEB* are equal, and since *BC* is parallel to *EF*, angles *LBE* and *CBP* are equal. Those angles are the complements of ϕ_I and ϕ_R; hence

$$\phi_I = \phi_R \tag{10-81}$$

Thus the angle of reflection is equal to the angle of incidence.

Since the direction of propagation of the incident wave set is in the plane of $\boldsymbol{a}_x$ and $\boldsymbol{a}_z$ but not parallel to either of these unit vectors, phase constants β_x and β_z must be derived if the traveling-wave functions are to have x and z for arguments. The phase velocities in those two coordinate directions may be found in terms of the velocity normal to the wave front ν and the angle of incidence. From Fig. 10-3 it may be seen that for the wave front GH to advance to JK, the phase velocities must be as follows:

$$\nu_x = \frac{\nu}{\sin \phi_I} \tag{10-82}$$

$$\nu_z = \frac{\nu}{\cos \phi_I} \tag{10-83}$$

The basic relation between phase velocity and phase function is, from Eq. 4-28,

$$\beta = \frac{\omega}{\nu}$$

Hence the phase functions for the $\boldsymbol{a}_x$ and $\boldsymbol{a}_z$ directions are

$$\beta_x = \beta \sin \phi_I \tag{10-84}$$

$$\beta_z = \beta \cos \phi_I \tag{10-85}$$

b. Boundary Conditions

Constraints of continuity are enjoined upon the electromagnetic fields by Maxwell's equations. The necessity for continuity of the tangential component of $\boldsymbol{E}$ and $\boldsymbol{H}$ was brought out in the earlier discussion of normal incidence (Sec. 10-2b). The same reasoning is applicable to the tangential components with oblique incidence, but mathematical statement of the constraint is more involved. For normal incidence the coordinate system could, without loss of generality, be rotated about the z axis to make the x axis parallel to the $\boldsymbol{E}$ field, and the y axis then would necessarily parallel the $\boldsymbol{H}$ field. For oblique incidence, as illustrated in Fig. 10-3, the x axis is parallel to the plane of incidence, and one must anticipate that each field may contain tangential components in both the x and y directions.

Furthermore, a normal component of either $\boldsymbol{E}$ or $\boldsymbol{H}$, or possibly both, will be present. The general boundary conditions in Sec. 9-3 require continuity of the normal components of $\boldsymbol{B}$ and $\boldsymbol{D}$.

(1) *Both Media Nonconducting.* The equations corresponding to Eqs. 10-47 and 10-48, applicable at the bounding surface ($z = 0$), for the tangential components are

$$\begin{aligned} E_{xI} + E_{xR} &= E_{xT} \\ E_{yI} + E_{yR} &= E_{yT} \end{aligned} \tag{10-86}$$

$$\begin{aligned} H_{xI} + H_{xR} &= H_{xT} \\ H_{yI} + H_{yR} &= H_{yT} \end{aligned} \tag{10-87}$$

Continuity of the normal components (z-directed) of $\boldsymbol{B}$ and $\boldsymbol{D}$ reduces to

$$\mu_1 H_{zI} + \mu_1 H_{zR} = \mu_2 H_{zT} \tag{10-88}$$

$$\epsilon_1 E_{zI} + \epsilon_1 E_{zR} = \epsilon_2 E_{zT} \tag{10-89}$$

(2) *Reflecting Surface of Infinite Conductivity.* In this case the following would prevail at the boundary surface: (a) the tangential component of $\boldsymbol{E}$ would vanish, (b) the normal component of $\boldsymbol{B}$ would vanish, (c) the tangential component of $\boldsymbol{H}$ in the conducting medium would be replaced by a surface-current-density sheet, and (d) the normal component of $\boldsymbol{D}$ would be terminated by a surface-charge-density sheet. The corresponding equations are

$$\begin{aligned} E_{xI} + E_{xR} &= 0 \\ E_{yI} + E_{yR} &= 0 \end{aligned} \tag{10-90}$$

$$\begin{aligned} \mu_1 H_{zI} + \mu_1 H_{zR} &= 0 \\ H_{zI} + H_{zR} &= 0 \end{aligned} \tag{10-91}$$

$$\begin{aligned} H_{xI} + H_{xR} &= -J_{yS} \\ H_{yI} + H_{yR} &= J_{xS} \end{aligned} \tag{10-92}$$

$$\epsilon_1 E_{zI} + \epsilon_1 E_{zR} = -\rho_S \tag{10-93}$$

The corresponding normal component of $\boldsymbol{J}_D$ is "absorbed" at the reflecting surface by the surface-current-density field. If $\boldsymbol{J}_S$ is regarded as a two-dimensional field, the divergence of $\boldsymbol{J}_S$ has the same units (amperes per square meter) as the impinging $\boldsymbol{J}_D$, and proves to be equal in magnitude. A positive value for $\nabla \cdot \boldsymbol{J}_S$ corresponds to displacement current flowing into the perfectly conducting surface. The resulting continuity of the composite current (conduction current and displacement current) will be demonstrated in examples in which the electric field has a component normal to such a surface. [See Sec. 10-3d(3)(b), Probs. 10-10 and 11-3, and Sec. 12-1b(2).]

c. Wave-Set Modes

The reflection coefficients of electromagnetic wave sets striking a surface at oblique incidence prove to be dependent on the polarization, or orientation, of the field-direction lines with respect to the surface; specifically, two modes of behavior exist, one for wave sets in which the electric field is parallel to the reflecting surface and another for wave sets in which the magnetic field is parallel to it. A plane electromagnetic wave set which is oriented so that neither its electric nor its magnetic field is parallel to the given surface may be resolved into two component wave sets, one of each type.

The terminology for designating the two wave-set modes has been complicated by separate developments of the subject by physicists working in optics and electrical engineers working with radio. Their viewpoints differed, with the result that physicists spoke of "perpendicular" polarization when referring to a wave set in which the $\boldsymbol{E}$ field is perpendicular to the plane of *incidence* and "parallel" polarization if the $\boldsymbol{E}$ field were parallel to the plane of incidence, whereas radio engineers spoke of "horizontal" polarization for the former case and "vertical" polarization for the latter. The frame of reference for "horizontal" and "vertical" is that of the level surface of the earth as the reflecting surface, with the incident fields coming from a dipole antenna which is either horizontal or vertical. The expressions "perpendicular" and "parallel" are less ambiguous and hence will be used in the remainder of this discussion.

(1) *Perpendicular Polarization.* A perpendicularly polarized incident wave set may be described in phasor form as follows, in accordance with Fig. 10-3:

$$\boldsymbol{E}_{I\perp}(x,z) = E_{IM} \exp[-j(\beta_{x1}x + \beta_{z1}z)]\boldsymbol{a}_y \tag{10-94}$$

$$\boldsymbol{H}_{I\perp}(x,z) = \frac{E_{IM}}{\eta_1}(-\cos\phi_I \boldsymbol{a}_x + \sin\phi_I \boldsymbol{a}_z)\exp[-j(\beta_{x1}x + \beta_{z1}z)] \tag{10-95}$$

The corresponding reflected wave set may be stated as follows in terms of the reflection coefficient $K_\perp$:

$$\boldsymbol{E}_{R\perp}(x,z) = K_\perp E_{IM}\exp[-j(\beta_{x1}x - \beta_{z1}z)]\boldsymbol{a}_y \tag{10-96}$$

$$\boldsymbol{H}_{R\perp}(x,z) = K_\perp \frac{E_{IM}}{\eta_1}(\cos\phi_I\boldsymbol{a}_x + \sin\phi_I\boldsymbol{a}_z)\exp[-j(\beta_{x1}x - \beta_{z1}z)] \tag{10-97}$$

(2) *Parallel Polarization.* The corresponding form for a parallel polarized incident wave set is

$$\boldsymbol{E}_{I\parallel}(x,z) = E_{IM}(\cos\phi_I\boldsymbol{a}_x - \sin\phi_I\boldsymbol{a}_z)\exp[-j(\beta_{x1}x + \beta_{z1}z)] \tag{10-98}$$

$$\boldsymbol{H}_{I\parallel}(x,z) = \frac{E_{IM}}{\eta_1}\exp[-j(\beta_{x1}x + \beta_{z1}z)]\boldsymbol{a}_y \tag{10-99}$$

The reflected wave set is

$$\boldsymbol{E}_{R\parallel}(x,z) = K_\parallel E_{IM}(-\cos\phi_I\boldsymbol{a}_x - \sin\phi_I\boldsymbol{a}_z)\exp[-j(\beta_{x1}x - \beta_{z1}z)] \tag{10-100}$$

$$\boldsymbol{H}_{R\parallel}(x,z) = K_\parallel\frac{E_{IM}}{\eta_1}\exp[-j(\beta_{x1}x - \beta_{z1}z)]\boldsymbol{a}_y \tag{10-101}$$

The angle $\theta_{K\parallel}$ of the reflection coefficient $K_\parallel$ has not been shown explicitly, but it should be noted that this angle is zero if E_{Rz} is in phase with E_{Iz}.

d. Reflection Coefficients and Transmitted Components

As was indicated earlier, the solutions for cases other than that of a surface of infinite conductivity are complicated, so general comments will be made and references cited.

(1) *Reflecting Medium of Zero Conductivity.* It will be assumed that $\mu_1 = \mu_2$ and that $\epsilon_1 < \epsilon_2$. The solution is straightforward, although lengthy.[2] The reflection coefficients for the two wave sets are as follows:

$$|K_\perp|\underline{/\theta_{K\perp}} = \frac{\cos\phi_I - \sqrt{(\epsilon_2/\epsilon_1) - \sin^2\phi_I}}{\cos\phi_I + \sqrt{(\epsilon_2/\epsilon_1) - \sin^2\phi_I}} \quad \begin{pmatrix}\text{perpendicular}\\ \text{polarization}\end{pmatrix} \tag{10-102}$$

$$\theta_{K\perp} = \pi \quad (\text{for } 0 \le \phi_I < \pi/2)$$

$$|K_\parallel|\underline{/\theta_{K\parallel}} = \frac{(\epsilon_2/\epsilon_1)\cos\phi_I - \sqrt{(\epsilon_2/\epsilon_1) - \sin^2\phi_I}}{(\epsilon_2/\epsilon_1)\cos\phi_I + \sqrt{(\epsilon_2/\epsilon_1) - \sin^2\phi_I}} \quad \begin{pmatrix}\text{parallel}\\ \text{polarization}\end{pmatrix} \tag{10-103}$$

The latter reflection coefficient vanishes at a critical angle of incidence, known as *Brewster's angle*, ϕ_{IB}:

$$\phi_{IB} = \tan^{-1}\sqrt{\frac{\epsilon_2}{\epsilon_1}} \tag{10-104}$$

$$\theta_{K\parallel} = 0 \quad (\text{for } 0 \le \phi_I < \phi_{IB})$$

$$\theta_{K\parallel} = \pi \quad (\text{for } \phi_{IB} < \phi_I < \pi/2)$$

The magnitudes of the two reflection coefficients are shown as functions of ϕ_I in Fig. 10-4 for $\epsilon_2/\epsilon_1 = 4$.

The transmitted, or refracted, wave set moves along rays inclined at angle ϕ_T, which is, according to Snell's law (see Prob. 10-6),

$$\frac{\sin\phi_T}{\sin\phi_I} = \sqrt{\frac{\epsilon_1}{\epsilon_2}} \tag{10-105}$$

(2) *Reflecting Medium of Finite, Nonzero Conductivity.* This situation is of importance in signal-propagation studies for radiobroadcasting. The reflection coefficients, which are complex, may be found by replacing ϵ_2 in

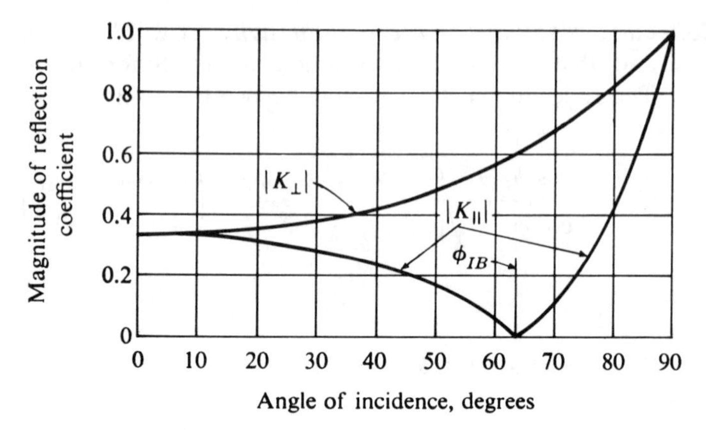

Figure 10-4. Reflection coefficients for waves obliquely incident upon nonconducting material; $\epsilon_2/\epsilon_1 = 4.0$.

Eqs. 10-102 and 10-103 with $\epsilon_2 + (\sigma_2/j\omega)$.[2, 3] Charts of computed values are given in Refs. 1, 2, and 4. Some qualitative observations concerning the differences between these results and those for the nonconducting reflecting medium are the following: (1) the reflection-coefficient magnitude for parallel polarization dips to a minimum at an angle analogous to Brewster's angle (pseudo-Brewster angle, in some writing) but does not go to zero, (2) the phase angles for the reflection coefficients of both modes vary continuously with the angle of incidence, and (3) the transmitted wave set is not strictly TEM, and the angle of refraction differs from that given in Eq. 10-105.

Results obtained from rigorous analyses of this general case justify the use of the simplified solutions for the infinite-conductivity case, which will be discussed in the next section, as valid approximations for the common physical situation of very high conductivity ($\sigma \ggg \omega\epsilon$).

(3) *Reflecting Medium of Infinite Conductivity.* The magnitudes of both reflection coefficients are unity for a surface of infinite conductivity, but the phase-shift angle is π radians for perpendicular polarization and zero for parallel polarization. This will reduce the field patterns in the z direction (normal to the reflecting surface) to pure standing waves. The resultant $\boldsymbol{E}$ and $\boldsymbol{H}$ fields may be found by adding the respective incident and reflected components and substituting identities paralleling Eqs. 6-41 or 6-42.

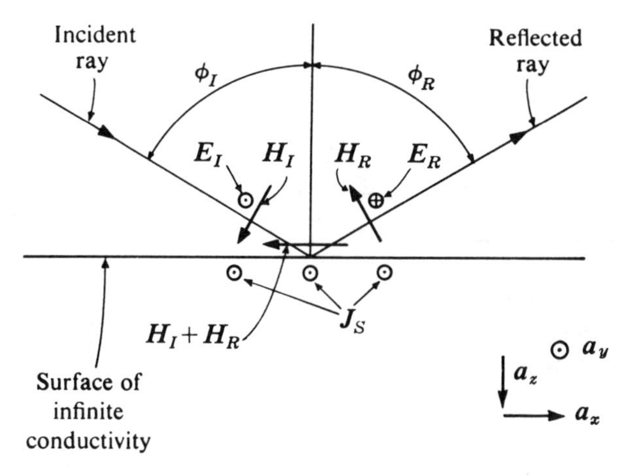

Figure 10-5. Perpendicularly polarized waves obliquely incident on ideal conductor.

(a) *Perpendicular Polarization*. (See Fig. 10-5 and Eqs. 10-94 through 10-97.)

$$\boldsymbol{E}_{I\perp}(x,z) + \boldsymbol{E}_{R\perp}(x,z) = -j2\boldsymbol{E}_{IM}\sin(\beta_{z1}z)\exp(-j\beta_{x1}x)\boldsymbol{a}_y \quad (10\text{-}106)$$

$$\boldsymbol{H}_{I\perp}(x,z) + \boldsymbol{H}_{R\perp}(x,z) = \frac{2E_{IM}}{\eta_1}[-\cos(\beta_{z1}z)\cos\phi_I\boldsymbol{a}_x$$

$$-j\sin(\beta_{z1}z)\sin\phi_I\boldsymbol{a}_z]\exp(-j\beta_{x1}x) \quad (10\text{-}107)$$

The surface-current density needed to maintain the difference between the sum of H_I and H_R above the reflecting surface, and no magnetic field below it, is, in accordance with Eq. 10-67,

$$\boldsymbol{J}_S(x) = \frac{2E_{IM}}{\eta_1}\cos\phi_I\exp(-j\beta_{x1}x)\boldsymbol{a}_y \quad (10\text{-}108)$$

(b) *Parallel Polarization*. (See Fig. 10-6 and Eqs. 10-98 through 10-101.)

$$\boldsymbol{E}_{I\parallel}(x,z) + \boldsymbol{E}_{R\parallel}(x,z) = -2E_{IM}[j\sin(\beta_{z1}z)\cos\phi_I\boldsymbol{a}_x$$

$$+\cos(\beta_{z1}z)\sin\phi_I\boldsymbol{a}_z]\exp(-j\beta_{x1}x) \quad (10\text{-}109)$$

$$\boldsymbol{H}_{I\parallel}(x,z) + \boldsymbol{H}_{R\parallel}(x,z) = \frac{2E_{IM}}{\eta_1}\cos(\beta_{z1}z)\exp(-j\beta_{x1}x)\boldsymbol{a}_y \quad (10\text{-}110)$$

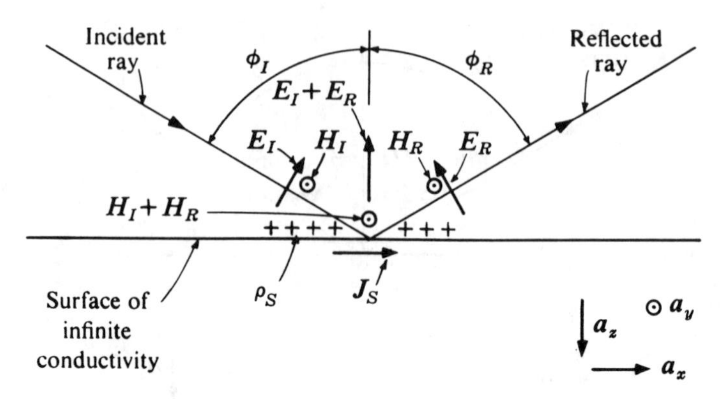

Figure 10-6. Parallel-polarized waves obliquely incident on ideal conductor.

The surface-current density is

$$\boldsymbol{J}_S(x) = \frac{2E_{IM}}{\eta_1} \exp(-j\beta_{x1}x)\boldsymbol{a}_x \tag{10-111}$$

Termination of the $\boldsymbol{D}$ flux impinging on the surface requires the following surface-charge density:

$$\rho_S(x) = 2E_{IM}\epsilon_1 \sin\phi_I \exp(-j\beta_{x1}x) \tag{10-112}$$

In accordance with the discussion in Sec. 10-3b(2) concerning continuity of current, the following may be derived from Eqs. 10-109 and 10-111:

$$\begin{aligned} \boldsymbol{J}_D(x,0) &= j\omega\epsilon[E_{zI}(x,0) + E_{zR}(x,0)]\boldsymbol{a}_z \\ &= -2j\omega\epsilon E_{IM} \sin\phi_I \exp(-j\beta_{x1}x)\boldsymbol{a}_z \end{aligned} \tag{10-113}$$

and

$$\begin{aligned} \nabla\cdot\boldsymbol{J}_S(x) &= \frac{\partial J_{xS}(x)}{\partial x} \\ &= \frac{-2j\beta_{x1}E_{IM}}{\eta_1} \exp(-j\beta_{x1}x) \end{aligned}$$

Substitution of Eqs. 10-84 and 10-23 reduces this to

$$\nabla\cdot\boldsymbol{J}_S(x) = -2j\omega\epsilon E_{IM} \sin\phi_I \exp(-j\beta_{x1}x) \tag{10-114}$$

Hence

$$\nabla\cdot\boldsymbol{J}_S(x) = J_{zD}(x,0) \tag{10-115}$$

10-4 PLANE WAVES TRAVELING PARALLEL TO HIGHLY CONDUCTING SURFACES

Metallic conducting surfaces are used abundantly to guide traveling electromagnetic fields. Field analyses are usually made under the assumption that the conductor surfaces are of infinite conductivity; then the $\boldsymbol{E}$ field tangential to those surfaces will be zero. In practice $\boldsymbol{E}$ will have a small but nonzero tangential component.[2]

a. Plane-Wave Fields between Parallel Planes — Effect of Finite Conductivity

The plane-wave mode, which was derived in terms of infinite space pervasiveness, is also applicable to the bounded space between two parallel planes of infinite conductivity if the $\boldsymbol{E}$ field is normal to those surfaces, and the direction of propagation and the $\boldsymbol{H}$ field are parallel to them, as suggested in Fig. 10-7. This geometrical arrangement is a convenient one in which to consider the effect of bounding surfaces of high, but finite, conductivity on TEM fields. A plane-wave set of fields in phasor form may be postulated. Let

$$\boldsymbol{E} \approx E_M \epsilon^{-j\beta z} \boldsymbol{a}_x \tag{10-116}$$

$$\boldsymbol{H} \approx \frac{E_M}{\eta_0} \epsilon^{-j\beta z} \boldsymbol{a}_y \tag{10-117}$$

The boundary condition related to an $\boldsymbol{H}$ field tangential to a highly conducting surface requires the presence of surface-current densities as follows:

$$\begin{aligned} \boldsymbol{J}_{SU} &= H_y \boldsymbol{a}_z \quad \text{(upper surface)} \\ \boldsymbol{J}_{SL} &= -H_y \boldsymbol{a}_z \quad \text{(lower surface)} \end{aligned} \tag{10-118}$$

In the limiting case of surfaces of infinite conductivity, expressions 10-116 and 10-117 would be equalities rather than approximations, and those equations plus 10-118 would suffice to describe the fields. A high but finite conductivity for the surface would require the presence of a longitudinal component of electric field E_z. Exact analyses indicate that $E_{\tan}$ and $|\boldsymbol{J}_S|$ are related, to a very good approximation, by the intrinsic impedance of the metal, even though that relationship was the result found for a

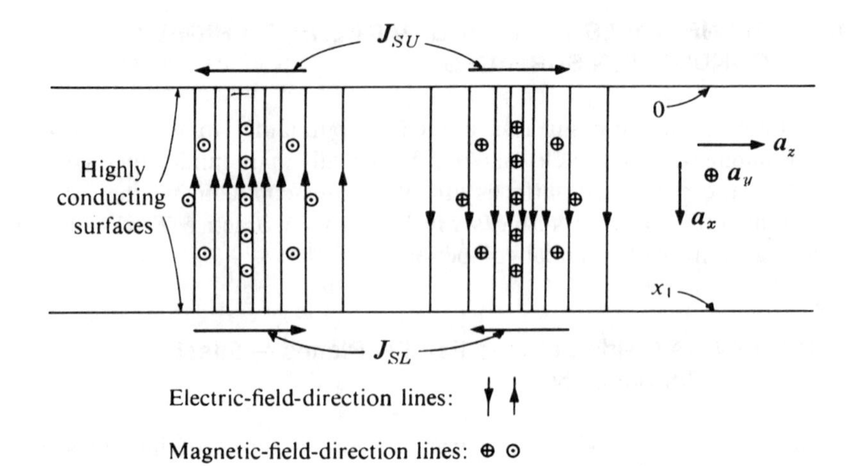

Figure 10-7. Parallel-plane conductors, traveling plane-wave mode.

normally incident wave set:

$$\eta_M = \sqrt{\frac{j\omega\mu}{\sigma}}$$

$$= \sqrt{\frac{\omega\mu}{\sigma}} \angle 45^\circ \tag{10-55}$$

$$E_z(0) \approx \frac{E_M}{\eta_0}\sqrt{\frac{\omega\mu}{\sigma}}\,\epsilon j^{r[(\pi/4)-\beta z]} \quad \text{(adjacent to upper surface)}$$

$$(10\text{-}119)$$

$$E_z(x_1) \approx \frac{-E_M}{\eta_0}\sqrt{\frac{\omega\mu}{\sigma}}\,\epsilon j^{r[(\pi/4)-\beta z]} \quad \text{(adjacent to lower surface)}$$

Interpolation of E_z for the region between the two planes will not be attempted here. Higher-order approximations would take account of the finite skin depth and the propagation rates in air and metal normal to the surface, and these would modify E_x and $\boldsymbol{H}$, but the approximations given here suffice for metallic conductors.

The maximum value of E_z is infinitesimal compared to that of E_x, even at microwave frequencies (see Prob. 10-9), but it should be noted that E_x

and E_z are not in time phase. Thus the magnitude of their resultant does not drop to zero but varies between a maximum and a minimum which differ enormously, while the space orientation of the vector field rotates continuously, but not uniformly, in the *xz* plane. This is a form of elliptical polarization (compare with Eq. 10-20).

Because of the longitudinal component of electric field, the electric and magnetic field-direction lines are not confined to planes transverse to the direction of propagation. The deviation from the true plane-wave pattern is slight, and since the plane wave is one configuration of the more general TEM mode, this lossy mode might appropriately be called a *quasi*-TEM mode. In practical usage it is called simply a TEM mode, even though noninfinite conductivity is involved.

b. Power Loss with Quasi-TEM Mode between Parallel Planes

Power dissipation in the conducting surfaces may be computed from the component of Poynting's vector normal to the surface:

$$P_x = -E_z H_y \tag{10-120}$$

Here E_z and H_y must be stated as explicit functions of time.

Substitutions will be made for the lower surface; the result for the upper surface will be the negative of that shown because it is oppositely directed:

$$P_x = \left(\frac{E_M}{\eta_0}\right)^2 \sqrt{\frac{\omega\mu}{\sigma}} \cos(\omega t - \beta z)\cos(\omega t + 45° - \beta z)$$

$$= \left(\frac{E_M}{\eta_0\sqrt{2}}\right)^2 \sqrt{\frac{\omega\mu}{\sigma}} \left[\cos 45° + \cos(2\omega t + 45° - 2\beta z)\right]$$

If this is time-averaged over an integral number of cycles, the result $P_{x\text{av}}$ is

$$P_{x\text{av}} = \left(\frac{E_m}{\eta_0\sqrt{2}}\right)^2 \sqrt{\frac{\omega\mu}{2\sigma}}$$

Equation 10-75 may be substituted:

$$P_{x\text{av}} = \left(\frac{E_M}{\eta_0 \sqrt{2}} \right)^2 R_S \qquad (10\text{-}121)$$

The squared quantity may be recognized as the rms value of the magnetic field tangent to the conducting surface (H_y), and that in turn as the rms value of the surface-current density.

10-5 CONCLUSIONS

Electromagnetic energy may be propagated in a homogeneous medium of infinite extent in the *plane-wave mode*, in which the electric and magnetic fields are both directed perpendicular to the direction of propagation, are uniform in planes transverse to the direction of propagation, and are perpendicular to each other. The propagation function for sinusoidal waves is analogous to that for a transmission line:

$$\alpha + j\beta = \sqrt{j\omega\mu(\sigma + j\omega\epsilon)} \qquad (10\text{-}10)$$

The magnitudes of sinusoidal $\boldsymbol{E}$ and $\boldsymbol{H}$ traveling-wave fields are related by the *intrinsic impedance* η:

$$\eta = \sqrt{\frac{j\omega\mu}{\sigma + j\omega\epsilon}} \qquad (10\text{-}13)$$

In a nonconducting medium these reduce to

$$\begin{aligned} \alpha &= 0 \\ \beta &= \omega\sqrt{\mu\epsilon} \\ \eta &= \sqrt{\mu/\epsilon} \end{aligned} \qquad (10\text{-}23)$$

The plane-wave set is a simple form of the *transverse electromagnetic*, or TEM, mode, which is defined as any mode in which the electric and magnetic fields are both perpendicular to the direction of propagation at all points.

Reflections occur when a traveling-wave set strikes the bounding surface between two media of differing intrinsic impedances. Boundary conditions which were derived in Chap. 9 relate the incident, reflected, and transmitted fields. For waves which are normally incident on the reflecting surface, the reflection phenomena correspond to those on a transmission line. Obliquely incident waves are reflected in a manner which depends on the polarization of the waves relative to the plane of incidence.

The limiting-case concept of a current sheet on a surface of infinite conductivity gives an acceptable approximation for the boundary effect of a metallic surface on the fields adjacent to it. A quasi-TEM mode may propagate between two parallel planes of highly conducting metal. Such a mode has a minute longitudinal component in its electric field and differs but slightly from a true TEM mode.

PROBLEMS

10-1. Compute the propagation constant, velocity of propagation, and intrinsic impedance at a frequency of 550 MHz for: (a) sea water (conductivity, 3.0 S/m; relative permittivity, 80), and (b) copper (conductivity, 5.8×10^7 S/m; assume relative permittivity unity). Assume the relative permeability to be unity for both media.

10-2. Find the divergence of the time-averaged Poynting vector for a plane wave, Eq. 10-40, and verify that this is equal to the time-averaged value of $|\boldsymbol{E}_1|^2\sigma$.

10-3. A 550-MHz plane wave moving through air is normally incident on seawater (see Prob. 10-1). Write expressions for the incident, transmitted, and reflected electric and magnetic fields, and the conduction-current density and displacement-current density, with numerical substitutions for all quantities except t and z. Assume the incident electric field has a maximum value of 1000 μV/m. Sketch a phasor diagram relating the various field magnitudes at the boundary surface. (This should correspond to Fig. 10-2.)

10-4. Repeat Prob. 10-3, assuming that the fields are normally incident on copper rather than seawater.

10-5. Compute the time-averaged power flow into a metallic surface from normally incident waves using Poynting's vector, and bring the result into agreement with Eq. 10-72.

10-6. Derive Snell's law for the angle of refraction of a plane wave obliquely incident upon a nonconducting medium, Eq. 10-105, by adding an equiphase line in medium 2 in Fig. 10-3.

10-7. Sketch a pattern in the plane of incidence of the direction lines of a perpendicularly polarized electromagnetic field set obliquely incident on a perfectly conducting plane surface. (*Hint*: The $\boldsymbol{H}$ field lines should be closed curves.)

10-8. Repeat Prob. 10-7 for a parallel-polarized field set.

10-9. Compute the ratio of the maximum value of E_z to the maximum value of E_x for a quasi-TEM mode between parallel-plane conductors at a frequency of 550 MHz, if the surfaces are copper. (Coordinate directions as in Fig. 10-7.)

10-10. Write an expression for the displacement-current density of a TEM mode between two parallel planes of infinite conductivity, beginning with Eq. 10-113. Show that continuity of the current field (displacement current and conduction current together) is maintained at the conductor surfaces.

REFERENCES

1. Burrows, C. R., Radio propagation over a plane earth, *Bell Syst. Tech. J.*, 16, 45, 1937.
2. Jordan, E. C. and Balmain, K. G., *Electromagnetic Waves and Radiating Systems*, 2nd ed., Prentice-Hall, New York, 1968.
3. Stratton, J. A., *Electromagnetic Theory*, McGraw-Hill, New York, 1941.
4. Terman, F. E., *Radio Engineers' Handbook*, McGraw-Hill, New York, 1943.

CHAPTER 11

Guided Fields: Arrays of Two or More Conductors

In the earlier chapters of this book, the inductance and capacitance parameters for a transmission line were postulated and used in the differential equations with no attempt to relate them to the cross-sectional dimensions or other physical properties of the line. It was mentioned that current is necessarily accompanied by a magnetic field and voltage by an electric field, and that those fields exist in the space between the conductors and possibly within them.

Electric and magnetic fields were discussed in the preceding two chapters, beginning with a derivation of Maxwell's equations and later considering the phenomena of sinusoidal traveling-wave trains propagating in continuous media and undergoing reflections at normal or oblique incidence. It was found that conduction current of finite density, describable as a vector field, is present whenever the propagation medium has a nonzero but finite conductivity.

Conducting surfaces parallel to the direction of field propagation influence the lateral extent of the field and other aspects of the process; in a sense all such conductors are waveguides, although usage often limits this term to hollow tubes of conducting material which enclose propagating fields.

The following conductor-set cross sections, each involving at least two electrically separated conductors, will be examined in some detail in this chapter: (1) the circularly symmetrical coaxial line, (2) the circular-wire parallel pair in space, (3) one wire above a highly conducting infinite plane, (4) a cable consisting of a circular outer conductor and an inner conductor which is offset from the center of the outer one, and (5) the circular-wire parallel pair with a highly conducting cylindrical shield.

11-1 COAXIAL LINE

A geometrically simple form of conductor-set cross section, one which yields algebraically simple expressions for the fields, is the coaxial cable, sketched in Fig. 11-1. The cylindrical coordinate system, with the $\boldsymbol{a}_z$ direction coincident with the axis of the conductor set, is an obvious choice here.

a. Fields for Traveling Waves

If one assumes (1) current flow in opposite directions in the two conductors at any given longitudinal location, (2) current flow uniformly distributed circumferentially, and (3) charge uniformly distributed circum-

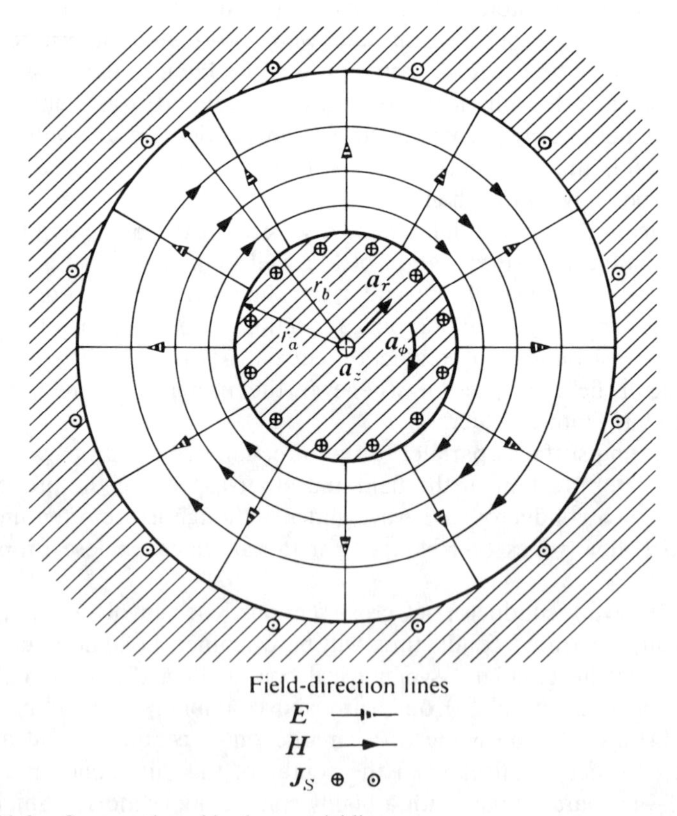

Figure 11-1. Cross section of lossless coaxial line.

ferentially on each conductor, one may surmise from symmetry that the resulting fields will be independent of ϕ and that one possible field configuration will have a purely radial $\boldsymbol{E}$ field and a purely circumferential $\boldsymbol{H}$ field. (This is the mode of principal importance for signal and energy propagation in a coaxial line, but it is not the only mode which is possible; see Sec. 13-3.) It will be further assumed, and subsequently verified, that the fields may propagate longitudinally as traveling waves.

Strictly, it is necessary that the conductors be lossless (infinite conductivity) for the $\boldsymbol{E}$ field to be purely radial, and this will be assumed for initial investigation of the fields between the conductors. (The fields within the conductors are nonzero when the conductivity is finite; these, and the resulting components of impedance, are investigated in Chap. 14.)

(1) *Solution for E and H Fields.* In accordance with the foregoing assumptions, let the following traveling-wave electric-field function be postulated:

$$\boldsymbol{E}_1 = E_{1r}(r)\cos(\omega t - \beta z)\boldsymbol{a}_r \tag{11-1}$$

The functional variation of $\boldsymbol{E}$ with respect to r may be determined with the aid of Maxwell's equation involving the divergence of $\boldsymbol{D}$; the accompanying $\boldsymbol{H}$ field and the parameter β may then be found with Maxwell's curl equations.

The space between the conductors will be assumed to be charge-free, so that, from Eq. 9-19,

$$\nabla \cdot \boldsymbol{D} = 0$$

The divergence of $\boldsymbol{D}$ may be found by multiplying Eq. 11-1 by ϵ and using Eq. A-36:

$$\frac{1}{r}\frac{\partial[r\epsilon E_{1r}(r)\cos(\omega t - \beta z)]}{\partial r} = 0$$

This reduces to

$$\frac{\partial[rE_{1r}(r)]}{\partial r} = 0 \tag{11-2}$$

Equation 11-2 may be integrated directly; a constant of integration A_1 is introduced thereby:

$$rE_{1r}(r) = A_1$$
$$E_{1r}(r) = \frac{A_1}{r} \tag{11-3}$$

Hence, by substituting in Eq. 11-1,

$$\boldsymbol{E}_1 = \frac{A_1}{r}\cos(\omega t - \beta z)\boldsymbol{a}_r \tag{11-4}$$

Thus the magnitude of the $\boldsymbol{E}$ field varies inversely with r.

One may continue by testing this function in Maxwell's curl equations:

$$\nabla \times \boldsymbol{E} = -\mu\frac{\partial \boldsymbol{H}}{\partial t} \tag{9-23}$$

The curl of $\boldsymbol{E}_1$ is obtained with Eq. A-37:

$$\nabla \times \boldsymbol{E}_1 = \frac{\beta A_1}{r}\sin(\omega t - \beta z)\boldsymbol{a}_\phi \tag{11-5}$$

Substitution of this into Eq. 9-23 and integration with respect to time yields

$$\boldsymbol{H}_1 = \frac{\beta A_1}{\omega r \mu}\cos(\omega t - \beta z)\boldsymbol{a}_\phi \tag{11-6}$$

Thus the $\boldsymbol{H}_1$ field is, at all times and locations, (1) perpendicular to $\boldsymbol{E}_1$, (2) directly proportional to it in magnitude, and (3) in time phase with it.

From Maxwell's equations

$$\nabla \times \boldsymbol{H} = \epsilon\frac{\partial \boldsymbol{E}}{\partial t} \tag{9-21}$$

Another expression for $\boldsymbol{E}_1$ may be found by taking the curl of $\boldsymbol{H}_1$, substituting in Eq. 9-21, and integrating

$$\nabla \times \boldsymbol{H}_1 = -\frac{\beta^2 A_1}{\omega r \mu}\sin(\omega t - \beta z)\boldsymbol{a}_r$$

$$\boldsymbol{E}_1 = \frac{\beta^2 A_1}{\omega^2 r \mu \epsilon}\cos(\omega t - \beta z)\boldsymbol{a}_r$$

If this result is equated to Eq. 11-4, β may be found in terms of the other parameters.

$$\frac{A_t}{r}\cos(\omega t - \beta z)\boldsymbol{a}_r = \frac{\beta^2 A_1}{\omega^2 r \mu \epsilon}\cos(\omega t - \beta z)\boldsymbol{a}_r$$

After cancelling like terms the latter reduces to

$$\beta = \omega\sqrt{\mu\epsilon} \tag{11-7}$$

This result is identical with Eq. 10-23 for infinite plane waves propagating in a lossless medium. As before, the velocity of propagation is

$$\nu = \frac{1}{\sqrt{\mu\epsilon}} \tag{10-24}$$

Substitution of Eq. 11-7 for the first β in Eq. 11-6 yields

$$\boldsymbol{H}_1 = \frac{A_1}{r}\sqrt{\frac{\epsilon}{\mu}}\cos(\omega t - \beta z)\boldsymbol{a}_\phi$$

Intrinsic impedance as defined for the infinite plane wave in a lossless medium is appropriate here, too:

$$\eta = \sqrt{\frac{\mu}{\epsilon}} \tag{10-23}$$

$$\boldsymbol{H}_1 = \frac{A_1}{\eta r}\cos(\omega t - \beta z)\boldsymbol{a}_\phi \tag{11-8}$$

(2) *Field Magnitudes in Proportion to Current.* The constant A_1 in Eqs. 11-4 and 11-8 for $\boldsymbol{E}_1$ and $\boldsymbol{H}_1$ is yet to be related to the current in the conductors. Ampère's law, Eq. 9-10, may be used for this purpose. Since for this assumed mode, $\boldsymbol{E}$, and hence $\boldsymbol{D}$, has no longitudinal component, no such component of displacement current is present. Hence only conduction current will be encircled if the path of integration is confined to a plane perpendicular to $\boldsymbol{a}_z$. If the path of integration is further confined to the space between the two conductors, the instantaneous transmission-line current $i_1(z,t)$ is obtained. The simplest path is a circle concentric with the conductor.

$$dL = r\,d\phi\,\boldsymbol{a}_\phi \tag{9-49}$$

$$\int \boldsymbol{H}_1 \cdot dL = \int_0^{2\pi} \frac{A_1}{\eta}\cos(\omega t - \beta z)\,d\phi$$

$$= \frac{2\pi A_1}{\eta}\cos(\omega t - \beta z)$$

Let

$$i_1(z,t) = I_{1M}\cos(\omega t - \beta z) \tag{11-9}$$

Equating the last two expressions yields

$$A_1 = \frac{I_{1M}\eta}{2\pi} \tag{11-10}$$

Equations 11-4 and 11-8 may be rewritten, after substituting Eq. 11-10,

$$\begin{aligned} \boldsymbol{E}_1 &= \frac{I_{1M}\eta}{2\pi r}\cos(\omega t - \beta z)\boldsymbol{a}_r \\ \boldsymbol{H}_1 &= \frac{I_{1M}}{2\pi r}\cos(\omega t - \beta z)\boldsymbol{a}_\phi \end{aligned} \tag{11-11}$$

(3) *Conduction-Current Fields.* A consequence of assuming infinite conductivity is that conduction current concentrates into sheets of infinitesimal thickness at the conductor surfaces, as discussed in Sec. 10-2d(1) and 10-4. Such a sheet maintains a magnetic-field difference between the field immediately above the conductor surface $[\boldsymbol{H}_1(r_a, z, t)$ or $\boldsymbol{H}_1(r_b, z, t)]$ and the field immediately below the surface, which is zero. The conduction-current field may be regarded as a pair of surface-current densities; the symbols $\boldsymbol{J}_{S1a}$ and $\boldsymbol{J}_{S1b}$ may be used for the inner and outer conductors, respectively:

$$\begin{aligned} \boldsymbol{J}_{S1a} &= \frac{I_{1M}}{2\pi r_a}\cos(\omega t - \beta z)\boldsymbol{a}_z \\ \boldsymbol{J}_{S1b} &= \frac{-I_{1M}}{2\pi r_b}\cos(\omega t - \beta z)\boldsymbol{a}_z \end{aligned} \tag{11-12}$$

Because the conduction current has no circumferential component, a longitudinal slit of infinitesimal width could be cut through the outer conductor without disturbing the field patterns. This is the basis of the *slotted line*, a useful experimental device for measuring relative field strength and thereby obtaining standing-wave data. An insulated probe connected to a crystal detector is used to sample the field. (For mechanical clearance the slot actually used is noninfinitesimal in width. An alternative cross section for slotted lines which is sometimes used in coaxial systems is the *parallel-slab line*, discussed in Appendix C.)

(4) *E and H Pair for Other Direction of Travel.* In Chap. 2 it was noted that a second voltage-and-current traveling-wave pair, moving in the negative direction, would also satisfy the transmission-line differential equations. Equations for the fields and current of this wave pair, corresponding to those just derived in Eqs. 11-1 through 11-2, are

$$\boldsymbol{H}_2 = \frac{I_{2M}}{2\pi r}\cos(\omega t + \beta z)\boldsymbol{a}_\phi$$

$$\boldsymbol{E}_2 = \frac{-I_{2M}\eta}{2\pi r}\cos(\omega t + \beta z)\boldsymbol{a}_r \tag{11-13}$$

$$i_2(z,t) = I_{2M}\cos(\omega t + \beta z) \tag{11-14}$$

b. Derivation of Transmission-Line Parameters

The field-configuration mode just examined for the coaxial line is, like the infinite plane wave in a uniform medium, an instance of transverse electromagnetic (TEM) propagation. When TEM-mode fields are bounded by conducting surfaces, as in this instance, the concept of voltage as the line integral of $\boldsymbol{E}$ may be used. The expressions already derived for electric field are directly proportional to current, and this will enable one to find the characteristic impedance in terms of the radii r_a and r_b and the properties of the medium between.

(1) *Voltage in Time-Varying TEM Fields.* The line integral of $\boldsymbol{E}$ was mentioned in the description of electric fields in Sec. 9-1b(2). Because the line integral between two conductor surfaces in a non-time-varying field is independent of path, it is a meaningful measure of the "resultant" strength of the intervening field. Known as *voltage*, it is a key element in the concept of the electric circuit as an approximate means of analyzing an electromagnetic-field problem.

The curl function was defined in Sec. A-1c(2) in terms of the line integral around a small closed path. Unless the curl of a given vector field vanishes throughout a given region, the line integral of that vector field will not be independent of path. According to Eq. 9-23 (one of Maxwell's equations), the curl of $\boldsymbol{E}$ does not vanish if the fields are time-varying, and hence the concept of voltage as a path-independent line integral is not, in general, realizable for the dynamic $\boldsymbol{E}$ field.

For the TEM mode, however, independence of path for the line integral exists on a limited basis. Specifically, if one confines the paths of integration to planes perpendicular to the direction of wave propagation, line integrals of $\boldsymbol{E}$ in each such plane are independent of path. The z component of the curl of $\boldsymbol{E}_1$ given in Eq. 11-11 vanishes, and that is all that is required if the path is restricted as just described. This is the basis for the voltage concept as commonly used in alternating-current circuit theory.

$$v_1(z,t) = -\int_{r_b}^{r_a} \boldsymbol{E}_1 \cdot (dr\,\boldsymbol{a}_r) \tag{11-15}$$

$$v_1(z,t) = -\int_{r_b}^{r_a} \frac{I_{1M}\eta}{2\pi r}\cos(\omega t - \beta z)\,dr$$

$$= \frac{I_{1M}\eta}{2\pi}\ln\left(\frac{r_b}{r_a}\right)\cos(\omega t - \beta z) \tag{11-16}$$

Similarly, for the wave traveling in the $-\boldsymbol{a}_z$ direction,

$$v_2(z,t) = \frac{-I_{2M}\eta}{2\pi}\ln\left(\frac{r_b}{r_a}\right)\cos(\omega t + \beta z) \tag{11-17}$$

(2) *Characteristic Impedance.* In Chap. 2 it was indicated that the voltage and current functions of a traveling-wave pair should be related by the characteristic impedance, a function which for a lossless line was purely resistive. Comparison of Eqs. 11-9 and 11-16 indicates that the characteristic impedance of a coaxial line is

$$Z_0 = \frac{\eta}{2\pi}\ln\left(\frac{r_b}{r_a}\right)$$

$$= \frac{1}{2\pi}\sqrt{\frac{\mu}{\epsilon}}\ln\left(\frac{r_b}{r_a}\right) \tag{11-18}$$

(3) *Maximum $|E|$, Minimum Radii, and Parasitic Modes.* In some applications such as the cables connecting transmitters to antennas, and power transmission cables, the maximum voltage is high, and the breakdown gradient of the insulation material imposes a minimum value for r_a. From Eqs. 11-11 and 11-16 the maximum values (as functions of time) of $|\boldsymbol{E}_1|$ and v_1 may be related as

$$E_{1M}(r) = \frac{v_{1M}}{r\ln(r_b/r_a)} \qquad r_a \le r \le r_b \tag{11-19}$$

As a function of r, E_{1M} will be a maximum at $r = r_a$. (The ratio r_b/r_a will be determined by the characteristic impedance, Eq. 11-18.) If the maximum allowable, or critical, gradient in the insulation material is designated by E_{CR}, the corresponding minimum value for r_a, r_{aCR}, is

$$r_{aCR} = \frac{V_{1M}}{E_{CR} \ln(r_b/r_a)} \tag{11-20}$$

Parasitic *propagating* waveguide modes are possible in a coaxial cable if the operating frequency is sufficiently high. The boundary conditions for such modes are detailed in Sec. 13-3; the cutoff (minimum) frequency for the dominant mode (the one with the lowest cutoff frequency) is approximately

$$f_c = \frac{1}{\pi(r_a + r_b)\sqrt{\mu\epsilon}} \tag{11-21}$$

(4) *Distributed Inductance and Capacitance.* Characteristic impedance and velocity of propagation are related to the inductance and capacitance per unit length of a lossless transmission line as

$$Z_0 = \sqrt{\frac{l}{c}} \tag{2-35}$$

$$\nu = \frac{1}{\sqrt{lc}} \tag{2-30}$$

Since expressions have been obtained for Z_0 and ν in terms of the physical properties of the cable (μ, ϵ, r_a, and r_b), it is appropriate to solve Eqs. 2-30 and 2-35 for l and c:

$$l = \frac{Z_0}{\nu} \tag{11-22}$$

$$c = \frac{1}{Z_0 \nu} \tag{11-23}$$

$$\nu = \frac{1}{\sqrt{\mu\epsilon}} \tag{10-24}$$

Substitution of Eqs. 10-24 for ν and 11-18 for Z_0 yields

$$l = \frac{\mu}{2\pi} \ln\left(\frac{r_b}{r_a}\right) \tag{11-24}$$

$$c = \frac{2\pi\epsilon}{\ln(r_b/r_a)} \tag{11-25}$$

Equations 11-24 and 11-25 define the distributed inductance and capacitances of a lossless coaxial cable in terms of its cross-sectional dimensions and the properties of the medium between the conductors. Equation 11-18 similarly defines the characteristic impedance. In each instance the cross-sectional dimensions appear solely in the form of the ratio r_b/r_a. A change in scale of the cross section leaves these derived parameters unchanged.

Problems 11-1 through 11-4 consider other aspects of the fields in a lossless coaxial line.

c. Parameters When Resistance is Nonzero

If the conductors have finite conductivity, the resistance parameter r is nonzero and, as noted in Secs. 1-2a and 4-2, and shown in Chap. 14, it and the inductance are frequency-dependent. An example of the calculation of distributed impedance in the high-frequency range is given in Sec. 14-3b.

11-2 CIRCULAR-WIRE PARALLEL PAIR

A set of two circular conductors parallel to each other and carrying equal currents in opposite directions, as shown in Fig. 11-2, is an arrangement commonly used in transmission-line practice. From the standpoint of field analysis, it is less symmetrical than the coaxial cable, but amenable nevertheless.

a. TEM Fields between Conductors

It will be assumed that the distance between the conductors is much less than one-quarter of the TEM wavelength of the highest frequency component of the transmitted signal. Under this restriction the fields may be assumed to be *quasistatic*, that is, in a given transverse plane they will have the same configurations as those of direct current and be in phase timewise. Superposition may be used to obtain a description of the fields for this conductor array from the results for the coaxial line, Eqs. 11-11.

Initially it will be assumed that the conductor centers are at $-h/2, 0$ and $h/2, 0$, and that the conductor radii are extremely small compared to the distance h. [As is shown in Sec. 11-2a(3), the results may be general-

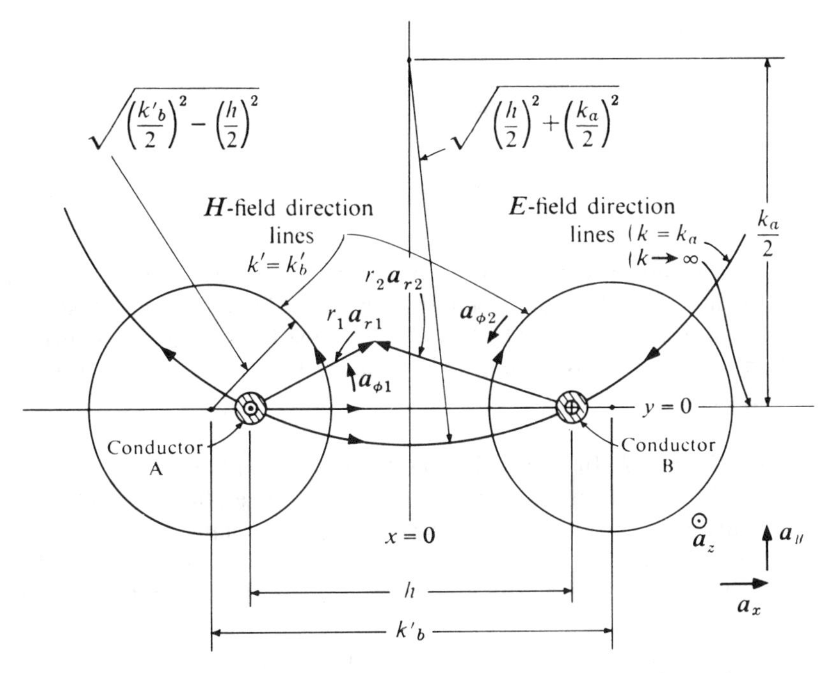

Figure 11-2. Field-direction lines for two widely spaced, parallel circular conductors.

ized to apply to conductors of any diameter, provided their centers are offset by appropriate distances from $-h/2, 0$ and $h/2, 0$.]

One may first visualize the fields which would exist between conductor A and a coaxial return path with a radius very much larger than h, with conductor B removed. A traveling-wave set moving in the $\boldsymbol{a}_z$ direction will be assumed, and will be designated by $\boldsymbol{E}_A$ and $\boldsymbol{H}_A$. Equations 11-11 will be applicable if r is replaced by r_1. If conductor B has an infinitesimal diameter and is unchanged, it may be replaced without disturbing fields $\boldsymbol{E}_A$ and $\boldsymbol{H}_A$. In like manner a circuit may be formed through conductor B and a coaxial return, and this may be assumed to carry a traveling-wave set also moving in the $\boldsymbol{a}_z$ direction, but in phase opposition to that between conductor A and its coaxial return. For these fields, which will be designated by $\boldsymbol{E}_B$ and $\boldsymbol{H}_B$, Eqs. 11-11 are applicable if r is replaced by r_2 and I_{1M} by $-I_{1M}$. Strictly the coaxial return paths would not be coincident (their centers are separated by h), but as their radii approach infinity, the resultant $\boldsymbol{H}$ field in their vicinity would approach zero (see Prob. 11-6), and hence they would effectively cancel each other. Thus the only conduction currents remaining would be those in conductors A and B, and the resultant fields would be

$$\boldsymbol{E} = \boldsymbol{E}_A + \boldsymbol{E}_B \tag{11-26}$$

$$\boldsymbol{H} = \boldsymbol{H}_A + \boldsymbol{H}_B \tag{11-27}$$

(The foregoing is essentially what is inferred in derivations given in terms of the fields of single, isolated conductors. Every current must have a return path, and for the TEM mode this must be a conduction current. Furthermore, both ends of nonclosing lines of the $\boldsymbol{D}$ field must terminate on charges.)

(1) *Electric Field.* In the TEM mode the current and magnetic and electric fields are all in time phase in any transverse plane, and occasion will not arise, in the derivation which follows, to obtain derivatives with respect to t or z. Accordingly $i_1(z, t)$ in Eqs. 11-11 may be replaced by $i_+(z, t)$, where this will be understood to represent any function of time traveling in the $+z$ direction.

The vector fields $\boldsymbol{E}_A$ and $\boldsymbol{E}_B$ at a point (x, y) due to a traveling-wave set moving in the $\boldsymbol{a}_z$ direction are, if one adapts Eq. 11-11 in the manner noted above,

$$\boldsymbol{E}_A = \frac{\eta i_+(z,t)}{2\pi r_1}\boldsymbol{a}_{r1} \tag{11-28}$$

$$\boldsymbol{E}_B = \frac{-\eta i_+(z,t)}{2\pi r_2}\boldsymbol{a}_{r2} \tag{11-29}$$

To put these expressions in terms of the common coordinate system shown in Fig. 11-2, a rectangular one, the following substitutions are needed:

$$r_1 = \sqrt{\left(x + \frac{h}{2}\right)^2 + y^2} \tag{11-30}$$

$$r_2 = \sqrt{\left(x - \frac{h}{2}\right)^2 + y^2} \tag{11-31}$$

The partial $\boldsymbol{E}$ fields will be resolved into components:

$$E_{Ax} = \frac{\eta i_+(z,t)[x + (h/2)]}{2\pi r_1^2} \tag{11-32}$$

$$E_{Ay} = \frac{\eta i_+(z,t) y}{2\pi r_1^2} \tag{11-33}$$

$$E_{Bx} = \frac{-\eta i_+(z,t)[x - (h/2)]}{2\pi r_2^2} \tag{11-34}$$

$$E_{By} = \frac{-\eta i_+(z,t) y}{2\pi r_2^2} \tag{11-35}$$

A helpful technique for describing a vector field is that of *direction lines*, which were introduced from the field-exploration point of view in Sec. 9-1a. Such lines are loci which are continuously tangent to the field. Thus at any point x, y in a transverse plane, the directional tangent of an incremental segment of such a line, or the ratio of Δy to Δx, is equal to the ratio of the y component of the field to the x component at that point. If the length of the segment is reduced to differential size, the ratio $\Delta y/\Delta x$ approaches, as a limit, the derivative dy/dx. Thus

$$\frac{dy}{dx} = \frac{E_y}{E_x} \tag{11-36}$$

In this instance Eq. 11-36 becomes

$$\frac{dy}{dx} = \frac{E_{Ay} + E_{By}}{E_{Ax} + E_{Bx}}$$

$$= \frac{(y/r_1^2) - (y/r_2^2)}{\{[x + (h/2)]/r_1^2\} - \{[x - (h/2)]/r_2^2\}}$$

Clearing the fractions within the numerator and denominator of the main expression and substituting Eqs. 11-30 and 11-31 reduces it to

$$\frac{dy}{dx} = \frac{2xy}{x^2 - y^2 - (h/2)^2}$$

This is a nonlinear differential equation of the first order. It is not one for which the variables may be separated, but it proves to be convertible into an exact differential:

$$2xy\,dx + \left[y^2 + (h/2)^2 - x^2\right]dy = 0 \tag{11-37}$$

Given a first-order differential equation of the form

$$M\,dx + N\,dy = 0 \tag{11-38}$$

The necessary and sufficient condition that it be an exact differential is

$$\frac{\partial M}{\partial y} = \frac{\partial N}{\partial x} \tag{11-39}$$

This condition is not met by Eq. 11-37 as it stands, but if that equation is multiplied by the integrating factor $1/y^2$, the resulting equation will meet the requirement just given:

$$\frac{2x\,dx}{y} + \left[1 + \left(\frac{h}{2}\right)^2 \frac{1}{y^2} - \left(\frac{x}{y}\right)^2\right]dy = 0$$

or

$$\frac{2x}{y} + \left[1 + \left(\frac{h}{2}\right)^2 \frac{1}{y^2} - \left(\frac{x}{y}\right)^2\right]\frac{dy}{dx} = 0 \tag{11-40}$$

The expression which, if differentiated with respect to x (considering y to be a function of x) would yield Eq. 11-40, is

$$\frac{x^2}{y} + y - \left(\frac{h}{2}\right)^2 \frac{1}{y} = k \tag{11-41}$$

Here k is an arbitrary constant. Equation 11-41 may be reduced to the usual form for the equation of a circle by (1) multiplying it by y,

(2) moving the two terms nearest the equality sign to their respective opposite sides, and (3) adding the quantity $(k/2)^2$ to both sides:

$$x^2 + y^2 - \left(\frac{h}{2}\right)^2 = ky$$

$$x^2 + y^2 - ky = \left(\frac{h}{2}\right)^2$$

$$x^2 + \left(y - \frac{k}{2}\right)^2 = \left(\frac{h}{2}\right)^2 + \left(\frac{k}{2}\right)^2 \tag{11-42}$$

Hence the direction lines of the $\boldsymbol{E}$ field are circles of radii $\sqrt{(h/2)^2 + (k/2)^2}$, with centers located on the Y axis but offset from the origin by amount $k/2$; the relation between the offset and the radius of each circle is such that every field line will pass through the points $(h/2, 0)$ and $(-h/2, 0)$. The arbitrary constant k may be assigned any value from $-\infty$ to $+\infty$, and each such value defines a direction line of the $\boldsymbol{E}$ field.

(2) *Magnetic Field.* Direction lines for the magnetic field may be found similarly:

$$\boldsymbol{H}_A = \frac{i_+(z,t)}{2\pi r_1}\boldsymbol{a}_{\phi 1} \tag{11-43}$$

$$\boldsymbol{H}_B = \frac{-i_+(z,t)}{2\pi r_2}\boldsymbol{a}_{\phi 2} \tag{11-44}$$

$$H_{Ax} = \frac{-i_+(z,t)y}{2\pi r_1^2} \tag{11-45}$$

$$H_{Ay} = \frac{i_+(z,t)[x + (h/2)]}{2\pi r_1^2} \tag{11-46}$$

$$H_{Bx} = \frac{i_+(z,t)y}{2\pi r_2^2} \tag{11-47}$$

$$H_{By} = \frac{-i_+(z,t)[x - (h/2)]}{2\pi r_2^2} \tag{11-48}$$

The differential equation for the direction lines is

$$\frac{dy}{dx} = \frac{H_{Ay} + H_{By}}{H_{Ax} + H_{Bx}} \tag{11-49}$$

The clearing of fractions after the appropriate terms have been substituted parallels the procedure used for the electric field and results in the following:

$$\frac{dy}{dx} = \frac{x^2 - y^2 - (h/2)^2}{-2xy}$$

$$\left[x^2 - y^2 - (h/2)^2\right] dx + 2xy\, dy = 0 \tag{11-50}$$

In this instance the integrating factor which will convert the differential equation into an exact differential is $1/x^2$. The result of integrating is

$$x + \frac{y^2}{x} + \left(\frac{h}{2}\right)^2 \frac{1}{x} = k' \tag{11-51}$$

Here k' is an arbitrary constant. This result may be changed to the standard form for a circle by multiplying by x, rearranging terms, and adding $(k'/2)^2$ to both sides:

$$x^2 + y^2 - k'x = -\left(\frac{h}{2}\right)^2$$

$$\left(x - \frac{k'}{2}\right)^2 + y^2 = \left(\frac{k'}{2}\right)^2 - \left(\frac{h}{2}\right)^2 \tag{11-52}$$

Direction lines of the $\boldsymbol{H}$ fields are circles, with centers on the X axis but offset from the origin by distances $k'/2$, which distances are necessarily greater than $h/2$. Each value of $k'/2$ defines a separate field-direction line. The radius of each circle is $\sqrt{(k'/2)^2 - (h/2)^2}$, a quantity which approaches zero for values of k' approaching h.

(3) ***General Properties—Applicability to Conductors of Nonzero Diameter.*** The $\boldsymbol{E}$ and $\boldsymbol{H}$ fields are mutually perpendicular; this may be demonstrated by showing that the scalar product $\boldsymbol{E} \cdot \boldsymbol{H}$ vanishes (see Prob. 11-6).

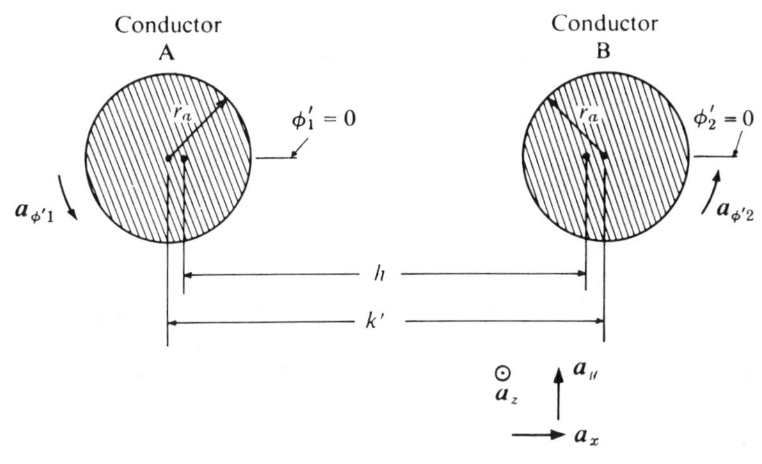

Figure 11-3. Closely spaced parallel conductors; $(h/2)^2 = (k'/2)^2 - r_a^2$.

As was true of the coaxial line, the mode of propagation is transverse-electromagnetic (TEM), and the concept of voltage as a path-independent line integral is applicable on a limited basis, namely, if one restricts each line integral of $\boldsymbol{E}$ to a single plane perpendicular to the direction of propagation.

Curves faired so as to be everywhere *normal to the electric-field direction lines* (but in a given transverse plane) have the property that the line integral of $\boldsymbol{E}$ along any such curve is zero. Because the line integral of $\boldsymbol{E}$ in a transverse plane is independent of path, its value from one such curve to another will be independent of the particular terminal points, and these curves are called *equipotential lines*. As has been noted, the $\boldsymbol{H}$ field-direction lines lie in transverse planes and are everywhere perpendicular to the $\boldsymbol{E}$ field-direction lines, hence they have the same geometric pattern as do equipotentials of the $\boldsymbol{E}$ field.

Thus, so far as the $\boldsymbol{E}$ field is concerned, the initially postulated conductors of infinitesimally small radii can be replaced by larger circular conductors, as indicated in Fig. 11-3, provided the center of each is displaced from the infinitesimal-conductor location so that the perimeter coincides with a locus (in this instance a circle) of constant potential. The field pattern between the new, larger conductors is the same as before, but the field within each conductor becomes zero.

(4) *Surface-Charge and Surface-Current Densities.* Surface-charge density is equal in magnitude to the normal component of $\boldsymbol{D}$ field touching the conductor and hence will not be uniform around the circumference of the conductor. It will be highest at the point closest to the other conductor and weakest at the diametrically opposite point.

Under the limiting condition of lossless conductors, as assumed here, the conduction current consists of a surface-current sheet, and no penetration of the conductor surface by magnetic flux will occur. The surface current will distribute itself over the perimeter of each conductor so that the resultant $\boldsymbol{H}$ field is tangent to the surface of each. The particular current distribution which will achieve this at each point on the perimeter is equal in magnitude to the strength of the $\boldsymbol{H}$ field immediately adjacent to it. Since $|\boldsymbol{H}|$ and $|\boldsymbol{E}|$ are in the same proportion throughout the field, the surface-current density is in direct proportion, with respect to circumferential position, to the surface-charge distribution just mentioned.[1] The resulting circumferential nonuniformity in surface-charge distributions and in current density is called *proximity effect*.*

Cylindrical coordinates that are coaxial with one of the conductors form a logical reference for describing the intensities of the surface-charge and surface-current fields. An equation will be written for the $\boldsymbol{E}$ field leaving conductor A in Fig. 11-3. Since the conductor surface was placed on an equipotential surface in the elementary field, the tangential component $E_{\phi'1}$ is zero. The radial component, considered positive when directed outwardly, will be designated as $E_{r'1}$. (The prime marks in the subscripts will indicate that the origin is at the center of the conductor of nonzero radius r_a rather than the geometric-line conductors.) The partial fields defined in Eqs. 11-28 through 11-31 will be used:

$$E_{r'1} = (E_{Ax} + E_{Bx})\cos\phi'_1 + (E_{Ay} + E_{By})\sin\phi'_1 \qquad \text{(11-54)}$$

*In a conductor of finite conductivity, proximity effect will increase the effective resistance above that which would be produced by skin effect with a circularly symmetrical current distribution. Sim[28] gives the following asymptotic approximation for the ratio of the AC resistance to the DC value for a two-conductor pair consisting of solid circular conductors of radius r_a, with a center-to-center spacing of k':

$$\frac{r_{\text{ac}}}{r_{\text{adc}}} \approx \frac{Bmr_a}{2\sqrt{2}} + \frac{B(2-B^2)}{4} + \frac{B(9-10B^2+4B^4)}{16\sqrt{2}\,mr_a} \qquad \text{(11-53)}$$

Here,

$$B = \frac{1}{\sqrt{1-4(r_a/k')^2}}$$

$$m = \sqrt{\omega\mu\sigma}$$

Coordinates for points on the cylinder radius r_a may be transformed by the equations

$$x + \frac{k'}{2} = r_a \cos \phi'_1 \tag{11-55}$$

$$y = r_a \sin \phi'_1 \tag{11-56}$$

After substitution of Eqs. 11-32 through 11-35, and 11-55 and 11-56, Eq. 11-54 may be reduced to

$$E_{r'1} = \frac{\eta i_+(z,t)}{2\pi}\left[\frac{r_a + (h/2 - k'/2)\cos \phi'_1}{r_1^2} - \frac{r_a - (h/2 + k'/2)\cos \phi'_1}{r_2^2}\right] \tag{11-57}$$

Substitution of Eqs. 11-55 and 11-56 in Eqs. 11-30 and 11-31 for the denominators yields

$$r_1^2 = r_a^2 + (h - k')r_a \cos \phi'_1 + \tfrac{1}{4}(h - k')^2 \tag{11-58}$$

$$r_2^2 = r_a^2 - (h + k')r_a \cos \phi'_1 + \tfrac{1}{4}(h + k')^2 \tag{11-59}$$

As noted after Eq. 11-52,

$$r_a^2 = \left(\frac{k'}{2}\right)^2 - \left(\frac{h}{2}\right)^2 \tag{11-60}$$

A common denominator may be found by multiplying r_1^2 by $k' + h$, and r_2^2 by $k' - h$; after substituting Eq. 11-60 it may be reduced to $r_a^2(2k' - 4r_a \cos \phi'_1)$. Equation 11-57 then simplifies to

$$E_{r'1} = \frac{\eta i_+(z,t)h}{2\pi r_a(k' - 2r_a \cos \phi'_1)} \tag{11-61}$$

Equation 11-60 may be substituted to eliminate h from Eq. 11-61.

$$E_{r'1} = \frac{\eta i_+(z,t)\sqrt{(k'/2)^2 - r_a^2}}{\pi r_a(k' - 2r_a \cos \phi'_1)} \tag{11-62}$$

For the surface-charge density:

$$\rho_{SA} = \epsilon_0 E_{r'1} \tag{11-63}$$

For the surface-current density:

$$\boldsymbol{J}_{SA} = \frac{1}{\eta} E_{r'1} \boldsymbol{a}_z \tag{11-64}$$

From Eqs. 11-62 and 11-64 a first-order approximation may be calculated of the increase in effective resistance at high frequencies due to proximity effect (see Prob. 11-8). These results also prove useful in the analyses in Secs. 11-3 and 11-4 on the effects of an ideal ground plane and of a cylindrical shield.

b. Derivation of Transmission-Line Parameters

The voltage function $v_+(z,t)$ may be found by integrating $-\boldsymbol{E}$ from the surface of conductor B to that of conductor A, restricting the path to the XY plane. The easiest path to use is along the X axis:

$$v_+(z,t) = -\int_{(k'/2)-r_a}^{-(k'/2)+r_a} (\boldsymbol{E}_A + \boldsymbol{E}_B) \cdot (dx\, \boldsymbol{a}_x) \tag{11-65}$$

At $y = 0$, E_{Ax} and E_{Bx} reduce to the forms shown in the integral

$$v_+(z,t) = -\frac{\eta i_+(z,t)}{2\pi} \int_{(k'/2)-r_a}^{-(k'/2)+r_a} \left[\frac{1}{x + (h/2)} - \frac{1}{x - (h/2)} \right] dx \tag{11-66}$$

Integration of the right-hand side, substitution of limits, and elimination of h by means of Eq. 11-60 will yield

$$v_+(z,t) = \frac{\eta i_+(z,t)}{\pi} \ln\left[\frac{k'}{2r_a} + \sqrt{\left(\frac{k'}{2r_a}\right)^2 - 1} \right] \tag{11-67}$$

The following identity may be used:

$$\cosh^{-1} u = \ln\left(u + \sqrt{u^2 - 1}\right) \tag{11-68}$$

Thus

$$v_+(z,t) = \frac{\eta i_+(z,t)}{\pi} \cosh^{-1}\left(\frac{k'}{2r_a}\right) \tag{11-69}$$

If $r_a \ll k'$, a series expansion may be used for the radical in Eq. 11-67:

$$\sqrt{\left(\frac{k'}{2r_a}\right)^2 - 1} = \frac{k'}{2r_a}\sqrt{1 - \left(\frac{2r_a}{k'}\right)^2}$$

$$\approx \frac{k'}{2r_a}\left[1 - \frac{1}{2}\left(\frac{2r_a}{k'}\right)^2\right] \tag{11-70}$$

Hence
$$v_+(z,t) \approx \frac{\eta i_+(z,t)}{\pi} \ln\left(\frac{k'}{r_a}\right) \tag{11-71}$$

The characteristic impedance, which relates the voltage and current of a traveling-wave pair, is, therefore, from Eq. 11-69 or 11-71,

$$Z_0 = \frac{\eta}{\pi} \cosh^{-1}\left(\frac{k'}{2r_a}\right) \tag{11-72}$$

$$Z_0 \approx \frac{\eta}{\pi} \ln\left(\frac{k'}{r_a}\right) \tag{11-73}$$

The intrinsic impedance of a lossless medium is

$$\eta = \sqrt{\frac{\mu}{\epsilon}} \tag{10-23}$$

Hence
$$Z_0 = \frac{1}{\pi}\sqrt{\frac{\mu}{\epsilon}} \cosh^{-1}\left(\frac{k'}{2r_a}\right) \tag{11-74}$$

$$Z_0 \approx \frac{1}{\pi}\sqrt{\frac{\mu}{\epsilon}} \ln\left(\frac{k'}{r_a}\right) \tag{11-75}$$

The distributed parameters l and c may be related to Z_0 in the same manner as for the TEM mode in the coaxial line:

$$l = \frac{Z_0}{\nu} \tag{11-22}$$

$$c = \frac{1}{Z_0 \nu} \tag{11-23}$$

The velocity of propagation ν for the TEM mode in a lossless medium was found to be

$$\nu = \frac{1}{\sqrt{\mu\epsilon}} \tag{10-24}$$

Substitution of Eqs. 11-74 or 11-75, and 10-24, in Eqs. 11-22 and 11-23 yields

$$l = \frac{\mu}{\pi}\cosh^{-1}\left(\frac{k'}{2r_a}\right) \tag{11-76}$$

$$\approx \frac{\mu}{\pi}\ln\left(\frac{k'}{r_a}\right) \tag{11-77}$$

$$c = \frac{\epsilon\pi}{\cosh^{-1}(k'/2r_a)} \tag{11-78}$$

$$\approx \frac{\epsilon\pi}{\ln(k'/r_a)} \tag{11-79}$$

These results for the distributed inductance and capacitance of a lossless two-parallel-wire line correspond to Eqs. 11-24 and 11-25 for the lossless coaxial line.

11-3 IMAGE-CONDUCTOR ANALYSIS OF GROUND PLANES AND SHIELDS

The geometrical analysis of the ***E*** and ***H*** fields of the two-parallel-conductor line may be applied to other high-conductivity-surface situations by the method of images. Specific cases which will be examined here are: (a) a single conductor with a lossless ground return,* (b) a cable with a single inner conductor which is not concentric with the shield, and (c) a cable with two identical parallel conductors which are surrounded by a shield.

*The physical earth has a conductivity of 10^{-1} to 10^{-3} S/m, compared to 5.8×10^7 S/m for copper. Substitution of $\sigma = 10^{-2}$ S/m in Eq. 10-79 will yield, for 60 Hz, a nominal skin depth of 650 m. For power-frequency studies a more thoroughgoing approach than that of images is generally required; results of such an analysis are summarized in Appendix D.

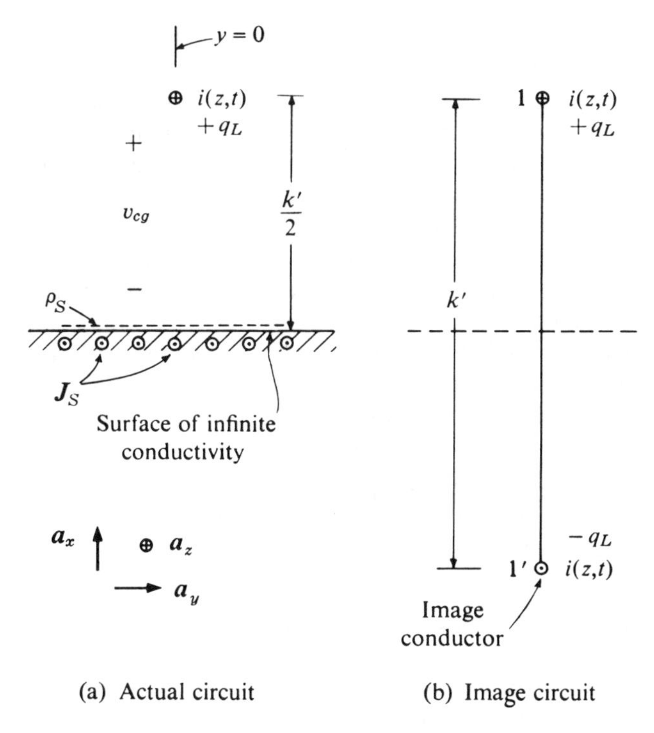

Figure 11-4. Single conductor with lossless ground return.

a. Single Conductor with Perfectly Conducting Ground Return

A simple example which will illustrate the technique of images is that of a single conductor with ground return of infinite conductivity as shown in Fig. 11-4(a). The boundary conditions at an infinitely conducting surface are the same as those derived for oblique reflection, namely: (1) that the tangential component of $\boldsymbol{E}$ vanish at the conducting surface; (2) that the normal component of $\boldsymbol{B}$ vanish at the conducting surface; (3) that a surface-current density, equal in magnitude to the tangential component of $\boldsymbol{H}$ at the conducting surface, flow at right angles to the $\boldsymbol{H}$ field in the direction required by the right-hand rule; and (4) that a surface-charge density, equal in magnitude to the normal component of $\boldsymbol{D}$ leaving the conducting surface, be present.

It may be seen from the symmetry of the equivalent image system in Fig. 11-4(b) that if line 1-1′ is perpendicular to the plane corresponding to ground, if the image conductor 1′ is at the same distance from that surface as the real conductor 1 is, and if the image conductor carries a current equal in magnitude but directed oppositely to that in the real conductor, the magnetic field at the midplane will be purely tangential. Likewise, if the charge per unit length of the image conductor is the negative of that on the real conductor, the electric field at the midplane will be purely normal. Thus conditions 1 and 2 have been satisfied, and the field configurations between the real conductor and the conducting surface are the same as those found in Sec. 11-2 for the space between two parallel circular conductors. From the results of that section, expressions may be derived for surface-current density and surface-charge density to satisfy conditions 3 and 4.

Thus Eqs. 11-46 and 11-48 may be added together, with x set equal to zero, to find the resultant H_y, after which the surface current density may be written in accordance with Eq. 10-67:

$$\boldsymbol{J}_S = \frac{-hi_+(z,t)}{2\pi\left[(h/2)^2 + y^2\right]}\boldsymbol{a}_z \quad \text{(amperes per meter)} \qquad (11\text{-}80)$$

It is assumed that r_a is much smaller than k'; hence, in accordance with Eq. 11-60, $h \approx k'$.

The surface-charge density may be found with the aid of Eqs. 11-32 and 11-34, with x set equal to zero:

$$\rho_S = \frac{-h\epsilon_0\eta i_+(z,t)}{2\pi\left[(h/2)^2 + y^2\right]} \quad \text{(coulombs per square meter)} \qquad (11\text{-}81)$$

The voltage from the conductor to the ground plane is one-half of that between the two conductors of Fig. 11-4(b). In the traveling-wave situation it is related to the current by the characteristic impedance:

$$v_{cg}(z,t) = \frac{i_+(z,t)\eta}{2\pi}\cosh^{-1}\left(\frac{k'}{2r_a}\right)$$

$$\approx \frac{i_+(z,t)\eta}{2\pi}\ln\left(\frac{k'}{r_a}\right) \quad \text{(volts)} \qquad (11\text{-}82)$$

The characteristic impedance, distributed inductance, and distributed capacitance for the conductor and its ground return are related to the corresponding values for the real-and-image conductor pair (see Eqs.

11-74 through 11-79) in the following manner:

$$Z_{0cg} = \frac{Z_{011'}}{2}$$

$$l_{cg} = \frac{l_{11'}}{2} \tag{11-83}$$

$$c_{cg} = 2c_{11'}$$

By the use of superposition, this analysis may be used to find the resultant surface-current density on a lossless ground caused by two or more overhead conductors. (See Prob. 11-9.)

b. Cable with Nonconcentric Inner Conductor

The field configuration developed in Fig. 11-2 is immediately applicable to the problem of a cable with an offset inner conductor. Consider conductor A as the inner conductor; it is shown as enclosed by a particular circular $\boldsymbol{H}$-field direction line. A perfectly conducting cylindrical surface (outer conductor or shield, of radius designated as r_s carrying current in the $-\boldsymbol{a}_z$ direction could be located on that circle without disturbing the fields between it and conductor A. In other words, conductor B would then be the image of conductor A in the resultant cable assembly. The foregoing is restated in Fig. 11-5 with additional dimensioning for this application. Let r_a be the radius of conductor A.

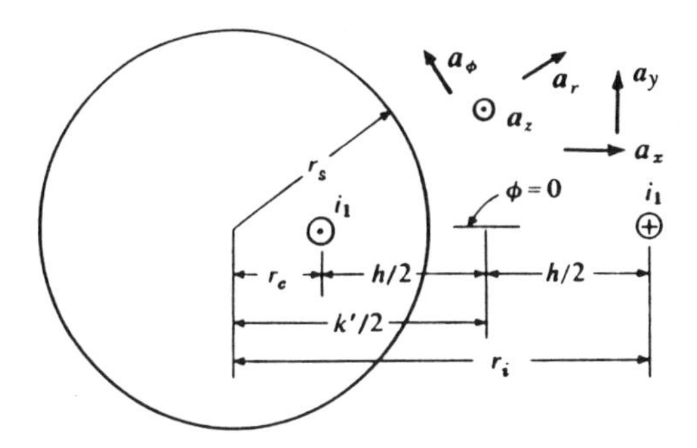

Figure 11-5. Real and image conductors for analysis of effect of nonconcentric return conductor or shield.

A formula for the displacement of the image conductor from the center of the outer conductor may be derived in the following manner:

$$r_c = \frac{k'}{2} - \frac{h}{2} \tag{11-84}$$

$$r_i = \frac{k'}{2} + \frac{h}{2} \tag{11-85}$$

The following equation corresponds to Eq. 11-60, although the conducting and nonconducting regions are interchanged:

$$r_s^2 = \left(\frac{k'}{2}\right)^2 - \left(\frac{h}{2}\right)^2 \tag{11-86}$$

If Eqs. 11-84 and 11-85 are multiplied together, the right-hand side may be substituted in Eq. 11-86:

$$r_i r_c = r_s^2 \tag{11-87}$$

The voltage which would accompany a traveling-wave current $i_{+A}(z,t)$, and thence the characteristic impedance, may be found by integrating, in a transverse plane, the $-\boldsymbol{E}$ field produced by $i_{+A}(z,t)$ in conductor A and its image, from the shield to the surface of conductor A. The intersection of that plane with the plane of the axes of the conductors is the simplest path from the standpoint of algebraic manipulation. From Eqs. 11-28 and 11-29, with $\boldsymbol{a}_{r1} = \boldsymbol{a}_x = \boldsymbol{a}_{r2}$, $r_1 = x - r_c$, and $r_2 = x - r_i$,

$$v_{+As}(z,t) = -\frac{\eta i_{+A}(z,t)}{2\pi} \int_{r_s}^{r_c + r_a} \left(\frac{1}{x - r_c} - \frac{1}{x - r_i} \right) dx$$

$$= \frac{\eta i_{+A}(z,t)}{2\pi} \ln\left(\frac{(r_s - r_c)(r_c + r_a - r_i)}{r_a(r_s - r_i)} \right) \tag{11-88}$$

The characteristic impedance is the ratio of v_{+As} to i_{+A}. Substitution of Eq. 10-23 for η and Eq. 11-87 for r_i yields the following (here r_a may be neglected in comparison with $r_i - r_c$):

$$Z_0 = \frac{1}{2\pi}\sqrt{\frac{\mu}{\epsilon}} \ln\left[\frac{(r_s - r_c)(r_s^2 - r_c^2)}{r_a(r_s^2 - r_s r_c)} \right]$$

$$= \frac{1}{2\pi}\sqrt{\frac{\mu}{\epsilon}} \ln\left(\frac{r_s^2 - r_c^2}{r_a r_s} \right) \tag{11-89}$$

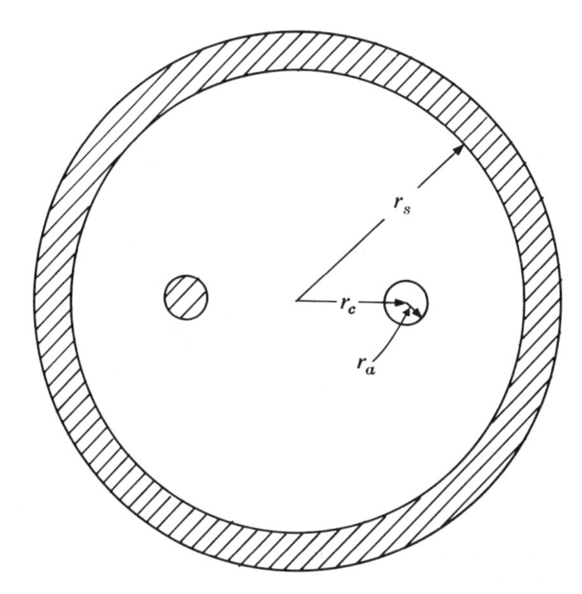

Figure 11-6. Two-wire shielded cable.

c. Two-Wire Shielded Cable[12, 14, 21]

The cross section of such a cable is shown in Fig. 11-6. The fields between the wires and the shield will be the same as those that would be produced by the wires and their images, provided the wire diameters are small compared to the spacing between the wires and the minimum distance to the shield. (The perimeter of a conductor establishes an equipotential surface, and hence the presence of one conductor distorts locally the field produced by the other and by the latter's image.)

The cross section is symmetrical with respect to the longitudinal mid-plane normal to the plane through the wires; accordingly two independent modes, or patterns of positive current directions and voltage polarities, are possible. The first, which is antisymmetric, is the one ordinarily used in applications and will be called the *balanced mode*; the second, equally interesting from the standpoint of electromagnetics, is symmetrical and will be called the *shield-return mode*.

(1) *Balanced Mode.* In this mode it is assumed that the currents in the two wires are equal in magnitude and oppositely directed; Fig. 11-7 shows the corresponding array of real and image conductors. By symmetry, the potential of the shield is midway between those of the wires.

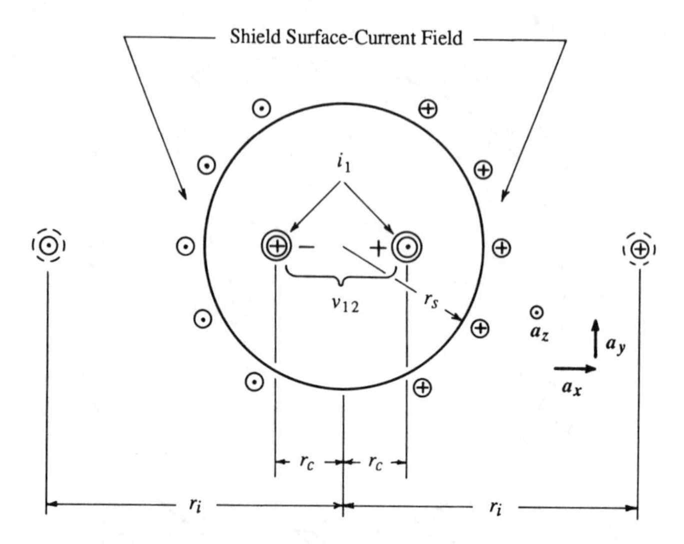

Figure 11-7. Shielded two-wire cable: balanced-mode current pattern. (Image currents are enclosed in dashed circles.)

Along the intersection of the transverse plane and the plane through the wires, if the origin is taken at the center of the assembly, the $\boldsymbol{E}$ field is

$$\boldsymbol{E}_{+}(z,t) = \frac{\eta i_{+1}(z,t)}{2\pi}\left(\frac{1}{x-r_c} - \frac{1}{x-r_i} - \frac{1}{x+r_c} + \frac{1}{x+r_i}\right)\boldsymbol{a}_x \tag{11-90}$$

The voltage from conductor 1 to the shield is

$$v_{+1s}(z,t) = -\int_{r_s}^{r_c+r_a} E_{+}(z,t)\,dx \tag{11-91}$$

Paralleling Eq. 11-88, the result is

$$v_{+1s}(z,t) = \frac{\eta i_{+1}(z,t)}{2\pi}\left\{\ln\left[\frac{(r_s-r_c)(r_c+r_a-r_i)}{r_a(r_s-r_i)}\right]\right.$$

$$\left. - \ln\left[\frac{(r_s+r_c)(r_c+r_a+r_i)}{(r_a+2r_c)(r_s+r_i)}\right]\right\} \tag{11-92}$$

This may be reduced by substituting Eq. 11-87 for r_i, and neglecting r_a in comparison with $r_i - r_c$:

$$v_{+1s}(z,t) = \frac{\eta i_{+1}(z,t)}{2\pi}\left\{\ln\left[\frac{(r_s - r_c)(r_s^2 - r_c^2)}{r_a(r_s^2 - r_s r_c)}\right] - \ln\left[\frac{(r_s + r_c)(r_s^2 + r_c^2)}{2r_c(r_s^2 + r_s r_c)}\right]\right\}$$

$$= \frac{\eta i_{+1}(z,t)}{2\pi}\left[\ln\left(\frac{r_s^2 - r_c^2}{r_a r_s}\right) - \ln\left(\frac{r_s^2 + r_c^2}{2r_c r_s}\right)\right] \tag{11-93}$$

$$= \frac{\eta i_{+1}(z,t)}{2\pi}\ln\left[\frac{2r_c(r_s^2 - r_c^2)}{r_a(r_s^2 + r_c^2)}\right] \tag{11-94}$$

The voltage from conductor 1 to conductor 2 is

$$v_{+12}(z,t) = 2v_{+1s}(z,t) \tag{11-95}$$

Division by $i_{+1}(z,t)$ and substitution of Eq. 10-23 for η yields

$$Z_{0\,\mathrm{bal}} = \frac{1}{\pi}\sqrt{\frac{\mu}{\epsilon}}\ln\left[\frac{2r_c(r_s^2 - r_c^2)}{r_a(r_s^2 + r_c^2)}\right] \tag{11-96}$$

Again, formulas for l' and c may be found by substitution of Eq. 11-96 in 11-22 and 11-23.

(2) *Shield-Return Mode.* In this mode it is assumed that the currents in the two wires have equal magnitudes and the same direction; see Fig. 11-8. Thus

$$i_{+\mathrm{shr}}(z,t) = 2i_{+1}(z,t) \tag{11-97}$$

The voltage from wire to shield may be found from Eq. 11-93 if one reverses the sign of the second logarithm:

$$v_{+\mathrm{shr}}(z,t) = \frac{\eta i_{+1}(z,t)}{2\pi}\ln\left(\frac{r_s^4 - r_c^4}{2r_a r_c r_s^2}\right) \tag{11-98}$$

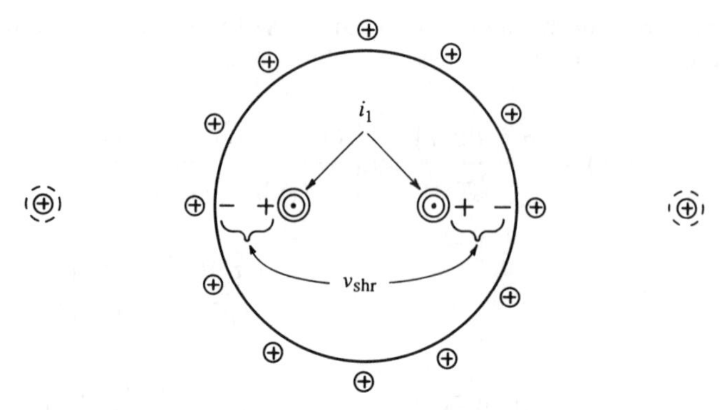

Figure 11-8. Shielded two-wire cable; shield-return-mode current pattern.

This reduces to

$$Z_{0\,\text{shr}} = \frac{1}{4\pi}\sqrt{\frac{\mu}{\epsilon}}\,\ln\left(\frac{r_s^4 - r_c^4}{2r_a r_c r_s^2}\right) \tag{11-99}$$

11-4 TRANSMISSION MODES OF MULTICONDUCTOR SYSTEMS

Some properties of transmission systems with more than two conductors should be noted.[2, 3, 29] The cable consisting of two wires and an enclosing shield just studied was a special case involving a high degree of symmetry. Because of the geometric symmetry with respect to a midplane, two modes could be readily visualized in which, magnitudewise, the corresponding field patterns were symmetrical with respect to the midplane. One mode was antisymmetric when account was taken of the directions of the fields and of the positive senses of current and voltage; the other mode was symmetric.

If the conductors or the insulating medium, or both, were lossy, one could anticipate, because of the differing paths of current and electric flux, that the propagation functions of the two modes would differ. But the allocation of currents and voltages to the two modes would be the same at all frequencies. (A line consisting of two wires at equal heights above the earth would also have those properties.*)

---

*A cable with three conductors arranged in an equilateral triangle, and surrounded by a circular sheath, would have (a) a distinct shield-return mode (equal currents in the conductors) and (b) two degenerate modes in which the currents in the three conductors sum to zero, but are not otherwise restricted.

A set of n conductors will have $n - 1$ modes. If the system is assumed to be lossless, the attenuation is zero and the velocity of propagation is independent of frequency and independent of how one might choose to constitute the modes. But the presence of loss will result, at any one frequency, in a unique set of modes, each with its own composition and propagation function. Except in geometrically symmetrical cases as described above, both the propagation function and the composition of each mode are frequency-dependent.

11-5 SOME ELECTROMAGNETIC PROPERTIES OF PRACTICAL LINES

The following topics which illustrate the application of field-propagation and boundary-condition concepts will be considered here briefly: (1) braided conductors—the effects of using a braided outer conductor on a coaxial cable or of using a braided shield—and (2) conductor-support techniques—the effects caused by various structures which are used to support the conductors of a line and to maintain the correct spacing.

a. Braided Conductors[5]

Mechanical flexibility is often desired in a cable, and this may be achieved by using stranded conductors at or near the center and braided conductors at greater radii. In some flexible coaxial cables one braid serves both as the outer conductor and as a shield: in other designs separate braids are used for these two purposes.*

Braids are formed by weaving together two sets of closely spaced wires which are wound helically in opposite directions. Ideally, strand-to-strand conduction should take place at every crossing of two oppositely spiralled wires. If this occurs with negligible contact resistance, the large-scale current pattern in the braid should be essentially the same as that which would flow in a continuous-sheet or tubular cylindrical conductor. However, a base metal will accumulate an oxide film, and this may introduce significant contact resistance, or it might be sufficient to inhibit strand-to-strand conduction altogether at the rather low voltages induced between strands.

---

*Two alternate forms of construction are (a) forming the shield with two thin copper strips which are spiral-wound in opposite directions, and (b) using a solid, corrugated outer conductor. These designs yield more effective shielding than braids.

Experimental measurements have indicated that the attenuation of coaxial cables is affected by flexing and that the shielding behavior of braids is sometimes erratic. Flexing could be expected to change, by differing amounts, the mechanical pressure between the strands at each of the many crossings in the affected portion. In order to sense whether the phenomenon suggested in the preceding paragraph could account for this, the ideal current patterns should be noted.

(1) *Coaxial Cable.* Equation 11-12 indicates that the current in a tubular outer conductor of a coaxial line is strictly longitudinal in direction and is of uniform intensity circumferentially. The corresponding filamentary path in a braid would be a "minimum-width zigzag"; in other words, it would pass, at every crossing, from the strand on which it arrived to the oppositely spiralled one. Variation in the contact resistances involved could reasonably be expected to cause local distortions in the current pattern and hence a change in the total losses and the attenuation of the cable. Goldberg and Slaughter[13] have examined the braid-attenuation problem in some detail.

(2) *Shielded Single-Pair Cable.* As indicated in Fig. 11-7, the surface-current-density field on a tubular shield is longitudinal but not of uniform intensity circumferentially; at every location on the inner surface of the shield the current direction is opposite to that of the current in the nearer wire.

In a braided shield the differential voltages in any one strand are oppositely directed at any two locations which differ by the longitudinal distance corresponding to half a turn of the spiral (assuming that this distance is short compared with a wavelength). If strand-to-strand conduction does not take place anywhere in the braid, essentially no shield currents will flow and the electromagnetic fields will extend uninterruptedly beyond the shield structure. Conduction at some strand-contacts but not at others would permit some current flow, but uniform containment of the fields would not be achieved.

b. Conductor-Support Techniques

Supports for the conductors are obviously necessary, and these often modify the assumption of line uniformity in the longitudinal direction. Because of their differing geometries, coaxial lines and parallel-wire lines will be considered separately. A modified form of coaxial cross section, "strip" line, will also be commented on.

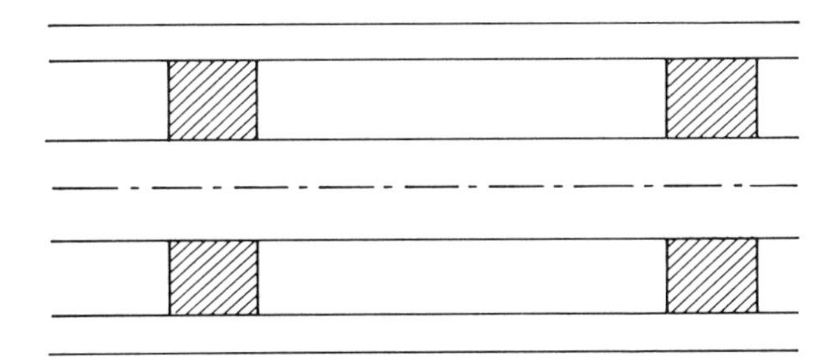

Figure 11-9. Disc-supported coaxial line.

(1) *Coaxial Lines.* The aspect of interest here is that of supporting the inner conductor in a centered position.

The simplest method, from the standpoint of electrical theory, is to fill the space between the conductors with a solid dielectric material. This technique is commonly used, especially in flexible cables, but dielectric losses become objectionably high at microwave frequencies.[17] A porous or "foam" type of dielectric yields a lower effective permittivity, and lower loss.

If the line is to consist of mechanically rigid sections, sufficient centering support and a great reduction in losses may be obtained by means of spaced circular discs or "beads" of various shapes. An alternative form of support consists of resonant stubs, or short-circuited sections of coaxial cable with nominal lengths of one-quarter wavelength. For semiflexible cables, a reduction in losses with acceptable mechanical properties may be had by using a continuous helical bar or rod of dielectric material between the two conductors. These methods complicate the wave-propagation phenomena.

(a) *Disc or Bead Supports*[25]—An elementary form of disc support is a short cylinder with flat ends, as illustrated in Fig. 11-9. Practical dielectric materials have permittivities two or more times that of air, and hence each such disc will constitute a short section of transmission line with a lower characteristic impedance than the air-filled sections. Each air-to-dielectric interface is a source of reflected waves with respect to waves incident from either direction. The minimum thickness of a disc is set by the need for mechanical rigidity, and this yields, at microwave frequencies, an electrical length which is sufficient to cause intolerably large standing waves.

Cancellation may be effected among the waves sent by two or more reflecting surfaces toward the generator by (1) making the thickness of each disc one half-wavelength (see Prob. 11-11), or (2) by fixing both the disc thickness and the spacings between discs according to some selected

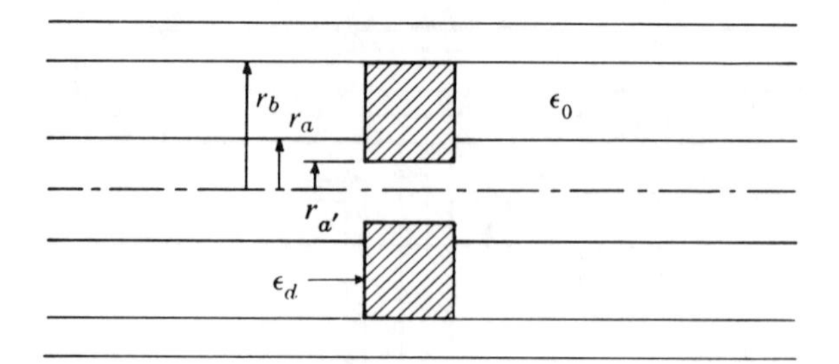

Figure 11-10. Undercut disc-supported coaxial line.

pattern of electrical distances. Such arrangements have the disadvantage of being frequency-sensitive.

Reflections might be avoided by making the characteristic impedance of the dielectric-filled sections the same as for the air-filled sections. It would seem that this could be done by undercutting the inner conductor as illustrated in Fig. 11-10, such that its radius $r_{a'}$ satisfied the following relationship (see Prob. 11-12):

$$r_{a'} = r_b\left(\frac{r_a}{r_b}\right)^p \tag{11-100}$$

where

$$p = \sqrt{\epsilon_d/\epsilon_0}$$

Abrupt changes in inner-conductor radius distort the fields from the TEM pattern however, and cause some reflections. These discontinuities are described in equivalent-circuit terminology as *fringing capacitances*.[32] If the undercut radius is made slightly smaller than the predicted $r_{a'}$, and the disc thickness is appropriately chosen, a workable broadband design may be arrived at.[8]

More elaborate disc supports have been devised to minimize standing waves,[24] but fabrication costs must be weighed against improvements in performance.

Dielectric supports in commercially produced line are often perforated discs or pegs, but the undercut-conductor principle is still used to obtain broadband uniformity.

(b) *Resonant Stubs*. The input admittance to a quarter-wavelength lossless short-circuited transmission line is zero. Thus it would appear that such a section of rigid coaxial line could be used as a support for the center conductor of the transmission line proper, as indicated in Fig. 11-11. Because the radius of the outer conductor is an appreciable fraction of a wavelength, and because the fields are distorted in the vicinity of the junction, the stub length from line center to short-circuiting plate which will yield resonance may be expected to differ slightly from $\lambda/4$.

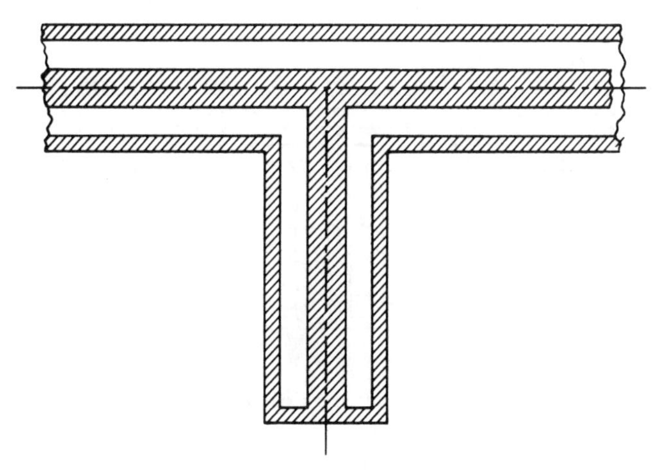

Figure 11-11. Resonant-stub-supported coaxial line.

Any deviation from the design frequency, as for example the side bands of a modulated signal, would change the electrical length of the stub and yield the following input admittance:

$$Y = \frac{1}{jZ_0} \cot \beta z_{\text{st}}$$

$$= \frac{1}{jZ_0} \cot\left(\omega \sqrt{lc}\, z_{\text{st}}\right) \tag{11-101}$$

Let

$$\omega = \omega_0 + \Delta\omega \qquad \left(\frac{\Delta\omega}{\omega_0} \ll 1\right)$$

$$\omega_0 \sqrt{lc}\, z_{\text{st}} = \frac{\pi}{2}$$

$$\omega \sqrt{lc}\, z_{\text{st}} = \frac{\pi}{2}\left(1 + \frac{\Delta\omega}{\omega_0}\right)$$

$$\cot\left(\omega \sqrt{lc}\, z_{\text{st}}\right) = -\tan\left(\frac{\pi\,\Delta\omega}{2\omega_0}\right)$$

$$\approx -\frac{\pi\,\Delta\omega}{2\omega_0}$$

$$Y \approx \frac{j\pi\,\Delta\omega}{Z_0 2\omega_0} \tag{11-102}$$

This simple form of stub is obviously frequency-sensitive.

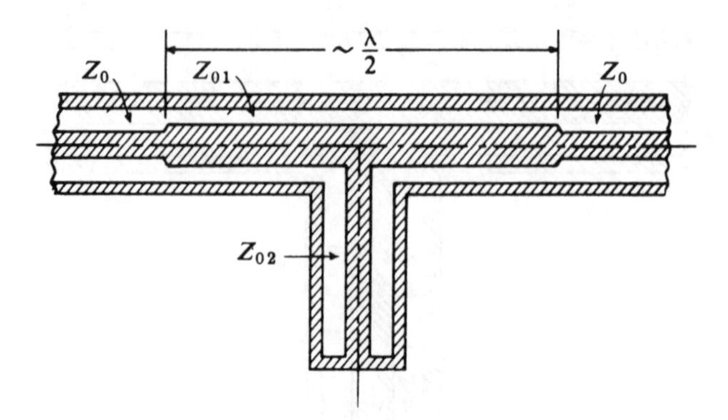

Figure 11-12. Broadbanded stub-supported coaxial line.

The design indicated in Fig. 11-12 may be made much less sensitive to frequency variations than that in Fig. 11-11 by suitable choices of Z_{01} and Z_{02} in proportion to Z_0.[25] (See Prob. 11-13.)

(c) *Helical Support*. The traveling-wave fields within a coaxial cable with a helical dielectric bar or rod have been analyzed by Griemsmann[15] in terms of two quasi-TEM modes of propagation. One of these follows the dielectric helix and the other follows a helical path at right angles to the dielectric helix. The path of the latter mode consists of alternate sections of air and solid dielectric, and, within certain frequency bands, propagation properties vary markedly. The behavior is comparable to that of a filter, and the cable will have frequencies of increased attenuation. The lowest such frequency has as its half-wavelength the helical distance between centers of the dielectric at the mean of radii r_a and r_b. At frequencies below about 35% of the lowest critical frequency, performance may be predicted to good approximation on the basis of an assumed TEM mode and a uniform dielectric medium of averaged permittivity.

(2) *Parallel-Wire Arrangements.* From the point of view of this section, two types of parallel-wire arrangements will be commented on: (1) open-wire-and-pole construction, and (2) plastic-embedded wire pairs without shielding. Parallel-wire sets are subject to more or less intense coupling with other circuits.[3]

(a) *Open-Wire-and-Pole Construction*. For telephone lines of this type the parallel conductors are bare wires supported by glass insulators at

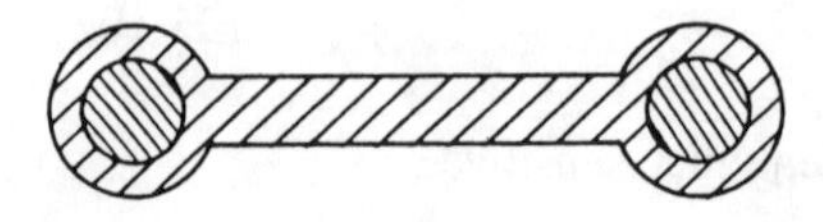

Figure 11-13. Plastic-embedded wire pair.

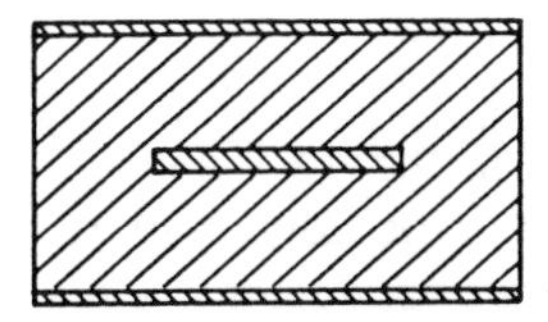

Figure 11-14. Strip line.

intervals of perhaps 100 ft. At 150 kHz the wavelength is 2000 m, or about 6500 ft.

Most open-wire carrier channels use frequencies below this value, and under these circumstances the lumped-discontinuity effect of the insulators is negligible. The insulators have the effect of increasing the distributed capacitance of the line slightly, and they are the principal cause of losses dependent upon voltage. Most of these losses are related to displacement-current effects and hence are dependent on frequency. The nominal value of distributed shunt conductance is selected to account for them.[33]

(*b*) *Plastic-Embedded Wire Pairs*. This form of line is convenient for short lead-in sections; a typical cross section is shown in Fig. 11-13. In the absence of a metallic shield over the plastic, the fields will extend into the air, where TEM waves would have a higher velocity of propagation than in the plastic. As a result the transmission mode will not be TEM but one in which both $\boldsymbol{E}$ and $\boldsymbol{H}$ may be expected to have longitudinal components.

(3) *"Strip Line."* The term *strip line* is commonly applied to assemblies of two or more conductors in the form of thin sheets or films, separated by solid dielectric material. The cross section of one type is sketched in Fig. 11-14. Such a line is simpler to fabricate than coaxial cable, and has almost as good shielding properties. The two outer metallic strips correspond to the outer conductor of the coaxial cable, much as if it were split longitudinally at two diametrically opposite locations and the segments then flattened. As with the plastic-embedded wire pair, the transmission mode would not be TEM, because the fields are not confined to the homogeneous dielectric slab. Analysis of the fields in this and similar cross sections is difficult[4, 16]; results have been presented in graphical form.[10, 26] Cohn,[6] Harvey,[18, 19] and Wheeler[31] list extensive bibliographies.

11-6 CONCLUSIONS

Propagation in the TEM mode is possible in a line composed of two or more parallel lossless conductors. The fields in the space between the conductors are accompanied by a surface-current-density field on each conductor, equal in magnitude at every point on the surfaces to that of the ***H*** field tangential there.

Voltage in a two-conductor set conveying energy in the TEM mode is defined as the line integral of ***E*** from one conductor to the other, with the path restricted to a transverse plane. Current in either conductor is equal to the line integral of ***H***, confined to a transverse plane, around the conductor.

The following transmission-line parameters were derived for the TEM mode from the electromagnetic fields, using the line integrals just mentioned.

For the coaxial line (Fig. 11-1),

$$Z_0 = \frac{1}{2\pi}\sqrt{\frac{\mu}{\epsilon}}\,\ln\left(\frac{r_b}{r_a}\right) \tag{11-18}$$

$$l = \frac{\mu}{2\pi}\ln\left(\frac{r_b}{r_a}\right) \tag{11-24}$$

$$c = \frac{2\pi\epsilon}{\ln(r_b/r_a)} \tag{11-25}$$

For the two-parallel-wire line (Fig. 11-3, $r_a \ll k'$),

$$Z_0 \approx \frac{1}{\pi}\sqrt{\frac{\mu}{\epsilon}}\,\ln\left(\frac{k'}{r_a}\right) \tag{11-75}$$

$$l \approx \frac{\mu}{\pi}\ln\left(\frac{k'}{r_a}\right) \tag{11-77}$$

$$c \approx \frac{\epsilon\pi}{\ln(k'/r_a)} \tag{11-79}$$

The method of *images* may be used to analyze the fields caused by the presence of an ideal ground plane or cylindrical shield.

A transmission line consisting of n parallel conductors will have $n - 1$ voltage-and-current modes.

Physical structures which support the conductors commonly cause discontinuities which, at the higher frequencies, may make the electromagnetic performance significantly different from that of a uniform line.

PROBLEMS

11-1. A coaxial line is terminated with a short-circuiting plate at $z = 0$. Combine Eqs. 11-11 and 11-13 for the traveling-wave fields, in accordance with appropriate boundary conditions. Sketch the resultant fields, as viewed in a longitudinal plane passing through the axis of the line, at the following values of ωt: (a) zero, (b) 45°, (c) 90°, and (d) 135°.

11-2. Write Poynting's vector $\boldsymbol{P}_1$ for the traveling-wave set in a coaxial line as defined by Eqs. 11-11. Evaluate $\iint \boldsymbol{P}_1 \cdot d\boldsymbol{S}$ over the transverse plane $z = z_1$. Find the time-averaged values of $\boldsymbol{P}_1$ and $\iint \boldsymbol{P}_1 \cdot d\boldsymbol{S}$. Compare the results for $\iint \boldsymbol{P}_1 \cdot d\boldsymbol{S}$ with the product $v_1(z,t)i_1(z,t)$ from Eqs. 11-9 and 11-16.

11-3. Write expressions for (a) the displacement-current density in the dielectric portion of a coaxial cable for the traveling-wave set described by Eqs. 11-11, and (b) the divergences of the surface-current densities given by Eqs. 11-12. By comparing these results, show that continuity of current (conduction or displacement or both) exists at the conductor surfaces.

11-4. Given a lossless coaxial cable with the following dimensions (see Fig. 11-1): $r_a = 2.00 \times 10^{-3}$ m; $r_b = 9.00 \times 10^{-3}$ m.

(a) If the characteristic impedance is 50.0 Ω and the relative permeability of the insulation is unity, what is the relative permittivity?

(b) The inner conductor has the following surface-current-density field:

$$\boldsymbol{J}_{SI}(z,t) = 2.70 \cos(\omega t - \beta z)\boldsymbol{a}_z \text{ A/m}$$

Find the surface-current-density field on the outer conductor, $\boldsymbol{J}_{SO}(z,t)$. Also find the magnetic intensity field (1) in the region $0 < r < r_a$, (2) in the region $r_a < r < r_b$, and (3) in the region $r_b < r$.

(c) What is the velocity of propagation?

11-5. Given the following expression for the traveling-wave $\boldsymbol{E}$ field in a coaxial line in which the conductors have infinite conductivity, but the insulating medium has a nonzero conductivity σ:

$$\boldsymbol{E} = \text{Re}\left\{ \frac{A}{2\pi r} \exp\left[-z\sqrt{j\omega\mu_0(\sigma + j\omega\epsilon_0)} + j\omega t \right] \boldsymbol{a}_r \right\} \qquad r_a < r < r_b$$

Write the corresponding expressions for (a) the conduction-current-density field (radial) $\boldsymbol{J}$ (amperes per square meter), (b) the displacement-current-density field $\boldsymbol{J}_D$ (amperes per square meter), and (c) the magnetic field $\boldsymbol{H}$.

11-6. Show that the $\boldsymbol{H}$ field of the two-parallel wire set, as described in Eqs. 11-27 and 11-43 through 11-48, is proportional to $1/x^2$ for $y = 0$ and $x \to \infty$.

11-7. Verify that $\boldsymbol{E} \cdot \boldsymbol{H}$ vanishes in the region between two parallel circular conductors, using Eqs. 11-32 through 11-35 and 11-45 through 11-48 for the field components.

11-8. Find, as a function of r_a/k', the ratio between the loss caused in a conductor of high conductivity for which the surface-current-density distri-

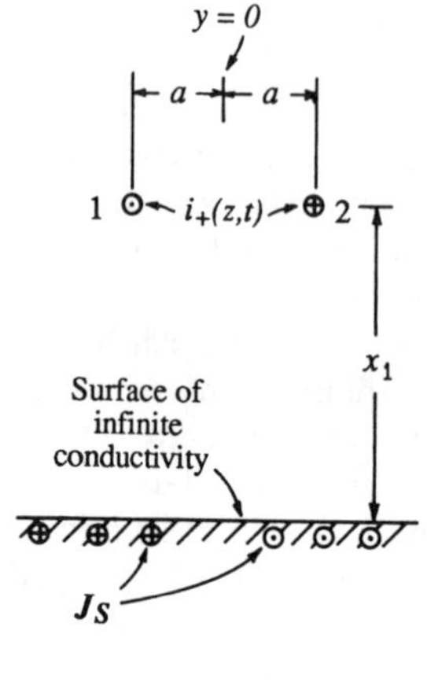

Figure 11-15. Two wire line above lossless ground.

bution found in Eqs. 11-64 and 11-62 may be assumed valid, and the loss caused by the same current when uniformly distributed around the conductor periphery. [Compare result with the first term in the approximation given in the footnote to Sec. 11-2a(4).]

11-9. Given the line shown in Fig. 11-15. Sketch the equivalent image array. By means of superposition and Eq. 11-80, write an expression for $\boldsymbol{J}_S$. Assuming that $i_+(z, t)$ is a traveling-wave mode, find the accompanying voltage with the image currents present, and find the corresponding characteristic impedance.

11-10. (a) For the cable with an offset inner conductor shown in Fig. 11-5, let $J_{SO}\boldsymbol{a}_z$ indicate the surface-current-density field on the inner surface of the outer conductor. Sketch a graph of J_{SO} versus ϕ.

(b) Sketch corresponding graphs for the shield surface-current-density fields of the two modes shown in Figs. 11-7 and 11-8 for the shielded pair of wires.

11-11. A solid-disc coaxial support like that shown in Fig. 11-9 has an electrical length of $\lambda/2$ at ω_0. Assume that the characteristic impedance of the air-filled sections is 50 Ω and that the relative permittivity of the disc is 2.25. Also assume that the right-hand line is terminated with its characteristic impedance.

(a) Sketch the standing-wave patterns for $\omega = \omega_0$, of $|V|$, $|I|$, $|E_r(r_a+)|$, $|D_r(r_a+)|$, and $|H_\phi(r_a+)|$. If $|V| = 0.80$ V and $|E_r(r_a+)| = 400$ V/m in the right-hand section of line, find the maximum and minimum numerical values for each of the functions in the sketch.

(b) Find the input impedance to the disc at its left-hand face as a function of $\Delta\omega/\omega_0$, where $\omega = \omega_0 + \Delta\omega$.

11-12. Verify Eq. 11-100 for the inner radius $r_{a'}$ of an undercut disc support for a coaxial line (Fig. 11-10). Find the ratios r_a/r_b and $r_{a'}/r_b$ for a line with a characteristic impedance of 50 Ω, in which the discs are made of polyethylene with a relative permittivity of 2.25.

11-13. Consider the broadbanded stub support shown in Fig. 11-12. Assume that the right-hand line is terminated with its characteristic impedance, that $Z_{02} = Z_0$, and that $Z_{01} = 0.8393Z_0$. (This combination is shown in ref. 25 to yield zero first-order variation in the input impedance for small values of $\Delta\omega/\omega_0$.) By using the Smith chart (1) show that the impedance at the input to the stub-support unit at the design frequency is equal to Z_0, and (2) find the corresponding impedance when the frequency is 10% above the design frequency. Also compute the latter result with a calculator, applying Eq. 7-9 to the stub and each section of line.

REFERENCES

1. Adams, E. P., The resistance of cylindrical conductors at high frequencies, *Proc. Am. Phil. Soc.*, 78, 271, 1937.
2. Amemiya, H., Time-domain analysis of multiple parallel transmission lines, *RCA Review*, 28, 241, 1967.
3. Babcock, W. C., Rentrop, E., and Thaeler, C. S., *Crosstalk on Open-Wire Lines*, Bell Telephone System Tech. Pubs., Monograph 2520, 1955.
4. Black, K. G. and T. J. Higgins, Rigorous determination of the parameters of microstrip transmission lines, *Trans. Inst. Radio Engrs., Microwave Theory and Techniques*, 3, no. 2, 93, 1955.
5. Blackband, W. T., Wire-braid screens for R.F. cables," *J. IEE (London) N.S.* 1, no. 6, 363, 1955.
6. Cohn, S. B., A reappraisal of strip transmission line, *Microwave J.*, 3, no. 3, 17, 1960.
7. *Communication System Engineering Handbook*, D. H. Hamsher, ed., McGraw-Hill, New York, 1967.
8. Cornes, R. W., A coaxial-line support for 0 to 4000 Mc, *Proc. IRE*, 37, 94, 1949.
9. *Electrical Transmission and Distribution Reference Book*, Central Station Engineers of the Westinghouse Electric Corporation, Pittsburgh, PA, 1950.
10. Freeman, R. L., *Reference Manual for Telecommunications Engineering*, Wiley, New York, 1985.
11. Freeman, R. L., *Telecommunication Transmission Handbook*, 2nd ed., Wiley, New York, 1982.
12. Gent, A. W., Capacitance of shielded-pair transmission line, *Elec. Commun.*, 33, 234, 1956.
13. Goldberg, J. L. and Slaughter, R. J., Braid construction and attenuation of coaxial cables at microwave frequencies, *Proc. IEE (London)*, 113, No. 6, 957, 1966.

14. Green, E. I., Leibe, F. A., and Curtis, H. E., The proportioning of shielded circuits for minimum high-frequency attenuation, *Bell Syst. Tech. J.*, 15, 248, 1936.
15. Griemsmann, J. W. E., An approximate analysis of coaxial line with a helical dielectric support, *IRE Trans. Microwave Theory Tech.*, MTT-4, 13, 1956.
16. Grivet, P., *The Physics of Transmission Lines at High and Very High Frequencies*, Vol. 1, Academic, New York, 1970.
17. Hannon, J. R., Factors affecting attenuation of solid-dielectric coaxial cables above 3000 megacycles, *Trans. IRE on Component Parts*, CP-3, no. 3, 99, 1956.
18. Harvey, A. F., Parallel-plate transmission systems for microwave frequencies, *Proc. IEE (London)*, 106, part B, 129, 1959.
19. Harvey, A. F., *Microwave Engineering*, Academic, London, 407, 1963.
20. Kimbark, E. W., *Electrical Transmission of Power and Signals*, Wiley, New York, 1949.
21. King, B. G., McKenna, J., and Raisbeck, G., Experimental check of formulas for capacitance of shielded balanced-pair transmission line, *Proc. IRE*, 46, 922, 1958.
22. Lowman, R. V., Transmission lines and waveguides, *Antenna Engineering Handbook*, 2nd ed., Chap. 42, R. C. Johnson and H. Jasik, eds., McGraw-Hill, New York, 1984.
23. Matick, R. E., Transmission line pulse transformers—Theory and applications, *Proc. IEEE*, 56, no. 1, 47, 1968.
24. Peterson, D. W., Notes on a coaxial line bead, *Proc. IRE*, 37, 1294, 1949.
25. Ragan, G. L., *Microwave Transmission Circuits*, McGraw-Hill, New York, 155, 1948.
26. *Reference Data for Radio Engineers*, 5th ed., Howard W. Sams & Co., New York, 1968.
27. Shugg, W. T., *Handbook of Electrical and Electronic Insulating Materials*, Van Nostrand Reinhold, New York, 1986.
28. Sim, A. C., New high-frequency proximity-effect formula, *Wireless Engr.*, 30, no. 8, 204, 1953; reprinted in *Elec. Commun.*, 131, no. 1, 66, 1954.
29. Wedepohl, L. M., Application of matrix methods to the solution of traveling-wave phenomena in polyphase systems, *IEE Proc.*, 110, 2200, 1963.
30. Wedepohl, L. M., Electrical characteristics of polyphase transmission systems with special reference to boundary-value calculations at power-line carrier frequencies, *IEE Proc.*, 112, 2103, 1965.
31. Wheeler, H. A., Transmission-line conductors of various cross sections, *IEEE Trans.*, MTT-28, 73, 1980. Extensive bibliography.
32. Whinnery, J. R., Jamieson, H. W., and Robbins, T. E., Coaxial-line discontinuities, *Proc. IRE*, 32, 695, 1944.
33. Wilson, L. T., A study of telephone line insulators, *Bell. Syst. Tech. J.*, 9, 697, 1930.
34. Wong, K. H., Using precision air dielectric transmission lines as calibration and verification standards, *Microwave J.*, 31, no. 12, 83, 1988.

CHAPTER 12

Fields in Hollow Rectangular Waveguides

The two preceding chapters have emphasized one particular type of electromagnetic wave propagation, the TEM mode, in which the electric and magnetic fields are everywhere perpendicular to the direction of propagation. Propagation is also possible in modes in which one field has a component parallel to that direction, and such modes are important in that they may convey energy through hollow tubes of conducting material.*

Numerous transverse cross sections for waveguides have been utilized; among these, three are amenable to closed form solution of the wave equation by the method of separable variables. Those cross sections are (a) rectangular, (b) circular, and (c) elliptical. The first of these will be examined in this chapter, and the latter two in Chap. 13. This analysis will be limited to waveguides which are uniform longitudinally; this is generally true of rigid waveguide structures. (By corrugating the tubing with circumferential or helical ridges and valleys, a waveguide may be made semiflexible, or by using articulated segments, additional flexibility may be realized.[17])

For the sake of mathematical simplicity, a particular traveling-wave mode for the waveguide with a rectangular cross section will be studied first. The same general technique employed for the coaxial cable, that of postulating a solution and testing it by substituting in Maxwell's equations,

---

*The possibility of electromagnetic wave propagation in hollow metallic tubes was analyzed (mode configurations and cutoff frequencies predicted for tubes of rectangular and circular cross sections) by Rayleigh[20] in 1897. Oscillators which would generate useful amounts of power in the gigahertz range were not developed until some years later, and applications in that frequency range were of limited scope until the 1930s. Research teams led by George C. Southworth[3, 25, 26] at Bell Telephone Laboratories, and by Wilmer L. Barrow[2] at the Massachusetts Institute of Technology, independently of each other, and unaware of Rayleigh's work, produced functioning waveguide assemblies. Both published their results in 1936. Packard[18] has compiled a detailed historical account.

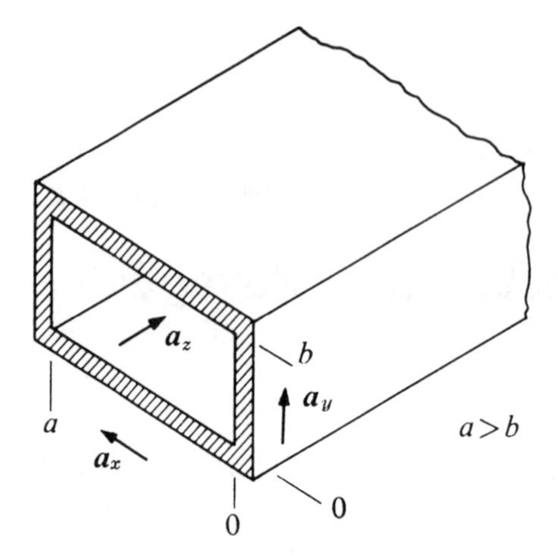

Figure 12-1. Dimensions and coordinates for rectangular waveguide.

will be used initially. Afterward a more general mathematical approach will be tried. Figure 12-1 shows the geometry of the rectangular waveguide.

12-1 TE_{m0} PROPAGATING MODES

The simplest and also the most commonly used mode in a rectangular waveguide has these properties: (1) the electric field is confined to planes perpendicular to the direction of propagation, whereas the magnetic field has both transverse and longitudinal components, and (2) the electric field is directed from one of the guide surfaces straight across to the opposite surface. Actually an infinite number of modes possess these properties in common; they are referred to as TE_{m0} modes by which the following is meant: (1) TE designates *transverse electric* (property 1 above), (2) the second subscript, 0, indicates an absence of variation in field configuration as a function of the shorter transverse direction (y direction in Fig. 12-1), and (3) the first subscript, m, which is an integer, designates the particular mode of this class. This third item will be described specifically in Sec. 12-1a(2), after the TE_{m0} field patterns have been derived.

a. Electric and Magnetic Fields

For the ordinarily used approximation of the electric and magnetic fields, the waveguide walls may be assumed to be of infinite conductivity, in which case a traveling wave should propagate without attenuation.

(1) *Traveling-Wave Solution.* In view of the success experienced in finding traveling-wave solutions in unbounded space and in a coaxial cable, such a solution will be postulated here. The phase function β will be regarded as an unknown and may well differ from that for the TEM mode. In accordance with the properties just listed for the TE_{m0} mode, the $\boldsymbol{E}$ field will be assumed to consist solely of a y-directed component, E_y. However, because the component of $\boldsymbol{E}$ tangential to a perfectly conducting surface must vanish (Sec. 10-2b) and, as shown in Fig. 12-1, such surfaces exist at $x = 0$ and at $x = a$, E_y will be assumed to be a function of x. The requirement that the divergence of $\boldsymbol{D}$ should vanish (Eq. 9-19, with $\rho = 0$) will be met if the $\boldsymbol{E}$ thus chosen is not a function of y. The trial solution may be stated as follows:

$$\boldsymbol{E}_1 = \mathrm{Re}\left[E_y(x)\epsilon^{j(\omega t - \beta z)}\right]\boldsymbol{a}_y \tag{12-1}$$

Sequential substitution of Eq. 12-1 into Maxwell's curl equations, 9-23 and 9-21, will be made with a view to fixing β:

$$\nabla \times \boldsymbol{E}_1 = \mathrm{Re}\left[j\beta E_y(x)\epsilon^{j(\omega t - \beta z)}\boldsymbol{a}_x + \frac{dE_y(x)}{dx}\epsilon^{j(\omega t - \beta z)}\boldsymbol{a}_z\right]$$

$$\boldsymbol{H}_1 = \mathrm{Re}\left[\frac{-\beta}{\omega\mu}E_y(x)\epsilon^{j(\omega t - \beta z)}\boldsymbol{a}_x - \frac{1}{j\omega\mu}\frac{dE_y(x)}{dx}\epsilon^{j(\omega t - \beta z)}\boldsymbol{a}_z\right] \tag{12-2}$$

$$\nabla \times \boldsymbol{H}_1 = \mathrm{Re}\left\{\left[\frac{j\beta^2}{\omega\mu}E_y(x) + \frac{1}{j\omega\mu}\frac{d^2E_y(x)}{dx^2}\right]\epsilon^{j(\omega t - \beta z)}\right\}\boldsymbol{a}_y$$

$$\boldsymbol{E}_1 = \mathrm{Re}\left\{\left[\frac{\beta^2}{\omega^2\mu\epsilon}E_y(x) + \frac{1}{(j\omega)^2\mu\epsilon}\frac{d^2E_y(x)}{dx^2}\right]\epsilon^{j(\omega t - \beta z)}\right\}\boldsymbol{a}_y \tag{12-3}$$

Comparing Eq. 12-3 with Eq. 12-1 indicates that $E_y(x)$ must satisfy the following differential equation:

$$\frac{1}{(j\omega)^2\mu\epsilon}\frac{d^2E_y(x)}{dx^2} + \left[\frac{\beta^2}{\omega^2\mu\epsilon} - 1\right]E_y(x) = 0 \tag{12-4}$$

Equation 12-4 may be solved as follows:

$$\frac{d^2E_y(x)}{dx^2} + (\omega^2\mu\epsilon - \beta^2)E_y(x) = 0$$

$$E_y(x) = A_1 \sin\left(\sqrt{\omega^2\mu\epsilon - \beta^2}\, x\right) + B_1 \cos\left(\sqrt{\omega^2\mu\epsilon - \beta^2}\, x\right) \quad (12\text{-}5)$$

Here A_1 and B_1 are arbitrary constants which are determined by boundary conditions, and the phase function β is also fixed by boundary conditions.

(2) *Boundary Conditions.* As was mentioned at the start of the derivation, a perfectly conducting surface will support no tangential component of the $\boldsymbol{E}$ field, so E_y must vanish along the surfaces at $x = 0$ and $x = a$. From Eq. 12-5,

$$E_y(0) = B_1$$

Therefore,

$$B_1 = 0$$

Then

$$E_y(a) = A_1 \sin\left(\sqrt{\omega^2\mu\epsilon - \beta^2}\, a\right)$$

This can equal zero only by choosing β so that

$$\sqrt{\omega^2\mu\epsilon - \beta^2}\, a = m\pi$$

Here m is any integer, and β may be subscripted as β_{m0} to emphasize its functional dependence on the assumed mode features:

$$\beta_{m0}^2 = \omega^2\mu\epsilon - \left(\frac{m\pi}{a}\right)^2 \quad (12\text{-}6)$$

Equation 12-5 may be written with these results substituted:

$$E_y(x) = A_1 \sin\left(\frac{m\pi x}{a}\right) \quad (12\text{-}7)$$

The constant A_1 has the same dimensions as E_y and is simply a measure of its magnitude.

Complete expressions for the $\boldsymbol{E}$ and $\boldsymbol{H}$ fields may now be written by substitution in Eqs. 12-1 and 12-2:

$$\boldsymbol{E}_1 = \operatorname{Re}\left\{A_1 \sin\left(\frac{m\pi x}{a}\right)\exp[j(\omega t - \beta_{m0} z)]\right\}\boldsymbol{a}_y$$

$$= A_1 \sin\left(\frac{m\pi x}{a}\right)\cos(\omega t - \beta_{m0} z)\boldsymbol{a}_y \tag{12-8}$$

$$\boldsymbol{H}_1 = \operatorname{Re}\left\{-\frac{A_1\beta_{m0}}{\omega\mu}\sin\left(\frac{m\pi x}{a}\right)\exp[j(\omega t - \beta_{m0} z)]\boldsymbol{a}_x - \frac{A_1 m\pi}{j\omega\mu a}\cos\left(\frac{m\pi x}{a}\right)\exp[j(\omega t - \beta_{m0} z)]\boldsymbol{a}_z\right\}$$

$$= -\frac{A_1\beta_{m0}}{\omega\mu}\sin\left(\frac{m\pi x}{a}\right)\cos(\omega t - \beta_{m0} z)\boldsymbol{a}_x - \frac{A_1 m\pi}{\omega\mu a}\cos\left(\frac{m\pi x}{a}\right)\sin(\omega t - \beta_{m0} z)\boldsymbol{a}_z \tag{12-9}$$

Equations 12-8 and 12-9, supplemented by Eq. 12-6, describe an electromagnetic traveling-wave system, or mode, which satisfies Maxwell's equations and appears to be capable of propagating energy in the rectangular waveguide. The field configuration for $m = 1$ is sketched in Fig. 12-2.

It has been assumed, in going from the first to the second forms of Eqs. 12-8 and 12-9, that β_{m0} is a real quantity. In accordance with the definition for β_{m0} in Eq. 12-6, it follows that, for the solution just given to be valid, the frequency must be greater than the following limiting value, which is known as the *cutoff frequency* and is designated by f_{cm0}:

$$f_{cm0} = \frac{1}{2\pi}\frac{m\pi}{a\sqrt{\mu\epsilon}}$$

$$= \frac{m}{2a\sqrt{\mu\epsilon}} \tag{12-10}$$

At frequencies below cutoff, β_{m0} is imaginary. The solution is then not of the traveling-wave type; appropriate equations are developed in Sec. 12-4. (It is shown there that such modes do not propagate but exist only in the vicinity of the guide discontinuity or other source which excites them.)

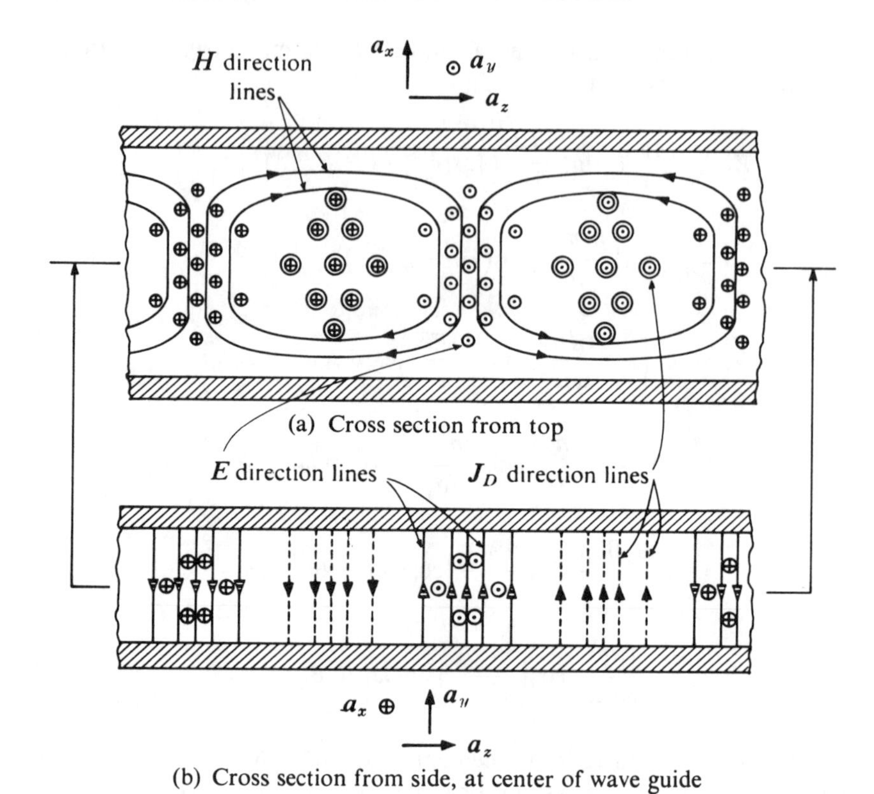

Figure 12-2. TE_{10} propagation mode in rectangular waveguide.

Since m may be any integer, Eqs. 12-8 and 12-9 actually describe a multiplicity of modes. For a given width a and with excitation at a stated frequency, only a limited number of these TE_{m0} modes will have cutoff frequencies which are below the excitation frequency, and only those modes (among those in the TE_{m0} class) may propagate. In accordance with the principle of superposition, however, all such possible modes may exist simultaneously without influencing one another. As a subscript in the mode designator, m indicates the number of half-period variations in the electric field along the longer transverse direction, the x direction in Fig. 12-1.

(3) *TE_{10} Mode as the Dominant Mode.* For a given size of waveguide, the TE_{10} mode has a lower cutoff frequency than any other TE_{m0} mode, and, as is shown in Sec. 12-3a and b, it has the lowest cutoff frequency of any possible mode. It can exist on a propagating basis at the same frequencies as any other modes can, and it can exist at a lower frequency than any of the others can; hence it is called the *dominant* mode. For most applications it is undesirable to have propagation taking place in more than one mode. Multimoding may be prevented by selecting the waveguide size for a particular installation such that only the TE_{10} mode has a cutoff frequency which is lower than the excitation frequency.

From the design point of view, if f_{ex} is the excitation frequency and the TE_{10} mode is to be a propagating mode but the TE_{20} and higher TE_{m0} modes are not, the width a should satisfy the following inequality, which is derivable from Eq. 12-10:

$$\frac{1}{f_{ex}\sqrt{\mu\epsilon}} > a > \frac{1}{2f_{ex}\sqrt{\mu\epsilon}} \tag{12-11}$$

This may be stated more conveniently in terms of the TEM-mode wavelength λ_{ex} of the frequency f_{ex}:

$$\lambda_{ex} = \frac{\nu_0}{f_{ex}}$$

$$= \frac{1}{f_{ex}\sqrt{\mu\epsilon}}$$

$$\lambda_{ex} > a > \frac{\lambda_{ex}}{2} \tag{12-12}$$

(4) *Parametric Equation of H-Field Direction Lines.* A parametric equation for the ***H***-field-direction lines (see the paragraph preceding Eq. 11-36) may be found by solving the following differential equation:

$$\frac{dx}{dz} = \frac{H_x}{H_z}$$

$$= \frac{-\beta_{m0} a \sin(m\pi x/a)\cos(\omega t - \beta_{m0} z)}{-m\pi \cos(m\pi x/a)\sin(\omega t - \beta_{m0} z)} \tag{12-13}$$

The separation of variables and integration of this equation is left as Prob. 12-2. The result is as follows:

$$\sin\left(\frac{m\pi x}{a}\right)\sin(\omega t - \beta_{m0} z) = C \tag{12-14}$$

Here C is an arbitrary constant. Inspection of the equation indicates that C must have a magnitude of unity or less in order to correspond to real values of x and z, but that each value of C within that range would represent a different field-direction line. For C slightly less than unity in magnitude, the field-direction lines are approximately elliptical.

b. Complementary Field Functions

The equations just given for $\boldsymbol{E}$ and $\boldsymbol{H}$ may be supplemented by expressions for displacement-current density, surface-current density, and power density. These help to unify the field and-circuit-theory aspects of waveguide propagation.

(1) *Displacement-Current Field.* Displacement current is present wherever the electric field is changing; it is defined mathematically as follows:

$$\boldsymbol{J}_D = \epsilon \frac{\partial \boldsymbol{E}}{\partial t} \tag{12-15}$$

Equation 12-8 may be substituted here:

$$\boldsymbol{J}_{D1} = -A_1 \omega \epsilon \sin\left(\frac{m\pi x}{a}\right) \sin(\omega t - \beta_{m0} z) \boldsymbol{a}_y \tag{12-16}$$

The following observations may be made from Fig. 12-2 and from the equations it illustrates: (1) the magnetic-field-direction lines encircle, in accordance with the right-hand rule, portions of the displacement-current field within the half-wavelength sections between displacement-current nulls; (2) the displacement-current density is maximum where the electric field is zero, and is so directed as to be changing the direction of the electric field from that of the electric-field maximum which has just passed the given location to that of the electric-field maximum which is next approaching; and (3) the transverse component of magnetic field is a maximum at the same locations as the electric field, and is so directed that the z component of the cross product $\boldsymbol{E} \times \boldsymbol{H}$ points in the direction of wave propagation.

(2) *Surface-Current Field.* Currents may be expected to flow in the side walls and in the top and bottom of the waveguide because: (1) continuity of the current field can be maintained only if the displacement-current field in the waveguide interior joins to a conduction-current field at the top and bottom of the waveguide, and (2) a difference in the magnetic-field strengths inside and outside the waveguide interior can be maintained only by means of currents in the waveguide conductor material. At waveguide frequencies (about 3×10^9 Hz and higher), skin effect is so pronounced that skin depth is negligible compared to the waveguide cross-sectional dimensions, and one may properly reason in terms of a surface-current density, the magnitude of which is everywhere equal to the magnitude of the $\boldsymbol{H}$ field tangent to the conducting surface in question and which flows at right angles to that field.

On the basis of the magnetic-field requirements the following expressions for surface-current density may be written; immediately afterward it will be shown that those expressions also satisfy the requirement of continuity between surface-current density in the conducting surfaces and displacement-current density in the waveguide interior:

$$\begin{aligned} \boldsymbol{J}_{S1}(0, y, z) &= -H_{1z}(0, y, z)\boldsymbol{a}_y \\ &= \frac{A_1 m\pi}{\omega\mu a} \sin(\omega t - \beta_{m0} z)\boldsymbol{a}_y \qquad \text{(sidewalls)} \qquad (12\text{-}17) \\ \boldsymbol{J}_{S1}(a, y, z) &= -(-1)^m \boldsymbol{J}_{S1}(0, y, z) \end{aligned}$$

$$\begin{aligned} \boldsymbol{J}_{S1}(x, 0, z) &= H_{1z}\boldsymbol{a}_x - H_{1x}\boldsymbol{a}_z \qquad \text{(bottom)} \\ \boldsymbol{J}_{S1}(x, b, z) &= -H_{1z}\boldsymbol{a}_x + H_{1x}\boldsymbol{a}_z \qquad \text{(top)} \end{aligned} \qquad (12\text{-}18)$$

Finding the parametric equation for the field-direction lines of the surface-current density on the top and bottom surfaces is left as Prob. 12-3.

Two important observations may be made concerning the surface-current patterns: (1) no current flows across the longitudinal centerlines of the top and bottom surfaces, and (2) no longitudinal component of current flows in the sidewalls. Item 1 is of particular value in measurement techniques in that a narrow longitudinal slit of any desired length may be cut through the middle of the top without interfering with propagation of the TE_{10} mode. A probe may be inserted in the slit, and with the aid of suitable detecting equipment, a series of measurements made of the relative strength of $E_y(z)$ at $x = a/2$. A *slotted-wave-guide* unit is functionally similar to the slotted coaxial line mentioned in Sec. 11-1a(3).

The divergence of surface-current density would have the units amperes per meter squared, the same as those of displacement-current density:

$$\nabla \cdot \boldsymbol{J}_{S1}(0, y, z) = 0 \qquad \text{(sidewall)}$$

$$\nabla \cdot \boldsymbol{J}_{S1}(a, y, z) = 0 \qquad \text{(sidewall)}$$

$$\nabla \cdot \boldsymbol{J}_{S1}(x, 0, z) = \frac{A_1(m\pi)^2}{\omega\mu a^2}\sin\left(\frac{m\pi x}{a}\right)\sin(\omega t - \beta_{m0}z) + \frac{A_1\beta_{m0}^2}{\omega\mu}\sin\left(\frac{m\pi x}{a}\right)\sin(\omega t - \beta_{m0}z) \qquad \text{(bottom)} \tag{12-19}$$

$$\nabla \cdot \boldsymbol{J}_{S1}(x, b, z) = -\nabla \cdot \boldsymbol{J}_{S1}(x, 0, z) \qquad \text{(top)}$$

Substitution of Eq. 12-6 reduces Eq. 12-19 to

$$\nabla \cdot \boldsymbol{J}_{S1}(x, 0, z) = A_1\omega\epsilon \sin\left(\frac{m\pi x}{a}\right)\sin(\omega t - \beta_{m0}z) \tag{12-20}$$

Comparison of these results with Eq. 12-16 for displacement-current density indicates an exact agreement in magnitude. The divergence of the surface-current-density function is positive when the function is *increasing* in strength because of a distributed source, and inflowing displacement-current flux has the nature of such a source with respect to the surface current. Hence the current density leaving the bottom surface [the negative of the divergence of $\boldsymbol{J}_{S1}(x, 0, z)$] corresponds to the upward-directed displacement-current density, and it in turn corresponds to that entering the top surface.

(3) *Poynting's Vector: Power Flow.* Poynting's vector for this traveling-wave set may be readily written from Eqs. 12-8 and 12-9:

$$\begin{aligned} \boldsymbol{P}_1 &= \boldsymbol{E}_1 \times \boldsymbol{H}_1 \\ &= \frac{A_1^2\beta_{m0}}{\omega\mu}\sin^2\left(\frac{m\pi x}{a}\right)\cos^2(\omega t - \beta_{m0}z)\boldsymbol{a}_z \\ &\quad - \frac{A_1^2 m\pi}{\omega\mu a}\sin\left(\frac{m\pi x}{a}\right)\cos\left(\frac{m\pi x}{a}\right)\sin(\omega t - \beta_{m0}z)\cos(\omega t - \beta_{m0}z)\boldsymbol{a}_x \\ &= \frac{A_1^2\beta_{m0}}{4\omega\mu}\left[1 - \cos\left(\frac{2m\pi x}{a}\right)\right][1 + \cos 2(\omega t - \beta_{m0}z)]\boldsymbol{a}_z \\ &\quad - \frac{A_1^2 m\pi}{4\omega\mu a}\sin\left(\frac{2m\pi x}{a}\right)\sin 2(\omega t - \beta_{m0}z)\boldsymbol{a}_x \end{aligned} \tag{12-21}$$

If this quantity is time-averaged over an integral number of cycles, the following is obtained:

$$\boldsymbol{P}_{1av}(x) = \frac{A_1^2 \beta_{m0}}{4\omega\mu}\left[1 - \cos\left(\frac{2m\pi x}{a}\right)\right]\boldsymbol{a}_z \qquad (12\text{-}22)$$

Average power flow through the waveguide may be found by taking the surface integral of $\boldsymbol{P}_{1av}(x)$ over a transverse cross section of the waveguide:

$$P_{1R\Sigma} = \int_0^b \int_0^a \boldsymbol{P}_{1av}(x) \cdot (dx\, dy\, \boldsymbol{a}_z)$$

$$= \frac{A_1^2 \beta_{m0} ab}{4\omega\mu} \qquad (12\text{-}23)$$

Examination of Eqs. 12-21 through 12-23 indicates (1) that power flow in the longitudinal direction is pulsating—it varies, at any given location, from zero up to twice the average value and back down to zero again in a half-cycle period; and (2) that power flow in the transverse direction is reactive in nature—it varies at any given location, from maximum in one direction to maximum in the opposite direction and back to maximum in the first direction in a half-cycle period.

c. Wave Set Traveling in Reverse Direction

With the two-wire transmission line it was found that waves could propagate with equal ease in either direction. One may surmise that the same should be true of waveguides, and this may be checked by postulating an $\boldsymbol{E}$ field which travels in the $-\boldsymbol{a}_z$ direction, and substituting in Maxwell's equations. Let

$$\boldsymbol{E}_2 = A_2 \sin\left(\frac{m\pi x}{a}\right)\cos(\omega t + \beta_{m0} z)\boldsymbol{a}_y \qquad (12\text{-}24)$$

The reader is urged to carry out the algebraic work for practice; the phase function β_{m0} proves to be the same as for the $+\boldsymbol{a}_z$-moving wave (Eq. 12-6), and the $\boldsymbol{H}$ field is as follows:

$$\boldsymbol{H}_2 = \frac{A_2 \beta_{m0}}{\omega\mu}\sin\left(\frac{m\pi x}{a}\right)\cos(\omega t + \beta_{m0} z)\boldsymbol{a}_x$$

$$- \frac{A_2 m\pi}{\omega\mu a}\cos\left(\frac{m\pi x}{a}\right)\sin(\omega t + \beta_{m0} z)\boldsymbol{a}_z \qquad (12\text{-}25)$$

If Poynting's vector is computed, the longitudinal component proves to be pointed in the $-\boldsymbol{a}_z$ direction, as one would expect.

12-2 DOMINANT-MODE STANDING WAVES — WAVEGUIDE IMPEDANCES

With expressions available for fields traveling in the two directions, one is prepared to examine simple discontinuities in the waveguide, namely, those which are uniform over a transverse cross section. Some examples are (1) a short-circuiting transverse plate, and (2) a discontinuity in ϵ at a transverse plane. (Fields will be postulated specifically in the TE_{10} mode in this discussion.)

These discontinuities will be found to yield results closely analogous to the reflection and standing-wave phenomena on two-conductor lines. The rms value of any component of the combination of the oppositely moving wave sets in the waveguide will vary with z, and the relative rms value of the resultant $E_y(z)$ along the centerline $x = a/2$ may be found experimentally with a slotted-waveguide-and-detector assembly such as that mentioned in Sec. 12-1b(2). The normalized input impedance along a transmission line proved to be directly related to the standing-wave ratio and to the locations of minima in the standing-wave plot; the same concepts and the same plan for determining impedance from standing-wave data are adaptable to the waveguide.

a. Fields in Short-Circuited Waveguide

In a guide with a short-circuiting end plate, the boundary conditions require that the $\boldsymbol{E}$ field should vanish over the surface of the plate. If one chooses the value of $z = 0$ for the location of the end plate, the following results:

$$\boldsymbol{E}_1(x,0,t) = -\boldsymbol{E}_2(x,0,t)$$

Hence

$$A_1 = -A_2 \tag{12-26}$$

If $\boldsymbol{E}_1$, Eq. 12-8, is added to $\boldsymbol{E}_2$, Eq. 12-24, and $\boldsymbol{H}_1$, Eq. 12-9, to $\boldsymbol{H}_2$, Eq. 12-25, expressions are obtained for the resultant fields. These may be reduced, by substituting Eq. 12-26, to the following complex exponential forms ($m = 1$):

$$\boldsymbol{E}_{\text{SC}} = \text{Re}\left[-2A_1 \sin\left(\frac{\pi x}{a}\right)\sin(\beta_{10} z) j\epsilon^{j\omega t}\boldsymbol{a}_y\right] \tag{12-27}$$

$$\boldsymbol{H}_{\text{SC}} = \text{Re}\left[\frac{-2A_1\beta_{10}}{\omega\mu}\sin\left(\frac{\pi x}{a}\right)\cos(\beta_{10} z)\epsilon^{j\omega t}\boldsymbol{a}_x + \frac{2A_1\pi}{\omega\mu a}\cos\left(\frac{\pi x}{a}\right)\sin(\beta_{10} z)\epsilon^{j\omega t}\boldsymbol{a}_z\right] \tag{12-28}$$

Important points concerning these results are (1) the $\boldsymbol{E}$ and $\boldsymbol{H}$ fields are in time quadrature with each other at every location in the waveguide, (2) the $\boldsymbol{E}$ field is always zero on the transverse planes located at half-wavelength intervals (π/β_{10}) from the end plate, (3) the longitudinal component of the $\boldsymbol{H}$ field is always zero on the transverse planes at half-wavelength intervals from the end plate, (4) the transverse component of the $\boldsymbol{H}$ field is always zero on the transverse planes at odd multiples of a quarter-wavelength from the endplate, (5) at every point between each adjacent pair of nulls the intensity of the $\boldsymbol{E}$ field rises and falls in synchronism, and (6) at every point between each adjacent pair of nulls the intensities of both components of the $\boldsymbol{H}$ field rise and fall in synchronism. These characteristics, with the exception of those aspects relating to the longitudinal component of the $\boldsymbol{H}$ field, correspond to those for the short-circuited coaxial cable.

The displacement-current density may also be found (Eq. 12-15):

$$J_{\text{DSC}} = \text{Re}\left[2A_1\omega\epsilon_0 \sin\left(\frac{\pi x}{a}\right)\sin(\beta_{10} z)\epsilon^{j\omega t}\boldsymbol{a}_y\right] \tag{12-29}$$

The displacement-current field has the same configuration in space as does the $\boldsymbol{E}$ field but is in time quadrature with it. The magnetic field is in time phase with the displacement-current field and encircles it spacewise in accordance with the right-hand rule. Problems 12-4 and 12-5 develop further the subject of standing-wave phenomena caused by a short-circuiting endplate.

The rms value of $E_{\text{SC}y}$, as given by Eq. 12-27, along the center plane $x = a/2$, is, in accordance with Eq. 4-33,

$$(E_{\text{SC}y})_{\text{rms}} = \sqrt{2}A_1 \sin\beta_{10} z \tag{12-30}$$

The corresponding standing-wave ratio (SWR) will approach infinity, as was the case for the short-circuited line. However, the distance between nulls is π/β_{10} meters rather than π/β_{TEM}, where

$$\beta_{\text{TEM}} = \omega\sqrt{\mu\epsilon} \tag{12-31}$$

As indicated by Eq. 12-6, $\beta_{10} < \beta_{\text{TEM}}$.

b. Reflection from Discontinuity in Waveguide Medium

As a more general illustration of boundary-value problems involving discontinuities which are transversely uniform, consider the following: A rectangular waveguide is filled with a dielectric medium of permittivity ϵ_1 for $z > 0$, and the permittivity is ϵ_0 for $z < 0$, $\epsilon_1 > \epsilon_0$. The frequency is such that the TE_{10} mode will propagate in both portions.

The following incident wave in the low-permittivity medium impinges on the discontinuity at $z = 0$:

$$\boldsymbol{E}_I = \text{Re}\left\{E_{IM}\sin\left(\frac{\pi x}{a}\right)\exp[j(\omega t - \beta_{10}z)]\right\}\boldsymbol{a}_y \tag{12-32}$$

Find the reflected and transmitted fields.

Continuity of the tangential components of the $\boldsymbol{E}$ and $\boldsymbol{H}$ fields, and of the normal components of the $\boldsymbol{D}$ and $\boldsymbol{B}$ fields, must be maintained across any boundary surface. For the given mode, the $\boldsymbol{D}$ field has no component normal to the surface in question. Locations on the two sides of the interface surface will be designated by $z = 0-$ and $z = 0+$:

$$[\boldsymbol{E}_I(x,0-,t)]_y + [\boldsymbol{E}_R(x,0-,t)]_y = [\boldsymbol{E}_T(x,0+,t)]_y$$

$$[\boldsymbol{H}_I(x,0-,t)]_x + [\boldsymbol{H}_R(x,0-,t)]_x = [\boldsymbol{H}_T(x,0+,t)]_x \tag{12-33}$$

$$[\mu\boldsymbol{H}_I(x,0-,t)]_z + [\mu\boldsymbol{H}_R(x,0-,t)]_z = [\mu\boldsymbol{H}_T(x,0+,t)]_z$$

The following forms for the reflected and transmitted $\boldsymbol{E}$-field components may be postulated; the angles θ_K and θ_T are introduced in the expectation that those fields might not be in time phase with each other nor with the incident field at the boundary:

$$\begin{aligned} \boldsymbol{E}_R &= \operatorname{Re}\left\{ E_{RM} \sin\left(\frac{\pi x}{a}\right) \exp[j(\omega t + \theta_K + \beta_{10} z)] \right\} \boldsymbol{a}_y \\ \boldsymbol{E}_T &= \operatorname{Re}\left\{ E_{TM} \sin\left(\frac{\pi x}{a}\right) \exp[j(\omega t + \theta_T - \beta'_{10} z)] \right\} \boldsymbol{a}_y \end{aligned} \tag{12-34}$$

Here

$$\begin{aligned} \beta_{10} &= \sqrt{\omega^2 \mu \epsilon_0 - \left(\frac{\pi}{a}\right)^2} \\ \beta'_{10} &= \sqrt{\omega^2 \mu \epsilon_1 - \left(\frac{\pi}{a}\right)^2} \end{aligned} \tag{12-35}$$

Substitution of Eqs. 12-32 and 12-34 in the first equation of set 12-33 yields the following, after cancelling the common factors: Re operator, $\sin(\pi x/a)$, and $\epsilon^{j\omega t}$:

$$E_{IM} + E_{RM} \exp(j\theta_K) = E_{TM} \exp(j\theta_T) \tag{12-36}$$

The ***H***-field components will not be given here in full, but by comparison with Eqs. 12-9, and 12-25 one may assemble the terms for the remaining two equations in set 12-33:

$$\frac{-E_{IM}\beta_{10}}{\omega\mu} + \frac{E_{RM}\beta_{10}}{\omega\mu} \exp(j\theta_K) = \frac{-E_{TM}\beta'_{10} \exp(j\theta_T)}{\omega\mu} \tag{12-37}$$

$$\frac{-E_{IM}\pi}{j\omega a} - \frac{E_{RM}\pi}{j\omega a} \exp(j\theta_K) = \frac{-E_{TM}\pi \exp(j\theta_T)}{j\omega a} \tag{12-38}$$

If the latter equation is multiplied by $-j\omega a/\pi$, it proves to be identical with Eq. 12-36 and hence is redundant. Substitution of Eq. 12-36 in Eq. 12-37 yields, after cancellation of like quantities and grouping terms,

$$E_{RM} \exp(j\theta_R) = \frac{\beta_{10} - \beta'_{10}}{\beta_{10} + \beta'_{10}} E_{IM} \tag{12-39}$$

$$E_{TM} \exp(j\theta_T) = \frac{2\beta_{10}}{\beta_{10} + \beta'_{10}} E_{IM} \tag{12-40}$$

Here $\epsilon_1 > \epsilon_0$, hence $\beta'_{10} > \beta_{10}$, and in this instance $\theta_R = \pi$, and $\theta_T = 0$.

c. Waveguide-Impedance Concept

A comparison of the results in Eqs. 12-39 and 12-40 with those in Chap. 3 for two lines in tandem (Eqs. 3-37 and 3-39) suggests that, for a consistent analogy, *waveguide characteristic impedance* (for the TE_{10} mode) should be defined as inversely proportional to β_{10}. Perhaps the simplest basis for defining absolute impedances in waveguides, a definition which meets this requirement and which may be applied consistently to other modes, is the following: Let $E_{trv}(z)\boldsymbol{a}_{Et}$ and $H_{trv}(z)\boldsymbol{a}_{Ht}$ represent the phasor forms of the transverse components of the respective fields, with the positive directions of the unit vectors so chosen that $\boldsymbol{a}_{Et}$, $\boldsymbol{a}_{Ht}$, and the direction to the load (in that cyclic order) form a right-hand set. Then let

$$Z_w(z) = \frac{E_{trv}(z)}{H_{trv}(z)} \tag{12-41}$$

For the traveling-wave fields in Eqs. 12-8 and 12-9, the direction to the load is $\boldsymbol{a}_z$; hence $\boldsymbol{a}_{Et} = \boldsymbol{a}_y$ and $\boldsymbol{a}_{Ht} = -\boldsymbol{a}_x$ will complete a right-handed set. The corresponding phasor terms are as follows.

$$E_{trv}(z) = E_y(z)$$

$$= A_1 \sin\left(\frac{\pi x}{a}\right)\exp(-j\beta_{10}z) \tag{12-42}$$

$$H_{trv}(z) = -H_x(z)$$

$$= \frac{A_1\beta_{10}}{\omega\mu}\sin\left(\frac{\pi x}{a}\right)\exp(-j\beta_{10}z) \tag{12-43}$$

The waveguide characteristic impedance is found by substituting Eqs. 12-42 and 12-43 in Eq. 12-41:

$$Z_{w0} = \frac{\omega\mu}{\beta_{10}} \tag{12-44}$$

The impedance in the short-circuited waveguide, as seen from the transverse plane at distance z from the termination, may be found from Eqs. 12-27 and 12-28. If the same coordinate plan is used as for Eqs. 12-42 through 12-44, the phasors of the transverse components are

$$E_{trv}(2) = -j2A_1 \sin\left(\frac{\pi x}{a}\right)\sin \beta_{10} z \tag{12-45}$$

$$H_{trv}(z) = \frac{2A_1\beta_{10}}{\omega\mu} \sin\left(\frac{\pi x}{a}\right)\cos \beta_{10} z \tag{12-46}$$

Equations 12-45 and 12-46 are substituted in Eq. 12-41:

$$Z_{w\,\mathrm{SC}}(z) = -j\frac{\omega\mu}{\beta_{10}} \tan \beta_{10} z \tag{12-47}$$

(If a_z represents the direction to the load, *negative* values of z correspond to locations in the energized waveguide; hence, as would be expected for a consistent analogy with transmission lines, $Z_{w\,\mathrm{SC}}$ is inductive in the first quarter-wavelength from the short circuit.)

In many instances a *normalized waveguide impedance* is more meaningful than an absolute impedance. The reflection coefficient and the standing-wave ratio are basically functions of the impedance of the load normalized with respect to the characteristic impedance of the line, rather than the absolute impedances of the load and the line. In the example in Sec. 12-2b, let the characteristic impedance of the section of waveguide in which $z > 0$ be represented by $Z_{w'0}$, and let this be regarded as a terminating impedance for the other section of waveguide ($z < 0$; characteristic impedance, Z_{w0}).

$$Z_{w'0} = \frac{\omega\mu}{\beta'_{10}} \tag{12-48}$$

The normalized waveguide impedance $\mathcal{Z}_w$ at the interface $z = 0$ is

$$\mathcal{Z}_w(0) = \frac{Z_{w'0}}{Z_{w0}} \tag{12-49}$$

Substitution of Eqs. 12-44 and 12-48 yields

$$\mathcal{Z}_w(0) = \frac{\beta_{10}}{\beta'_{10}} \tag{12-50}$$

With the aid of the Smith chart, the normalized waveguide impedance as seen from any transverse plane for which $z < 0$ (toward the generator from the discontinuity) may be found as indicated in Fig. 12-3. (See Prob. 12-7.)

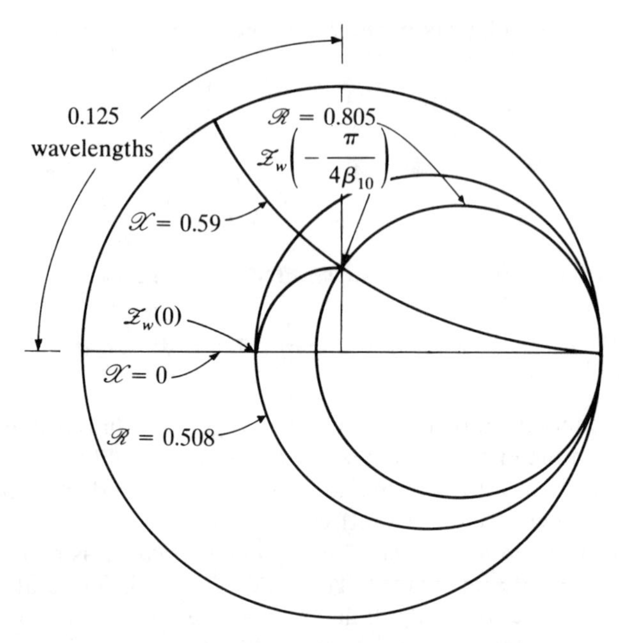

Figure 12-3. Smith chart solution for waveguide impedance at one eighth-waveguide from air–polyethylene interface (see Prob. 12-7).

12.3 GENERAL SOLUTION FOR TE AND TM MODES

A more comprehensive approach to the finding of waveguide modes* is that of reducing Maxwell's equations to a general wave equation. The two equations involving the curl function form the starting point:

$$\nabla \times \boldsymbol{H} = \epsilon \frac{\partial \boldsymbol{E}}{\partial t} \tag{12-51}$$

$$\nabla \times \boldsymbol{E} = -\mu \frac{\partial \boldsymbol{H}}{\partial t} \tag{9-23}$$

*The TEM mode, one in which both $\boldsymbol{E}$ and $\boldsymbol{H}$ are confined to planes transverse to the direction of propagation, cannot exist in a uniconductor waveguide. Because the divergence of $\boldsymbol{B}$ is everywhere zero (Eq. 9-20), the $\boldsymbol{H}$-field lines in an isotropic medium of finite extent necessarily form closed curves, and the line integral around the loop formed by any such curve is equal to the surface integral of the current density through the loop (Eq. 9-10). If $\boldsymbol{H}$ is confined to transverse planes, the current density function must include a longitudinal component. With no interior carrier for conduction current (as in a coaxial cable) a longitudinal component of displacement-current density and hence of $\boldsymbol{D}$ and $\boldsymbol{E}$ would be essential, and then the mode would be in the TM rather than the TEM class.

Equation 12-51 corresponds to Eq. 9-21 with conductivity σ set equal to zero. $\boldsymbol{H}$ may be eliminated from this set by differentiating Eq. 12-51 with respect to time, taking the curl of Eq. 9-23, and substituting:

$$\nabla \times (\nabla \times \boldsymbol{E}) = -\mu\epsilon \frac{\partial^2 \boldsymbol{E}}{\partial t^2} \tag{12-52}$$

a. Reduction to Scalar Wave Equations

Equation 12-52 is a *vector wave equation*; it may be reduced to a set of scalar wave equations by means of the following formal identity from vector analysis:

$$-\nabla \times (\nabla \times \boldsymbol{E}) + \nabla(\nabla \cdot \boldsymbol{E}) = \nabla^2 \boldsymbol{E} \tag{12-53}$$

This is actually the definition of the *Laplacian* of a *vector field*. For rectangular coordinates, a routine substitution of Eqs. A-25, A-26, A-28, and A-29 into the left-hand side of Eq. 12-53 (see Prob. 12-8) will yield the following:

$$\nabla^2 \boldsymbol{E} = \nabla^2 E_x \, \boldsymbol{a}_x + \nabla^2 E_y \, \boldsymbol{a}_y + \nabla^2 E_z \, \boldsymbol{a}_z \tag{12-54}$$

(This simple result cannot be generalized intuitively to other coordinate systems; see Sec. 13-1a regarding the outcome in cylindrical coordinates.)

Since the space within the waveguide is assumed to be free of electric charge, the divergence of $\boldsymbol{E}$ vanishes, and Eq. 12-52 may be replaced by three scalar equations, each representing the electric-field magnitude in one of the coordinate directions:

$$\begin{aligned}
\nabla^2 E_x &= \mu\epsilon \frac{\partial^2 E_x}{\partial t^2} \\
\nabla^2 E_y &= \mu\epsilon \frac{\partial^2 E_y}{\partial t^2} \\
\nabla^2 E_z &= \mu\epsilon \frac{\partial^2 E_z}{\partial t^2}
\end{aligned} \tag{12-55}$$

Each of these may be written in expanded form to show the functional dependence on x, y, and z explicitly. Thus for the equation in E_z,

$$\frac{\partial^2 E_z}{\partial x^2} + \frac{\partial^2 E_z}{\partial y^2} + \frac{\partial^2 E_z}{\partial z^2} = \mu\epsilon \frac{\partial^2 E_z}{\partial t^2} \tag{12-56}$$

Wave equations for the Cartesian components of $\boldsymbol{H}$ may be obtained by taking the curl of Eq. 12-51, differentiating Eq. 9-23 with respect to time, and making substitutions paralleling those of Eqs. 12-52, 12-53, and 12-54. The expanded form for the wave equation in H_z is

$$\frac{\partial^2 H_z}{\partial x^2} + \frac{\partial^2 H_z}{\partial y^2} + \frac{\partial^2 H_z}{\partial z^2} = \mu\epsilon \frac{\partial^2 H_z}{\partial t^2} \tag{12-57}$$

Equations 12-56 and 12-57 were written for the z components because that is the longitudinal direction of the waveguide shown in Fig. 12-1, and the longitudinal components of field prove to be useful as reference functions.

b. Interrelationships Among Field Components — Mode Types

Expressions may be derived for E_x, E_y, H_x, and H_y in terms of E_z and H_z. To find such an equation for H_x, take the y component of Eq. 12-51 and the x component of Eq. 9-23:

$$\frac{\partial H_x}{\partial z} - \frac{\partial H_z}{\partial x} = \epsilon \frac{\partial E_y}{\partial t} \tag{12-58}$$

$$\frac{\partial E_z}{\partial y} - \frac{\partial E_y}{\partial z} = -\mu \frac{\partial H_x}{\partial t} \tag{12-59}$$

If the first of these is differentiated with respect to z and the second with respect to t, E_y may be eliminated:

$$\frac{\partial^2 H_x}{\partial z^2} - \mu\epsilon \frac{\partial^2 H_x}{\partial t^2} = \frac{\partial}{\partial z}\left(\frac{\partial H_z}{\partial x}\right) + \epsilon \frac{\partial}{\partial t}\left(\frac{\partial E_z}{\partial y}\right) \tag{12-60}$$

This equation may be greatly simplified if the variation of each field component with respect to z and t is assumed to be that of a sinusoidal traveling wave:

$$\begin{aligned} H_x &= \mathrm{Re}\left[H_{0x}(x, y)\epsilon^{j(\omega t - \beta z)}\right] \\ H_z &= \mathrm{Re}\left[H_{0z}(x, y)\epsilon^{j(\omega t - \beta z)}\right] \\ E_z &= \mathrm{Re}\left[E_{0z}(x, y)\epsilon^{j(\omega t - \beta z)}\right] \end{aligned} \tag{12-61}$$

It should be anticipated that $H_{0x}(x, y)$, $H_{0z}(x, y)$, and $E_{0z}(x, y)$ may be complex.

Equations 12-61 may be substituted into Eq. 12-60 and the Re operator and the factor $\epsilon^{j(\omega t - \beta z)}$, which are common to every term, cancelled:

$$(-\beta^2 + \mu\epsilon\omega^2)H_{0x}(x, y) = -j\beta \frac{\partial H_{0z}(x, y)}{\partial x} + j\omega\epsilon \frac{\partial E_{0z}(x, y)}{\partial y} \quad (12\text{-}62)$$

Corresponding equations for the other components are

$$(-\beta^2 + \mu\epsilon\omega^2)H_{0y}(x, y) = -j\beta \frac{\partial H_{0z}(x, y)}{\partial y} - j\omega\epsilon \frac{\partial E_{0z}(x, y)}{\partial x}$$

$$(-\beta^2 + \mu\epsilon\omega^2)E_{0x}(x, y) = -j\beta \frac{\partial E_{0z}(x, y)}{\partial x} - j\omega\mu \frac{\partial H_{0z}(x, y)}{\partial y} \quad (12\text{-}63)$$

$$(-\beta^2 + \mu\epsilon\omega^2)E_{0y}(x, y) = -j\beta \frac{\partial E_{0z}(x, y)}{\partial y} + j\omega\mu \frac{\partial H_{0z}(x, y)}{\partial x}$$

Wave equations 12-56 and 12-57 separate E_z from H_z and, in accordance with Eqs. 12-62 and 12-63, H_x, H_y, E_x, and E_y may each be divided into two separate functions, one proportional to a derivative of E_z and the other proportional to a derivative of H_z. Thus, by the principle of superposition, a general sinusoidal traveling-wave field may be split into two parts, one of which has an E_z component but no H_z component, and the other of which has an H_z component but no E_z component. Functions of the first type are known as *transverse-magnetic* (TM) modes, because the magnetic field is confined to planes perpendicular to the direction of propagation; similarly, functions of the second type are known as *transverse electric* (TE) modes.

c. Solution by Product-Function Method

Substitution of the appropriate parts of Eq. 12-61 in Eqs. 12-56 and 12-57 gives the following:

$$\frac{\partial^2 E_{0z}(x, y)}{\partial x^2} + \frac{\partial^2 E_{0z}(x, y)}{\partial y^2} = (\beta^2 - \mu\epsilon\omega^2)E_{0z}(x, y) \quad (12\text{-}64)$$

$$\frac{\partial^2 H_{0z}(x, y)}{\partial x^2} + \frac{\partial^2 H_{0z}(x, y)}{\partial y^2} = (\beta^2 - \mu\epsilon\omega^2)H_{0z}(x, y) \quad (12\text{-}65)$$

The product-function method may be used to obtain the functions $E_{0z}(x, y)$ and $H_{0z}(x, y)$. Let

$$E_{0z}(x, y) = X_{Ez}(x) Y_{Ez}(y) \tag{12-66}$$

$$\frac{d^2 X_{Ez}(x)}{dx^2} Y_{Ez}(y) + X_{Ez}(x) \frac{d^2 Y_{Ez}(y)}{dy^2} = (\beta^2 - \mu\epsilon\omega^2) X_{Ez}(x) Y_{Ez}(y)$$

Variables may be separated by dividing this by $X_{Ez}(x)Y_{Ez}(y)$:

$$\frac{d^2 X_{Ez}(x)}{dx^2} \frac{1}{X_{Ez}(x)} + \frac{1}{Y_{Ez}(y)} \frac{d^2 Y_{Ez}(y)}{dy^2} = \beta^2 - \mu\epsilon\omega^2 \tag{12-67}$$

This must be satisfied for independently chosen values of x and y, and such can result only if (1) the first term, which is not a function of y, but is the only part of the equation which is a function of x, is equal to a constant, say, $-C_1^2$; and (2) the second term, which is not a function of x, but is the only part of the equation which is a function of y, is equal to another constant, say, $-C_2^2$:

$$\frac{d^2 X_{Ez}(x)}{dx^2} \frac{1}{X_{Ez}(x)} = -C_1^2 \tag{12-68}$$

$$\frac{d^2 Y_{Ez}(y)}{dy^2} \frac{1}{Y_{Ez}(y)} = -C_2^2 \tag{12-69}$$

Equation 12-67 imposes the following constraint on C_1 and C_2:

$$-C_1^2 - C_2^2 = \beta^2 - \mu\epsilon\omega^2 \tag{12-70}$$

Equations 12-68 and 12-69 yield the following, in which the A_is are arbitrary constants:

$$X_{Ez}(x) = A_1 \sin C_1 x + A_2 \cos C_1 x \tag{12-71}$$

$$Y_{Ez}(y) = A_3 \sin C_2 y + A_4 \cos C_2 y \tag{12-72}$$

Similarly,

$$H_{0z}(x, y) = X_{Hz}(x) Y_{Hz}(y) \tag{12-73}$$

$$X_{Hz}(x) = A_5 \sin C_3 x + A_6 \cos C_3 x \tag{12-74}$$

$$Y_{Hz}(y) = A_7 \sin C_4 y + A_8 \cos C_4 y \tag{12-75}$$

where

$$-C_3^2 - C_4^2 = \beta^2 - \mu\epsilon\omega^2 \tag{12-76}$$

d. Application of Boundary Conditions

The permissible values for the Cs and some of the As are fixed by the boundary condition that the tangential component of the electric field vanish along the conductor surfaces. The phase function is then fixed by Eq. 12-70 or 12-76.

(1) *TM Modes.* The TM case may be examined more directly, since E_z is parallel to the surfaces of the waveguide and hence must vanish at $x = 0$ and $x = a$ and at $y = 0$ and $y = b$. In accordance with Eq. 12-66,

$$E_{0z}(x, y) = (A_1 \sin C_1 x + A_2 \cos C_1 x)(A_3 \sin C_2 y + A_4 \cos C_2 y) \tag{12-77}$$

Setting x equal to zero,

$$E_{0z}(0, y) = A_2(A_3 \sin C_2 y + A_4 \cos C_2 y)$$

Therefore,

$$A_2 = 0$$

After substituting this in Eq. 12-77, x is set equal to a, where $\boldsymbol{E}$ must also vanish:

$$E_{0z}(a, y) = A_1 \sin C_1 a (A_3 \sin C_2 y + A_4 \cos C_2 y)$$

Therefore,

$$C_1 a = m\pi$$

Here m may be any integer. Substituting this in Eq. 12-77 gives

$$E_{0z}(x, y) = A_1 \sin\left(\frac{m\pi x}{a}\right)(A_3 \sin C_2 y + A_4 \cos C_2 y)$$

Similarly,

$$E_{0z}(x,0) = A_1 A_4 \sin\left(\frac{m\pi x}{a}\right)$$

Therefore,

$$A_4 = 0$$

$$E_{0z}(x,b) = A_1 A_3 \sin\left(\frac{m\pi x}{a}\right)\sin C_2 b$$

Therefore,

$$C_2 b = n\pi$$

Here n may be any integer.

The product $A_1 A_3$ may be replaced by a single constant C_{mn}, which corresponds to the particular mode defined by a given choice of m and n:

$$E_{0z}(x,y) = C_{mn} \sin\left(\frac{m\pi x}{a}\right)\sin\left(\frac{n\pi y}{b}\right) \tag{12-78}$$

From Eq. 12-70, the phase function is as follows:

$$\beta_{mn}^2 = \mu\epsilon\omega^2 - \left(\frac{m\pi}{a}\right)^2 - \left(\frac{n\pi}{b}\right)^2 \tag{12-79}$$

Let

$$h^2 = \left(\frac{m\pi}{a}\right)^2 + \left(\frac{n\pi}{b}\right)^2 \tag{12-80}$$

The functions of x, y for the remaining field components may be written after substituting in Eqs. 12-62 and 12-63:

$$\begin{aligned}
H_{0x}(x,y) &= \frac{j\omega\epsilon n\pi}{h^2 b} C_{mn} \sin\left(\frac{m\pi x}{a}\right)\cos\left(\frac{n\pi y}{b}\right) \\
H_{0y}(x,y) &= \frac{-j\omega\epsilon m\pi}{h^2 a} C_{mn} \cos\left(\frac{m\pi x}{a}\right)\sin\left(\frac{n\pi y}{b}\right) \\
E_{0x}(x,y) &= \frac{-j\beta_{mn} m\pi}{h^2 a} C_{mn} \cos\left(\frac{m\pi x}{a}\right)\sin\left(\frac{n\pi y}{b}\right) \\
E_{0y}(x,y) &= \frac{-j\beta_{mn} n\pi}{h^2 b} C_{mn} \sin\left(\frac{m\pi x}{a}\right)\cos\left(\frac{n\pi y}{b}\right)
\end{aligned} \tag{12-81}$$

One may verify by inspection that E_{0x}, which is tangential to the waveguide at $y = 0$ and $y = b$, vanishes there, and that E_{0y} vanishes at

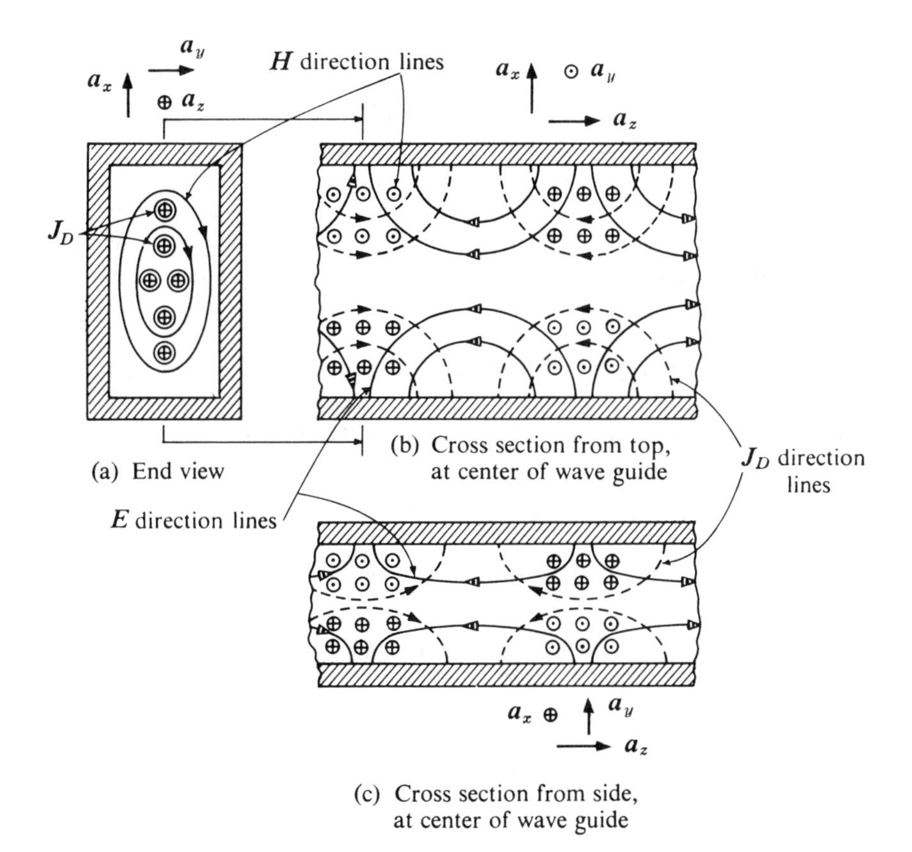

Figure 12-4. TM_{11} propagating mode in rectangular waveguide.

$x = 0$ and $x = a$. One may also note that the normal component of $\boldsymbol{H}$, and hence $\boldsymbol{B}$, vanishes at the conductor surface.

The TM_{11} mode is illustrated in Fig. 12-4.

An equation for the cutoff frequency f_{cmn} may be found from Eq. 12-79:

$$\omega_{cmn}^2 \mu \epsilon = \left(\frac{m\pi}{a}\right)^2 + \left(\frac{n\pi}{b}\right)^2$$

$$f_{cmn} = \frac{1}{2\pi\sqrt{\mu\epsilon}} \sqrt{\left(\frac{m\pi}{a}\right)^2 + \left(\frac{n\pi}{b}\right)^2} \tag{12-82}$$

For the $\boldsymbol{H}$-field to have direction lines which are closed curves, both H_x and H_y functions are needed, so m and n must both be nonzero. Comparison of Eq. 12-82 with Eq. 12-10 indicates that the cutoff frequencies of all TM modes are higher than that of the TE_{10} mode.

In usual applications of rectangular waveguides, transmission in the TE_{10} mode alone is sought, and the guide dimensions are chosen with respect to frequency so as to ensure this. Hence transverse-magnetic modes will exist only in evanescent form in the vicinity of guide discontinuities.

(2) *TE Modes.* For general analysis of TE modes one substitutes Eqs. 12-74 and 12-75 in Eq. 12-73, and that in turn in Eq. 12-63, so that the $\boldsymbol{E}$-field components parallel to the guide surfaces may be found:

$$\begin{aligned} E_{0x}(x, y) &= \frac{-j\omega\mu C_4}{-\beta^2 + \mu\epsilon\omega^2}(A_5 \sin C_3x + A_6 \cos C_3x) \\ &\quad \cdot (A_7 \cos C_4y - A_8 \sin C_4y) \\ E_{0x}(x, y) &= \frac{j\omega\mu C_3}{-\beta^2 + \mu\epsilon\omega^2}(A_5 \cos C_3x - A_6 \sin C_3\chi) \\ &\quad \cdot (A_7 \sin C_4y + A_8 \cos C_4y) \end{aligned} \tag{12-83}$$

The boundary condition that the tangential components of $\boldsymbol{E}$ vanish along the guide surfaces fixes most of these constants:

$$E_{0x}(x, 0) = 0 \quad \text{therefore } A_7 = 0$$

$$E_{0x}(x, b) = 0 \quad \text{therefore } C_4 b = n\pi$$

$$E_{0y}(0, y) = 0 \quad \text{therefore } A_5 = 0$$

$$E_{0y}(a, y) = 0 \quad \text{therefore } C_3 a = m\pi$$

From Eq. 12-76,

$$\beta_{mn}^2 = \mu\epsilon\omega^2 - \left(\frac{m\pi}{a}\right)^2 - \left(\frac{n\pi}{b}\right)^2 \tag{12-84}$$

This is the same as the equation for the phase function of TM modes, Eq. 12-79; Eq. 12-82 for cutoff frequencies of TM modes is also applicable here for TE modes.

Equations 12-79 and 12-84 may be restated in shortened form:

$$\beta_{mn} = \sqrt{\mu\epsilon(\omega^2 - \omega_{cmn}^2)} \tag{12-85}$$

The results just found may be combined with Eqs. 12-70, 12-83, 12-73, 12-62, and 12-63 to yield expressions for the field components.

Let $$A_6 A_8 = A_{mn}$$

$$\begin{aligned}
E_{0x}(x, y) &= \frac{j\omega\mu n\pi}{h^2 b} A_{mn} \cos\left(\frac{m\pi x}{a}\right)\sin\left(\frac{n\pi y}{b}\right) \\
E_{0y}(x, y) &= \frac{-j\omega\mu m\pi}{-h^2 a} A_{mn} \sin\left(\frac{m\pi x}{a}\right)\cos\left(\frac{n\pi y}{b}\right) \\
H_{0z}(x, y) &= A_{mn} \cos\left(\frac{m\pi x}{a}\right)\cos\left(\frac{n\pi y}{b}\right) \\
H_{0x}(x, y) &= \frac{j\beta_{mn} m\pi}{h^2 a} A_{mn} \sin\left(\frac{m\pi x}{a}\right)\cos\left(\frac{n\pi y}{b}\right) \\
H_{0y}(x, y) &= \frac{j\beta_{mn} n\pi}{h^2 b} A_{mn} \cos\left(\frac{m\pi x}{a}\right)\sin\left(\frac{n\pi y}{b}\right)
\end{aligned} \tag{12-86}$$

It should be noted that either m or n (but not both) may be set equal to zero for TE modes and still yield nontrivial results, but that both m and n must be nonzero for TM modes. Setting n equal to zero reduces the TE solution just given to the form developed in Eqs. 12-8 and 12-9. TE_{0n}-mode patterns resemble the TE_{m0} ones, except that they are turned 90° in space (in terms of Fig. 12-1) such that the ***E***-field variation takes place in the shorter transverse direction.

Except for a square waveguide, the TE_{10} mode has a lower cutoff frequency than the TE_{01} mode; in any case it has a lower cutoff frequency than any other TE mode, and, as noted in Sec. 12-4d(2), than any TM mode. Hence it is properly called the dominant mode.

e. Waveguide Characteristic Impedances

The definition of waveguide impedance given in Eq. 12-41 and the comments preceding it may be applied here to find the characteristic impedances of TE and TM modes in general. Since ***E*** and ***H*** in the wave sets developed in Sec. 12-4d both have x and y components, a more general representation of the transverse components is necessary than that used in Eqs. 12-42 and 12-43 for the analysis of TE_{m0} modes. For the traveling-wave fields described in Eqs. 12-78, 12-81, and 12-86, the direction to the load is $+\boldsymbol{a}_z$. Let

$$E_{trv}(z)\boldsymbol{a}_{Et} = E_y(z)\boldsymbol{a}_y + E_x(z)\boldsymbol{a}_x$$
$$H_{trv}(z)\boldsymbol{a}_{Ht} = [-H_x(z)](-\boldsymbol{a}_x) + H_y(z)\boldsymbol{a}_y \tag{12-87}$$

On this basis, the requirement for a right-handed set with $\boldsymbol{a}_z$ will be met with respect to the first unit vector on the right-hand side in each equation and with respect to the second unit vector on the right-hand side of each equation. The expressions in Eqs. 12-78, 12-81, and 12-86 are equivalent to traveling-wave-component phasors all divided by the common factor $\exp(-j\beta_{mn}z)$; furthermore the following is true of both the TM and the TE sets:

$$\frac{E_{0y}(x, y)}{-H_{0x}(x, y)} = \frac{E_{0x}(x, y)}{H_{0y}(x, y)} \tag{12-88}$$

The ratio in Eq. 12-88 is equal to the characteristic impedance; it has different values for the two types of modes:

$$Z_{w0} = \frac{\beta_{mn}}{\omega\epsilon} \quad \text{(TM modes)} \tag{12-89}$$

$$Z_{w0} = \frac{\omega\mu}{\beta_{mn}} \quad \text{(TE modes)} \tag{12-90}$$

Equation 12-44, for the TE_{m0} modes, is a special case of Eq. 12-90.

12-4 EVANESCENT MODES AND GENERAL DISCONTINUITIES

Many types of discontinuities which are not uniform over the waveguide cross section are present in waveguide equipment, and these lack the symmetry and simplicity which the examples in Sec. 12-2 possessed. Among the discontinuities of interest are bends in the waveguide, irises (transverse conducting sheets extending over a part of the waveguide cross section—one type is shown in Fig. 12-5), junctions with waveguide stubs (Fig. 12-6), and antenna structures to couple the waveguide to a coaxial cable. At such discontinuities the boundary conditions cannot be satisfied by means of one reflected and one transmitted traveling wave set, each of the same mode as the incident traveling wave. Rather, a multiplicity of modes, each of appropriate magnitude proportion and phase relationship

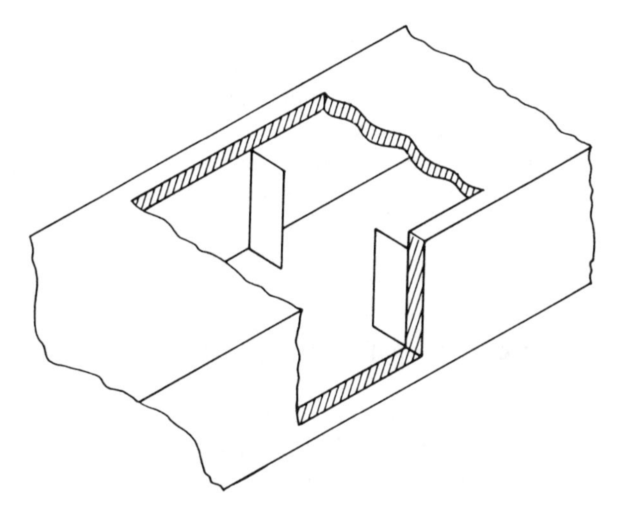

Figure 12-5. Vertical iris in rectangular waveguide.

with respect to the incident wave, is needed. In a rectangular waveguide of the usual proportions these additional, parasitic modes will all have cutoff frequencies which are above that of the mode of the incident wave, and presumably above the excitation frequency; hence they will be nonpropagating or *evanescent*. Accordingly it is desirable to investigate briefly the structure and properties of nonpropagating modes. For algebraic simplicity the TE_{m0}-mode class will be chosen as the subject for detailed study.

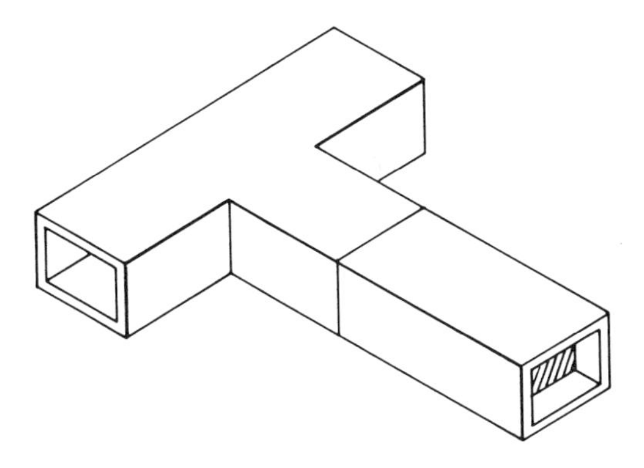

Figure 12-6. Shunt-tee junction with sort-circuiting plunger unit attached.

The term *port*, which was introduced at the beginning of Chap. 8 as a synonym for terminal-pair in the analysis of lumped-parameter networks, may be generalized to serve in waveguide theory as the designation for any opening through which electromagnetic energy may flow. A nonbranching section of waveguide, such as that shown in Fig. 12-5 containing an iris, is a two-port device, whereas the T in Fig. 12-6 has three ports. The reflection and transmission properties with respect to the dominant mode may be summarized, for an n-port discontinuity, in a *scattering matrix*.

a. Properties of Nonpropagating TE$_{m0}$ Modes

The phase function β_{m0} as given in Eq. 12-6, which relates it to the waveguide width and frequency, will be imaginary if the frequency is below the cutoff value given in Eq. 12-10. For this case let

$$\beta_{m0} = -j\alpha_{m0} \tag{12-91}$$

With this substitution the two forms of Eqs. 12-8 and 12-9 may be rewritten as follows:

$$\begin{aligned} \boldsymbol{E}_1 &= \mathrm{Re}\left[A_1 \sin\left(\frac{m\pi x}{a}\right)\exp(-\alpha_{m0}z + j\omega t)\right]\boldsymbol{a}_y \\ &= A_1 \sin\left(\frac{m\pi x}{a}\right)\exp(-\alpha_{m0}z)\cos\omega t\,\boldsymbol{a}_y \end{aligned} \tag{12-92}$$

$$\begin{aligned} \boldsymbol{H}_1 &= \mathrm{Re}\left[\frac{A_1 j\alpha_{m0}}{\omega\mu}\sin\left(\frac{m\pi x}{a}\right)\exp(-\alpha_{m0} + j\omega t)\boldsymbol{a}_x \right. \\ &\qquad \left. - \frac{A_1 m\pi}{j\omega\mu a}\cos\left(\frac{m\pi x}{a}\right)\exp(-\alpha_{m0}z + j\omega t)\boldsymbol{a}_z\right] \\ &= \frac{-A_1\alpha_{m0}}{\omega\mu}\sin\left(\frac{m\pi x}{a}\right)\exp(-\alpha_{m0}z)\sin\omega t\,\boldsymbol{a}_x \\ &\qquad - \frac{A_1 m\pi}{\omega\mu a}\cos\left(\frac{m\pi x}{a}\right)\exp(-\alpha_{m0}z)\sin\omega t\,\boldsymbol{a}_z \end{aligned} \tag{12-93}$$

This, like the one described by Eqs. 12-27 and 12-28, is a standing-wave pattern, since (1) $\boldsymbol{E}$ and $\boldsymbol{H}$ are in time quadrature at every location, (2) the $\boldsymbol{E}$ field changes in synchronism at every point, and (3) the $\boldsymbol{H}$ field changes in synchronism at every point. In this pattern neither field has any nulls with respect to z; rather each diminishes in magnitude with increase in distance in the $+\boldsymbol{a}_z$ direction. The magnitude appears to increase indefinitely for increasingly negative values of z, but in a physical problem the mode would exist only because it was excited by a discontinuity having mode-conversion characteristics and would be defined only for locations on the $+\boldsymbol{a}_z$ side of the discontinuity.

Equation 12-91 may be substituted in Eq. 12-6 to show the variation of α with frequency:

$$\alpha_{m0}^2 = \left(\frac{m\pi}{a}\right)^2 - \omega^2\mu\epsilon \tag{12-94}$$

Thus the farther the driving frequency is below the cutoff frequency of the given mode, the larger α is and hence the more pronounced the diminution of field strength with distance.

A second solution may be obtained by making the following substitution in Eqs. 12-8 and 12-9:

$$\beta_{m0} = j\alpha_{m0} \tag{12-95}$$

$$\boldsymbol{E}_2 = A_2 \sin\left(\frac{m\pi x}{a}\right)\exp(\alpha_{m0} z)\cos \omega t\, \boldsymbol{a}_y \tag{12-96}$$

$$\boldsymbol{H}_2 = \frac{A_2\alpha_{m0}}{\omega\mu} \sin\left(\frac{m\pi x}{a}\right)\exp(\alpha_{m0} z)\sin \omega t\, \boldsymbol{a}_x - \frac{A_2 m\pi}{\omega\mu a} \cos\left(\frac{m\pi x}{a}\right)\exp(\alpha_{m0} z)\sin \omega t\, \boldsymbol{a}_z \tag{12-97}$$

This pair of equations also describes a standing-wave pattern, but in this case the magnitude decreases exponentially in the $-a_z$ direction. Thus evanescent modes may exist in both the transmitted and reflected directions of any given incident wave.

This analysis may be applied to the other TE modes and to the TM modes by substituting Eq. 12-91 in Eqs. 12-61, 12-81, and 12-86.

b. Vertical Iris — Local and Distant Fields

The iris sketched in Fig. 12-5 possesses a high degree of symmetry, yet the fields surrounding it are far more difficult to compute than those of the problem considered in Sec. 12-2b. The boundary requirements are (1) that E_y vanish on both sides of the iris sheet, and (2) that continuity of E_y and H_x be maintained in the plane of the iris over the area between the conducting surfaces. Because the iris structure is uniform in the y direction, it does not tend to induce E_x or E_z field components when a TE_{10} mode is incident upon it. Hence reflected and transmitted fields are limited to the TE_{m0} type. Furthermore, because it is symmetrical with respect to the center of the guide in the x direction, it does not tend to induce even-ordered TE_{m0} modes (they are antisymmetric with respect to that axis).

In spite of restriction to odd-ordered TE_{m0} modes when the incident field is in the TE_{10} mode, an infinite number of modes within that class may be expected in the reflected and transmitted fields. A complete solution for the magnitude and relative phase of each would involve an infinite number of simultaneous linear equations. In waveguide applications the aspect of primary interest is the effect on the TE_{10} mode itself—the magnitude and phase of its reflected and transmitted components. Approximate solutions have been derived,[4, 9, 11, 16, 24] and these indicate that the iris is analogous to a shunt inductive reactance, if E_y is considered analogous to voltage.

The resultant field which a slotted-waveguide probe would sense includes, of course, the traveling-wave modes and all evanescent modes of the type just described. But because of the exponential decrease in amplitude of each evanescent mode with distance from the plane of the iris, those modes may be expected to have comparatively little effect on the rms value of the resultant E_y, unless readings are taken within, say, one wavelength from the iris. Thus one may anticipate a standing-wave plot similar to that shown in Fig. 7-5 (disregarding the particular location of the termination shown there); this plot could be extrapolated, on the basis of half-wavelength repetition, to the plane of the iris. The analytical techniques developed in Chap. 7 for determining an equivalent impedance or admittance at a given electrical distance from a minimum in the standing-wave plot may be applied.

c. Shunt-Tee Junction

Figure 12-6 shows a waveguide analogy to the shunt-connected short-circuited stub of Fig. 7-12 (E_y analogous to voltage). It would probably be installed either close to a load or close to a feed point in order that the impedance match might be improved and the transfer of power thereby increased. Movement of the short-circuiting plunger changes the electrical length of the stub and, in accordance with Eq. 12-47, the impedance as seen at a selected reference plane.

The junction itself is a discontinuity which is less symmetrical than those considered thus far. Because of its longitudinal extent when viewed from any of the three ports, a "junction point" or "junction plane" could be defined only on an arbitrary basis. The flange surfaces of a physical unit of this type may serve as reference input planes for the respective ports. Analyses of this unit on a dominant-mode equivalent circuit basis have been made.[16] As a means of consolidating data on reflection and transmission properties, the scattering matrix is convenient.

d. Scattering-Matrix Concept

A lumped-parameter network linking two transmission lines, such as that discussed in Sec. 3-3, provides an easily visualized example for the derivation of the scattering matrix.

(1) *Derivation of Scattering Matrix for Resistive Network.* The relationship between departing and arriving voltage waves at a junction was summarized in matrix form in Eq. 3-33, and the reflection-and-transmission-coefficients matrix was shown in expanded from in Eq. 3-34. The corresponding expanded form of the equations is the following:

$$v_{1D} = \rho_{11}v_{1I} + \rho_{12}v_{2I} \tag{3-33a}$$

$$v_{2D} = \rho_{21}v_{1I} + \rho_{22}v_{2I} \tag{3-33b}$$

In the situation considered there, the characteristic impedances of the lines differed, although both were purely resistive. In order to specify from Eqs. 3-33a and b what portion of the power which is incident on one port (p_{mI}) is transmitted out the other port (p_{mD}), we must know the values of those impedances, or at least the value of their ratio. For example, assuming that $v_{2I} = 0$:

$$p_{1I} = v_{1I}i_{1I}$$

$$= \frac{v_{1I}^2}{Z_{01}} \tag{12-98}$$

$$p_{2D} = v_{2D}i_{2D}$$

$$= \frac{v_{2D}^2}{Z_{02}} \tag{12-99}$$

From Eq. 3-33b,

$$v_{2D} = \rho_{21} v_{1I} \tag{12-100}$$

Equation 12-100 may be squared and divided by Z_{02}:

$$\frac{v_{2D}^2}{Z_{02}} = \rho_{21}^2 \frac{v_{1I}^2}{Z_{02}}$$

This may be factored into the following form by multiplying and dividing by Z_{01}:

$$\frac{v_{2D}^2}{Z_{02}} = \rho_{21}^2 \frac{Z_{01}}{Z_{02}} \cdot \frac{v_{1I}^2}{Z_{01}} \tag{12-101}$$

Thus, from Eqs. 12-98, 12-99, 12-101,

$$p_{2D} = \rho_{21}^2 \frac{Z_{01}}{Z_{02}} p_{1I} \tag{12-102}$$

Similarly, if $v_{1I} = 0$,

$$\frac{v_{1D}^2}{Z_{01}} = \rho_{12}^2 \frac{Z_{02}}{Z_{01}} \cdot \frac{v_{2I}^2}{Z_{02}} \tag{12-103}$$

The power-ratio property of Eqs. 12-101, 12-102, and 12-103 may be incorporated into Eqs. 3-33a and b if the voltage variables are *modified* or "$\sqrt{Z_0}$-normalized" as follows:

$$\frac{v_{1D}}{\sqrt{Z_{01}}} = \rho_{11} \frac{v_{1I}}{\sqrt{Z_{01}}} + \rho_{12} \sqrt{\frac{Z_{02}}{Z_{01}}} \frac{v_{2I}}{\sqrt{Z_{02}}} \tag{12-104}$$

$$\frac{v_{2D}}{\sqrt{Z_{02}}} = \rho_{21} \sqrt{\frac{Z_{01}}{Z_{02}}} \frac{v_{1I}}{\sqrt{Z_{01}}} + \rho_{22} \frac{v_{2I}}{\sqrt{Z_{02}}} \tag{12-105}$$

Let the scattering matrix for this all-resistive, two-port network be defined as follows:

$$[\mathscr{S}] = \begin{bmatrix} \rho_{11} & \rho_{12}\sqrt{\dfrac{Z_{02}}{Z_{01}}} \\ \rho_{21}\sqrt{\dfrac{Z_{01}}{Z_{02}}} & \rho_{22} \end{bmatrix} \tag{12-106}$$

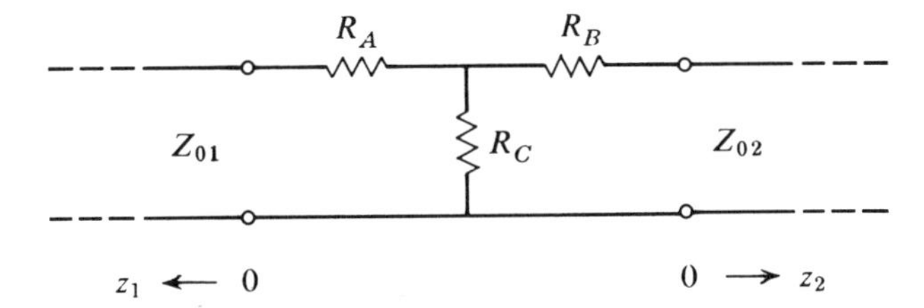

Figure 12-7. General resistive two-port network.

(The script capital $\mathscr{S}$ has been adopted here to avoid possible confusion with the differential surface vector $d\boldsymbol{S}$.)

Equations 12-104 and 12-105 may be restated in matrix form.

$$\left[\frac{v_D}{\sqrt{Z_0}}\right] = [\mathscr{S}]\left[\frac{v_I}{\sqrt{Z_0}}\right] \tag{12-107}$$

The reciprocity property may be considered next, after which the concept of $[\mathscr{S}]$ will be generalized to make it applicable to waveguide junctions.

(2) *Reciprocity Property of Transmission Coefficients.* A general resistive T network joining two transmission lines is shown in Fig. 12-7; any given passive resistive two-terminal-pair network could be reduced to this equivalent form by standard network transformations. Let R_{1T} be the equivalent terminating resistance viewed by the left-hand line:

$$R_{1T} = \frac{(Z_{02} + R_B)R_C}{Z_{02} + R_B + R_C} + R_A \tag{12-108}$$

The output voltage v_{2D} may be found in terms of the incident voltage v_{1I} by means of the reflection coefficient ρ_{11} and appropriate impedance ratios:

$$\rho_{11} = \frac{R_{1T} - Z_{01}}{R_{1T} + Z_{01}} \tag{12-109}$$

$$v_{2D} = v_{1I}(1 + \rho_{11})\frac{R_{1T} - R_A}{R_{1T}} \cdot \frac{Z_{02}}{Z_{02} + R_B} \tag{12-110}$$

This may be reduced to the following:

$$\rho_{21} = \frac{v_{2D}}{v_{1I}}$$

$$= \frac{2R_C Z_{02}}{Z_{02}(R_A + R_C) + Z_{01}(R_B + R_C) + R_A R_B + R_B R_C + R_C R_A + Z_{01} Z_{02}} \qquad (12\text{-}111)$$

The transmission coefficient for the opposite direction, ρ_{12}, may be found from Eq. 12-111 by interchanging Z_{01} and Z_{02}, and R_A and R_B. The denominators of the two coefficients prove to be identical; hence

$$\rho_{21}\sqrt{\frac{Z_{01}}{Z_{02}}} = \rho_{12}\sqrt{\frac{Z_{02}}{Z_{01}}} \qquad (12\text{-}112)$$

By comparing corresponding elements in the matrix in Eq. 12-106, one concludes that

$$\mathscr{S}_{21} = \mathscr{S}_{12} \qquad (12\text{-}113)$$

This analysis may be extended in piecemeal fashion to a network of n ports by examining each pair of ports for the corresponding transmission properties:

$$\mathscr{S}_{ij} = \mathscr{S}_{ji} \qquad (12\text{-}114)$$

where

$$\mathscr{S}_{ij} = \rho_{ij}\sqrt{\frac{Z_{0j}}{Z_{0i}}} \qquad (12\text{-}115)$$

(3) *Reference-Plane and Electrical-Length Effects.* If one is interested solely in the AC steady-state response, the foregoing may be extended to reactive networks simply by replacing the Rs with complex impedances, and using phasors (V_I, V_D) rather than instantaneous functions of time. The $\mathscr{S}$-matrix coefficients will, in general, then be complex:

$$\left[\frac{V_D}{\sqrt{Z_0}}\right] = [\mathscr{S}]\left[\frac{V_I}{\sqrt{Z_0}}\right] \qquad (12\text{-}116)$$

In every instance in which the term "lumped parameter" is used, it is inferred that the physical dimensions of that part of the circuit (the resistive network in Fig. 12-7, for example) are very small compared to wavelength, so that propagation time in the traveling-wave sense is negligible. The scattering matrix as set forth in Eq. 12-116 was in terms of modified voltages at the network terminals, but for steady-state AC conditions the coefficients may be restated for reference locations $\beta_1 z_1$ and $\beta_2 z_2$ (with coordinates as indicated in Fig. 12-7). At these locations the respective incident waves are advanced with respect to their phase at the terminals, whereas the departing waves are retarded. Let the matrix $[\mathscr{S}']$ be defined as follows:

$$\begin{bmatrix} \dfrac{V_{1D}}{\sqrt{Z_{01}}} \exp(-j\beta_1 z_1) \\ \dfrac{V_{2D}}{\sqrt{Z_{02}}} \exp(-j\beta_2 z_2) \end{bmatrix} = [\mathscr{S}'] \begin{bmatrix} \dfrac{V_{1I}}{\sqrt{Z_{01}}} \exp(j\beta_1 z_1) \\ \dfrac{V_{2I}}{\sqrt{Z_{02}}} \exp(j\beta_2 z_2) \end{bmatrix} \tag{12-117}$$

This will be consistent with Eq. 12-116 if $[\mathscr{S}']$ is related to $[\mathscr{S}]$ in the following manner:

$$[\mathscr{S}'] = \begin{bmatrix} \exp(-j2\beta_1 z_1)\mathscr{S}_{11} & \exp[-j(\beta_1 z_1 + \beta_2 z_2)]\mathscr{S}_{12} \\ \exp[-j(\beta_1 z_1 + \beta_2 z_2)]\mathscr{S}_{21} & \exp(-j2\beta_2 z_2)\mathscr{S}_{22} \end{bmatrix} \tag{12-118}$$

$$\text{or} \quad [\mathscr{S}'] = \begin{bmatrix} \exp(-j\beta_1 z_1) & 0 \\ 0 & \exp(-j\beta_2 z_2) \end{bmatrix} \times [\mathscr{S}] \begin{bmatrix} \exp(-j\beta_1 z_1) & 0 \\ 0 & \exp(-j\beta_2 z_2) \end{bmatrix} \tag{12-119}$$

(4) *Application to Waveguides.* In the uniform transverse-plane interface example which was discussed in Sec. 12-2b the reflection and transmission coefficients were found explicitly in Eqs. 12-39 and 12-40, and the characteristic impedances are given in Eqs. 12-44 and 12-48. Thus if one regards the transverse plane at $z = 0-$ as port 1, and the transverse plane at $z = 0+$ as port 2, the scattering matrix is as follows:

$$[\mathscr{S}] = \begin{bmatrix} \beta_{10} - \beta'_{10} & 2\sqrt{\beta_{10}\beta'_{10}} \\ 2\sqrt{\beta_{10}\beta'_{10}} & \beta'_{10} - \beta_{10} \end{bmatrix} \frac{1}{\beta_{10} + \beta'_{10}} \tag{12-120}$$

For more complicated discontinuities, and for intermode coupling, such as (1) between a waveguide and a coaxial line, (2) between two waveguides of different sizes, or (3) between two propagating modes in a single waveguide (see Sec. 13-2c on circular waveguides), the scattering-matrix concept may be extended by expressing the incident-wave or departing-wave power as the surface integral of Poynting's vector over a transverse plane. As in Eq. 12-23, the symbol Σ may be used in the subscript; thus

$$p_{1I\Sigma} = \int\int \mathbf{P}_{1I\,\mathrm{av}} \cdot d\mathbf{S}_1 \quad \text{etc.} \tag{12-121}$$

In many instances some constraints on possible values of the $\mathscr{S}$ coefficients may be discerned from geometrical reasoning in terms of symmetry.* For example, if ports 1 and 2 for the inductive iris in Fig. 12-5 are assumed to be at transverse planes equidistant from the iris, one may conclude that $\mathscr{S}_{11} = \mathscr{S}_{22}$.

For *lossless* units, the power leaving the junction must equal that entering it; hence the following constraint may be derived from Eqs. 12-98, 12-99, and 12-107:

$$\sum_{i=1}^{n} |\mathscr{S}_{ij}^2| = 1 \tag{12-122}$$

Experimental procedures for determining the numerical values of the scattering-matrix coefficients have been detailed in the literature.[7, 21]

12-5 VELOCITIES AND SIGNAL PROPAGATION

Velocity is an important measure of wave-propagation phenomena. For the TEM mode in a lossless medium, this was found to be equal to $1/\sqrt{\mu\epsilon}$ for all frequencies, but for waveguide modes (TE and TM) velocity is a more complicated function. More than one concept of "velocity of the fields" will prove useful. The simplest of these is the *phase velocity*, or apparent velocity of a given sinusoidal wave, and it proves to be dependent on the ratio between the wave frequency and the cutoff frequency.

As was mentioned in the introductory portion of Chap. 2, actual communication signals cover the frequency spectrum. Variation of phase

*In general the reciprocity property, Eq. 12-114, is applicable if the waveguide function is composed of isotropic media. Anisotropic media may cause *nonreciprocal* behavior, a characteristic that has been exploited in various devices.[5, 9, 10, 19]

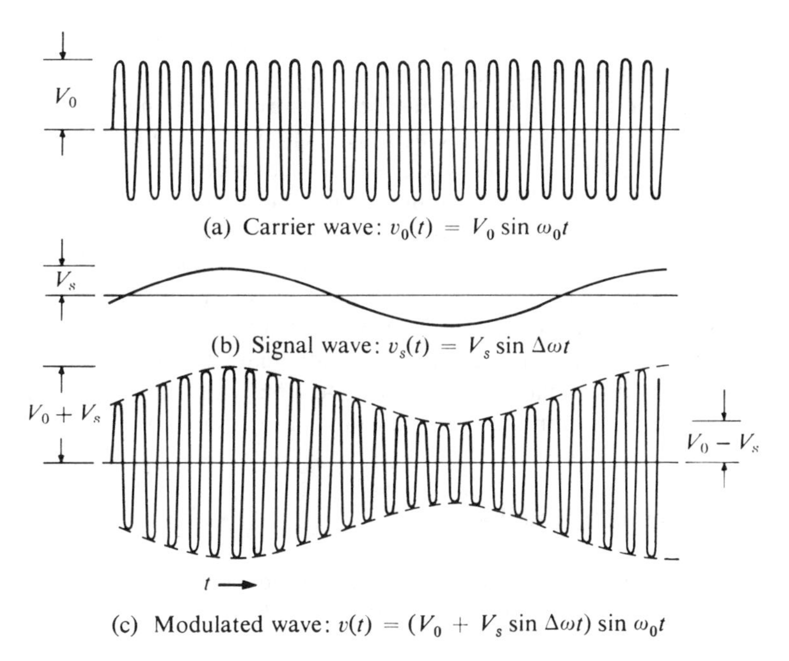

Figure 12-8. Amplitude modulation with sinusoidal signal.

velocity with frequency, which occurs in a waveguide even if the guide is lossless, causes *phase distortion*, or relative shifting of the sinusoidal components. (In the TEM mode in a two-conductor line, phase distortion does not occur if the line is lossless.) However, it was also mentioned that the intelligence put into a signal is in practice recoverable if distortion is kept small within some limited band of frequencies. With due regard for the restrictions thereby imposed, a waveguide can be used efficiently for signal transmission.

Modulation is a technique for adapting a signal so as to take best advantage of the characteristics of the transmission system. Several types of modulation have been exploited; *amplitude modulation*, illustrated in Fig. 12-8, is perhaps the simplest to analyze and will serve as the example in the discussion which follows. The signal contained in an amplitude-modulated wave may be sensed graphically by means of the *envelope*, which is a pair of curves faired so that one touches all the positive maxima of the modulated wave and the other touches all the negative maxima. This is indicated by the dashed lines in Fig. 12-8(c). Insight as to the "performance" of a transmission system may be obtained by studying particular envelopes for apparent speed of movement (*group velocity*) and for any progressive change in shape.

Two approaches to the "signal-transmitting performance" of a waveguide will be examined. An elementary analysis is possible if amplitude modulation by a uniform sinusoidal signal is assumed; in audio terminology this signal would be an uninterrupted tone of constant pitch and loudness. Such a signal would not convey any intelligence. In contrast to this, the step function and the isolated rectangular pulse embody much of the essential nature of practical communication, but mathematical analysis of the waveguide propagation of waves modulated by those signals is difficult.[8, 14, 15, 23] In keeping with the mathematical level of this text, a compromise will be made; the steady-state sinusoidal-signal case and the propagation of an impressed step-function wave front[6] will be derived.

a. Phase-Velocity Function

As has been noted in Chap. 4, the phase velocity of a sinusoidal traveling wave (steady state) is related to the phase function as follows:

$$\nu = \frac{\omega}{\beta} \tag{4-28}$$

The phase function for both TE and TM modes in a rectangular waveguide is given by Eq. 12-85. This may be rewritten with the subscripts m and n deleted:

$$\beta = \sqrt{\mu\epsilon\left(\omega^2 - \omega_c^2\right)} \tag{12-123}$$

Substituting Eq. 12-123 in Eq. 4-28 yields

$$\begin{aligned} \nu &= \frac{\omega}{\sqrt{\mu\epsilon\left(\omega^2 - \omega_c^2\right)}} \\ &= \frac{1}{\sqrt{\mu\epsilon\left[1 - (\omega_c/\omega)^2\right]}} \end{aligned} \tag{12-124}$$

At frequencies much greater than cutoff, ν approaches $1/\sqrt{\mu\epsilon}$, the same as for TEM waves, but as frequency is reduced toward the cutoff value, ν increases without limit. Thus ν may exceed $1/\sqrt{\mu_0\epsilon_0}$, the speed of light in vacuum. (If $\mu = \mu_0$ and $\epsilon = \epsilon_0$, ν is never less than the speed of light in vacuum.) This appears to contradict the principles of relativity, but an examination of the energy distribution and movement in a waveguide mode in comparison to that in the TEM mode will relieve the difficulty.

Let the *energy-density functions* w_ϵ and w_μ, introduced in connection with Poynting's vector in Sec. 10-1e, be used here:

$$w_\epsilon = \frac{\epsilon|\boldsymbol{E}|^2}{2} \tag{10-33}$$

$$w_\mu = \frac{\mu|\boldsymbol{H}|^2}{2} \tag{10-34}$$

Let

$$w = w_\epsilon + w_\mu \tag{12-125}$$

$$\boldsymbol{P} = \boldsymbol{E} \times \boldsymbol{H} \tag{10-38}$$

(1) *Energy Distribution and Poynting Vector for TEM Mode.* Equations 10-7 and 10-15 for the $\boldsymbol{E}$ and $\boldsymbol{H}$ fields in an unbounded medium may be adapted to the lossless case by setting α equal to zero and η equal to $\sqrt{\mu/\epsilon}$:

$$\boldsymbol{E} = E_{xM}\cos(\omega t - \beta z)\boldsymbol{a}_x \tag{12-126}$$

$$\boldsymbol{H} = E_{xM}\sqrt{\frac{\epsilon}{\mu}}\cos(\omega t - \beta z)\boldsymbol{a}_y \tag{12-127}$$

Substituting these into Eqs. 10-33 and 10-34 yields the following:

$$w = \frac{\epsilon}{2}E_{xM}^2\cos^2(\omega t - \beta z) + \frac{\mu}{2}E_{xM}^2\frac{\epsilon}{\mu}\cos^2(\omega t - \beta z)$$

$$= \epsilon E_{xM}^2\cos^2(\omega t - \beta z) \tag{12-128}$$

Similarly, substitution in Eq. 10-38 yields

$$\boldsymbol{P} = E_{xM}^2\sqrt{\frac{\epsilon}{\mu}}\cos^2(\omega t - \beta z)\boldsymbol{a}_z \tag{12-129}$$

The total energy density w is zero over the transverse planes corresponding to $\omega t - \beta z$ equal to an odd multiple of $(\pi/2)$, and the same is true of Poynting's vector. Furthermore Poynting's vector contains only a

z-directed component. To maintain the given energy distribution with respect to the $\boldsymbol{E}$ and $\boldsymbol{H}$ field patterns, the energy must move at the same speed as the fields do, or $1/\sqrt{\mu\epsilon}$.

(2) *Energy Distribution and Poynting Vector for the TE_{10} Mode.* The TE_{10} waveguide mode is of practical interest and is also the easiest to illustrate in sketches, so it will be used here. The second forms of Eqs. 12-8 and 12-9, with m set equal to unity, may be used for the waveguide fields:

$$\boldsymbol{E} = A\sin\left(\frac{\pi x}{a}\right)\cos(\omega t - \beta_{10}z)\boldsymbol{a}_y \tag{12-130}$$

$$\boldsymbol{H} = -\frac{A\beta_{10}}{\omega\mu}\sin\left(\frac{\pi x}{a}\right)\cos(\omega t - \beta_{10}z)\boldsymbol{a}_x - \frac{A\pi}{\omega\mu a}\cos\left(\frac{\pi x}{a}\right)\sin(\omega t - \beta_{10}z)\boldsymbol{a}_z \tag{12-131}$$

As before, the field expressions are substituted into Eqs. 10-33 and 10-34:

$$w = \frac{\epsilon}{2}A^2\sin^2\left(\frac{\pi x}{a}\right)\cos^2(\omega t - \beta_{10}z) + \frac{A^2\beta_{10}^2}{2\omega^2\mu}\sin^2\left(\frac{\pi x}{a}\right)\cos^2(\omega t - \beta_{10}z) + \frac{A^2\pi^2}{2\omega^2\mu a^2}\cos^2\left(\frac{\pi x}{a}\right)\sin^2(\omega t - \beta_{10}z) \tag{12-132}$$

From Eq. 12-6

$$\beta_{10}^2 = \omega^2\mu\epsilon - \left(\frac{\pi}{a}\right)^2 \tag{12-133}$$

The terms in Eq. 12-132 may be rearranged by substituting Eq. 12-133:

$$w = A^2\left[\epsilon - \frac{\pi^2}{2\omega^2\mu a^2}\right]\sin^2\left(\frac{\pi x}{a}\right)\cos^2(\omega t - \beta_{10}z) + \frac{A^2\pi^2}{2\omega^2\mu a^2}\cos^2\left(\frac{\pi x}{a}\right)\sin^2(\omega t - \beta_{10}z) \tag{12-134}$$

The function w is zero at the following locations:

$$\left.\begin{array}{l} x = 0 \\ x = a \end{array}\right\} \qquad \omega t - \beta_{10} z = k\pi \quad (k \text{ is any integer})$$

$$x = \frac{a}{2} \qquad \omega t - \beta_{10} z = \frac{(k+1)}{2}\pi \tag{12-135}$$

It has maxima at the following locations:

$$x = \frac{a}{2} \qquad \omega t - \beta_{10} z = k\pi \qquad (w_{M1})$$

$$\left.\begin{array}{l} x = 0 \\ x = a \end{array}\right\} \qquad \omega t - \beta_{10} z = \frac{(k+1)}{2}\pi \qquad (w_{M2}) \tag{12-136}$$

These are indicated in Fig. 12-9. If desired, the form of the scalar field w could be shown in greater detail by means of contours, but such lines tend to be confusing in a diagram in which magnetic-field lines have been drawn.

Poynting's vector is

$$\boldsymbol{P} = \frac{A^2\beta_{10}}{\omega\mu}\sin^2\left(\frac{\pi x}{a}\right)\cos^2(\omega t - \beta_{10} z)\boldsymbol{a}_z - \frac{A^2\pi}{\omega\mu a}\sin\left(\frac{\pi x}{a}\right)\cos\left(\frac{\pi x}{a}\right)\sin(\omega t - \beta_{10} z)\cos(\omega t - \beta_{10} z)\boldsymbol{a}_x \tag{12-137}$$

Arrows indicating the direction of Poynting's vector at a number of locations in the field are shown in Fig. 12-9. By comparing Poynting vector directions and the location of maximum energy densities at consecutive time intervals, it may be seen that the outward components of energy movement in the range $0 < (\omega t - \beta_{10} z) < \pi/2$ weaken the trailing portion of the energy "hump" in the center of the guide and strengthen the leading portions of the humps at the sidewalls, whereas the inward components of energy movement in the range $-\pi/2 < (\omega t - \beta_{10} z) < 0$ weaken the trailing portion of the humps at the sidewalls and strengthen the leading portion of the hump at the center. The result is apparent longitudinal movement at the phase velocity.

The limiting situation occurs when the frequency approaches the cutoff value; then β approaches zero and the phase velocity infinity. Under these

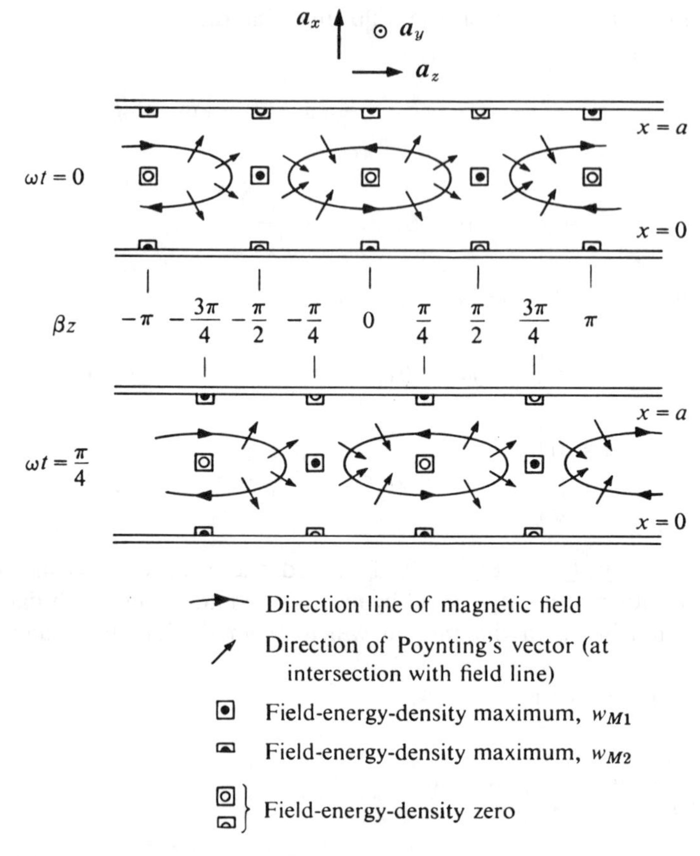

Figure 12-9. Locations of maxima and zeros of field-energy-density function relative to Poynting's vector field for TE_{10} mode (see Eqs. 12-130 through 12-137).

circumstances the energy movement is purely transverse and the field pattern changes simultaneously at all values of z (see Prob. 12-10).

b. Traveling Envelope of Sinusoidally Modulated Wave

An amplitude-modulated wave may be resolved mathematically into uniform sinusoidal components of differing frequencies which travel in a waveguide at differing phase velocities. If the frequency of the modulating signal is small compared to the carrier frequency, the envelope which characterizes the resultant wave at the sending end remains identifiable as the wave travels along the waveguide. The envelope appears to travel at a lower speed than do its sinusoidal components, and furthermore its shape is gradually but progressively changed with increase in the distance traveled.

The analysis which follows is applicable to any waveguide mode, and for convenience the modulated wave will be described in terms of an output voltage $v(z, t)$ (assuming a suitable termination-and-conversion device at the particular location in z). For TE modes in a rectangular waveguide, for example, $v(z, t)$ would be proportional to $|\boldsymbol{E}(x, y, z, t)|$.

(1) ***Equations for Modulated Wave.*** In accordance with Fig. 12-8, let $v_0(t)$ be the carrier wave and $v_s(t)$ be the signal by which it is amplitude-modulated:

$$v_0(t) = V_0 \sin \omega_0 t \tag{12-138}$$

$$v_s(t) = V_s \sin \Delta\omega\, t \tag{12-139}$$

Let

$$M = \frac{V_s}{V_0} \tag{12-140}$$

The modulated wave at $z = 0$ is

$$v(0, t) = V_0(1 + M \sin \Delta\omega\, t) \sin \omega_0 t$$

This may be changed by means of a trigonometric identity into the following:

$$v(0, t) = V_0\left[\sin \omega_0 t + \frac{M}{2}\cos(\omega_0 t - \Delta\omega\, t) - \frac{M}{2}\cos(\omega_0 t + \Delta\omega\, t)\right]$$

Each of the three frequency components will travel at its respective phase velocity, and the corresponding phase constants will be designated by β_0, β_-, and β_+. As a general function of z and t the traveling modulated wave is

$$v(z, t) = V_0\left[\sin(\omega_0 t - \beta_0 z) + \frac{M}{2}\cos(\omega_0 t - \Delta\omega\, t - \beta_- z) - \frac{M}{2}\cos(\omega_0 t + \Delta\omega\, t - \beta_+ z)\right] \tag{12-141}$$

(2) *Group Velocity.* If it is assumed that $\Delta\omega \ll \omega_0$, β_- and β_+ may be related to β_0 by Taylor's series and the principal effects of this variation brought into clear focus:

$$\beta_+ = \beta_0 + \Delta\omega\left(\frac{d\beta}{d\omega}\right)_{\omega_0} + \frac{(\Delta\omega)^2}{2}\left(\frac{d^2\beta}{d\omega^2}\right)_{\omega_0} + \cdots \qquad (12\text{-}142)$$

Let

$$\Delta\beta = \Delta\omega\left(\frac{d\beta}{d\omega}\right)_{\omega_0} \qquad (12\text{-}143)$$

and

$$\Delta'\beta = \frac{(\Delta\omega)^2}{2}\left(\frac{d^2\beta}{d\omega^2}\right)_{\omega_0} \qquad (12\text{-}144)$$

Then

$$\begin{aligned} \beta_+ &\approx \beta_0 + \Delta\beta + \Delta'\beta \\ \beta_- &\approx \beta_0 - \Delta\beta + \Delta'\beta \end{aligned} \qquad (12\text{-}145)$$

Substitution of Eq. 12-145 in Eq. 12-131 gives the following:

$$\begin{aligned} v(z,t) = V_0\Big[&\sin(\omega_0 t - \beta_0 z) + \frac{M}{2}\cos(\omega_0 t - \beta_0 z - \Delta'\beta\, z - \Delta\omega\, t + \Delta\beta z) \\ &- \frac{M}{2}\cos(\omega_0 t - \beta_0 z - \Delta'\beta\, z + \Delta\omega\, t - \Delta\beta z)\Big] \end{aligned}$$

The cosine terms may be combined to yield the following:

$$\begin{aligned} v(z,t) = V_0[&\sin(\omega_0 t - \beta_0 z) \\ &+ M\sin(\Delta\omega\, t - \Delta\beta z)\sin(\omega_0 t - \beta_0 z - \Delta'\beta z)] \end{aligned} \qquad (12\text{-}146)$$

For locations z for which $\Delta'\beta z \ll 1$ (see below), this may be simplified to

$$v(z,t) \approx V_0\left[1 + M\sin\Delta\omega\left(t - \frac{\Delta\beta}{\Delta\omega}z\right)\right]\sin(\omega_0 t - \beta_0 z) \qquad (12\text{-}147)$$

Let

$$\begin{aligned} \nu_g &= \left(\frac{\Delta\omega}{\Delta\beta}\right)_{\Delta\omega\to 0} \\ &= 1\Big/\left(\frac{d\beta}{d\omega}\right)_{\omega_0} \end{aligned} \qquad (12\text{-}148)$$

The bracketed terms in Eq. 12-147 describe the envelope of the traveling wave; it appears to move, without change of shape, at velocity ν_g. This is known as the *group velocity*. Differentiation of Eq. 12-123 for β gives

$$\frac{d\beta}{d\omega} = \frac{\omega\mu\epsilon}{\sqrt{\mu\epsilon\left(\omega^2 - \omega_c^2\right)}} \tag{12-149}$$

Substitution of Eq. 12-149 in Eq. 12-148 yields

$$\nu_g = \frac{\sqrt{\mu\epsilon\left(\omega_0^2 - \omega_c^2\right)}}{\omega_0\mu\epsilon}$$

This may be simplified by substituting the phase velocity of the carrier frequency ω_0 as given by Eq. 12-124:

$$\nu_g = \frac{1}{\nu\mu\epsilon}$$

The TEM velocity ν_0 may also be substituted (see Eq. 10-24):

$$\nu_g = \frac{\nu_0^2}{\nu} \tag{12-150}$$

(3) *Envelope Distortion.* The effect of neglecting $\Delta'\beta\, z$ in Eq. 12-147 should be investigated. Equation 12-149 may be differentiated to obtain $d^2\beta/d\omega^2$, which is related to $\Delta'\beta\, z$ by Eq. 12-144:

$$\frac{d^2\beta}{d\omega^2} = \frac{\mu\epsilon}{\sqrt{\mu\epsilon\left(\omega^2 - \omega_c^2\right)}} - \frac{\omega^2\mu^2\epsilon^2}{\left[\mu\epsilon\left(\omega^2 - \omega_c^2\right)\right]^{3/2}}$$

$$= \frac{-\omega_c^2\mu^2\epsilon^2}{\left[\mu\epsilon\left(\omega^2 - \omega_c^2\right)\right]^{3/2}}$$

$$\Delta'\beta = -\frac{(\Delta\omega)^2}{2}\frac{\omega_c^2\sqrt{\mu\epsilon}}{\left(\omega_0^2 - \omega_c^2\right)^{3/2}} \tag{12-151}$$

For given values of ω_0 and $\Delta\omega$, $\Delta'\beta\, z$ will exceed any stated finite magnitude if z is made sufficiently large. The result is a change in envelope shape, which will be demonstrated in an approximate manner.

The last term of Eq. 12-146 may be expanded as follows:

$$\sin(\omega_0 t - \beta_0 z - \Delta'\beta z) = \sin(\omega_0 t - \beta_0 z)\cos \Delta'\beta z - \cos(\omega_0 t - \beta_0 z)\sin \Delta'\beta z$$

Equation 12-146 may be rewritten as follows:

$$v(z,t) = V_0\{[1 + M\sin(\Delta\omega\, t - \Delta\beta z)\cos \Delta'\beta z]\sin(\omega_0 t - \beta_0 z) - M\sin(\Delta\omega\, t - \Delta\beta z)\sin \Delta'\beta z\cos(\omega_0 t - \beta_0 z)\} \quad (12\text{-}152)$$

The sine and cosine functions of $\omega_0 t - \beta_0 z$ may be combined by use of the following identity:

$$a\sin\theta + b\cos\theta = \sqrt{a^2 + b^2}\,\sin\left[\theta + \tan^{-1}\left(\frac{b}{a}\right)\right] \quad (12\text{-}153)$$

Direct substitution of the terms of Eq. 12-152 into Eq. 12-153 would yield a bulky and unilluminating expression. If M is assumed to be small compared to unity, an approximate solution may be obtained:

$$\sqrt{a^2 + b^2} \approx a + \left(\frac{b^2}{2a}\right) \approx a \qquad b \ll a \quad (12\text{-}154)$$

$$\tan^{-1}\left(\frac{b}{a}\right) \approx \frac{b}{a} \qquad b \ll a \quad (12\text{-}155)$$

$$v(z,t) \approx V_0[1 + M\sin(\Delta\omega\, t - \Delta\beta\, z)\cos \Delta'\beta z]\sin(\omega_0 t - \beta_0 z - \phi) \quad (12\text{-}156)$$

$$\phi \approx M\sin(\Delta\omega\, t - \Delta\beta z)\sin \Delta'\beta\, z \quad (12\text{-}157)$$

Thus the envelope appears to travel at the group velocity v_g defined in Eq. 12-148, but the envelope changes shape meanwhile. The signal amplitude is $M\cos \Delta'\beta z$ rather than M, and at those values of z at which $\Delta'\beta z$ is an odd multiple of $\pi/2$, the envelope shows no modulation.

If higher-order terms had been carried in approximation 12-142, additional modes of variation would have been brought into the description of the moving envelope.

c. Step-Function Excitation by Current Sheet in Infinite Waveguide

To study the propagation of a wave front, suppose that a transverse-plane current sheet, described by the following equation, is suddenly impressed at $z = 0$ in a rectangular waveguide (Fig. 12-1) which extends infinitely far in both directions:

$$\boldsymbol{J}_S(t) = 2H_{00} \sin\left(\frac{\pi x}{a}\right)\boldsymbol{a}_y U(t) \tag{12-158}$$

This will tend to induce, in each semiinfinite section, a TE_{10} mode traveling away from the current sheet. In the derivation of the concept of surface current [Sec. 10-2d(1)] it was indicated that $\boldsymbol{J}_S$ supports a tangential discontinuity in the magnetic field equal to $\boldsymbol{J}_S \times \boldsymbol{a}_n$. By symmetry, in this instance, a magnetic field of magnitude $H_{00} \sin(\pi x/a)$ would be created on each side. The magnitude field on the $+\boldsymbol{a}_z$ side will be studied here; its component parallel to the sheet has the $+\boldsymbol{a}_x$ direction.

(1) *Derivation of Transient Field Functions.* The transform of H_x as a traveling-wave field would have the following form:

$$H_x(z, s) = \frac{H_{00}}{s} \sin\left(\frac{\pi x}{a}\right)\epsilon^{-\gamma(s)z} \tag{12-159}$$

The operational propagation function may be derived from Eq. 12-6, which may be restated as

$$\beta_{10}^2 = -\left(\frac{\pi}{a}\right)^2 - (j\omega)^2\mu\epsilon$$

In a lossless waveguide,

$$\begin{aligned}\gamma_{10}^2 &= -\beta_{10}^2 \\ &= \left(\frac{\pi}{a}\right)^2 + (j\omega)^2\mu\epsilon\end{aligned} \tag{12-160}$$

The operational function may be obtained by replacing $j\omega$ by s:

$$\begin{aligned}\gamma(s) &= \left[\left(\frac{\pi}{a}\right)^2 + \mu\epsilon s^2\right]^{1/2} \\ &= \sqrt{\mu\epsilon}\sqrt{s^2 + \frac{\pi^2}{a^2\mu\epsilon}}\end{aligned} \tag{12-161}$$

From Eq. 12-10, the radian cutoff frequency ω_{c10} is

$$\omega_{c10} = \frac{\pi}{a\sqrt{\mu\epsilon}} \tag{12-162}$$

Equation 12-162 may be substituted in Eq. 12-161 as an identity to simplify it:

$$\gamma(s) = \sqrt{\mu\epsilon}\sqrt{s^2 + \omega_{c10}^2} \tag{12-163}$$

Substitution of this in Eq. 12-159 yields

$$H_x(z, s) = \frac{H_{00}}{s}\sin\left(\frac{\pi x}{a}\right)\exp\left(-z\sqrt{\mu\epsilon}\sqrt{s^2 + \omega_{c10}^2}\right) \tag{12-164}$$

The following equation may also be substituted as an identity.

$$t_z = z\sqrt{\mu\epsilon} \tag{12-165}$$

Hence

$$H_x(z, s) = \frac{H_{00}}{s}\sin\left(\frac{\pi x}{a}\right)\exp\left(-t_z\sqrt{s^2 + \omega_{c10}^2}\right) \tag{12-166}$$

The transforms for the other components of the field set, H_z and E_y, may be found with the aid of Eqs. 12-8 and 12-9. In terms of phasors,

$$H_z(z) = \frac{\pi}{j\beta_{10}a}H_x(z)\cot\left(\frac{\pi x}{a}\right)$$

$$= \frac{\pi}{\gamma_{10}a}H_x(z)\cot\left(\frac{\pi x}{a}\right) \tag{12-167}$$

$$E_y(z) = -\frac{\omega\mu}{\beta_{10}}H_x(z)$$

$$= -\frac{j\omega\mu}{\gamma_{10}}H_x(z) \tag{12-168}$$

The corresponding transforms, after substitution of Eq. 12-163, are

$$H_z(z,s) = \frac{H_{00}}{sa\sqrt{\mu\epsilon}\sqrt{s^2+\omega_{c10}^2}}\cos\left(\frac{\pi x}{a}\right)\epsilon^{-\gamma(s)z}$$

$$= \frac{H_{00}\omega_{c10}}{s\sqrt{s^2+\omega_{c10}^2}}\cos\left(\frac{\pi x}{a}\right)\exp\left(-t_z\sqrt{s^2+\omega_{c10}^2}\right) \tag{12-169}$$

$$E_y(z,s) = \frac{-H_{00}}{\sqrt{s^2+\omega_{c10}^2}}\sqrt{\frac{\mu}{\epsilon}}\sin\left(\frac{\pi x}{a}\right)\exp\left(-t_z\sqrt{s^2+\omega_{c10}^2}\right) \tag{12-170}$$

The inverse transform of H_x may be found from the following tabulated pair[22]:

$$\mathscr{L}^{-1}\left[\exp(-bs) - \exp\left(-b\sqrt{s^2+a^2}\right)\right] = \frac{abJ_1\left(a\sqrt{t^2-b^2}\right)}{\sqrt{t^2-b^2}}U(t-b) \tag{12-171}$$

The following operational form will make it applicable to Eq. 12-166:

$$\mathscr{L}^{-1}[F(s)] = \int_0^t\{\mathscr{L}^{-1}[sF(s)]\}\,d\tau \tag{8-63}$$

$$H_x(z,t) = H_{00}\sin\left(\frac{\pi x}{a}\right)\left[1 - \int_{t_z}^{t}\frac{t_z\omega_{c10}J_1\left(\omega_{c10}\sqrt{\tau^2-t_z^2}\right)}{\sqrt{\tau^2-t_z^2}}\,d\tau\right]U(t-t_z) \tag{12-172}$$

The inverse transform for E_y is given directly in tables[22]:

$$\mathscr{L}^{-1}\left[\frac{\exp\left(-b\sqrt{s^2+a^2}\right)}{\sqrt{s^2+a^2}}\right] = J_0\left(a\sqrt{t^2-b^2}\right)U(t-b) \tag{12-173}$$

Hence

$$E_y(z,t) = -H_{00}\sqrt{\frac{\mu}{\epsilon}}\,\sin\left(\frac{\pi x}{a}\right)J_0\left(\omega_{c10}\sqrt{t^2 - t_z^2}\right)U(t - t_z)$$

(12-174)

Equation 8-67 may be applied to Eq. 12-173 to find the inverse transform for H_z:

$$H_z(z,t) = H_{00}\omega_{c10}\cos\left(\frac{\pi x}{a}\right)\left[\int_{t_z}^{t} J_0\left(\omega_{c10}\sqrt{\tau^2 - t_z^2}\right)d\tau\right]U(t - t_z)$$

(12-175)

(2) *Analysis of Wave Front.* Equations 12-172, 12-174, and 12-175 all indicate a delay of t_z before the fields become nonzero. In view of the definition of that quantity (Eq. 12-165), this corresponds to wave-front propagation at the velocity of the TEM mode:

$$\nu_{\text{TEM}} = \frac{1}{\sqrt{\mu\epsilon}} \qquad (10\text{-}24)$$

The magnitudes of the various components at the wave front and their behavior for t slightly greater than t_z may be found by replacing each Bessel function with the first term or first two terms of its series expansion.

$$J_1(p) \approx \frac{p}{2} \qquad p \ll 1 \qquad (12\text{-}176)$$

$$H_x(z,t) \approx H_{00}\sin\left(\frac{\pi x}{a}\right)\left[1 - \int_{t_z}^{t}\frac{t_z\omega_{c10}^2}{2}\,d\tau\right]U(t - t_z)$$

$$\approx H_{00}\sin\left(\frac{\pi x}{a}\right)\left[1 - \frac{(t - t_z)t_z\omega_{c10}^2}{2}\right]U(t - t_z) \qquad (12\text{-}177)$$

$$J_0(p) \approx 1 - \left(\frac{p}{2}\right)^2 \qquad p \ll 1 \qquad (12\text{-}178)$$

$$E_y(z,t) \approx -H_{00}\sqrt{\frac{\mu}{\epsilon}}\,\sin\left(\frac{\pi x}{a}\right)\left[1 - \frac{\omega_{c10}^2(t^2 - t_z^2)}{4}\right]U(t - t_z) \qquad (12\text{-}179)$$

Only the first term in the brackets in Eq. 12-179 will be integrated for H_z:

$$H_z(z,t) \approx H_{00}\omega_{c10}\cos\left(\frac{\pi x}{a}\right)(t - t_z)U(t - t_z) \qquad (12\text{-}180)$$

Thus the transverse components E_x and H_y are, at the wave front, in the proportion of $\sqrt{\mu/\epsilon}$, the intrinsic impedance of the medium, and they do not diminish with the distance traveled. Immediately after the wave front passes a given location, the abruptly established fields begin to decrease.

(3) *Asymptotic Behavior at Large Values of Time.* At large values of time the first term of the asymptotic expansion for J_0 may be substituted in Eq. 12-174 for E_y:

$$J_0(p) \approx \sqrt{\frac{2}{\pi p}}\cos\left(p - \frac{\pi}{4}\right) \qquad p \gg 1 \qquad (12\text{-}181)$$

$$E_y(z,t) \approx -H_{00}\sin\left(\frac{\pi x}{a}\right)\sqrt{\frac{2\mu}{\pi\omega_{c10}\sqrt{t^2 - t_z^2}\epsilon}}\cos\left(\omega_{c10}\sqrt{t^2 - t_z^2} - \frac{\pi}{4}\right) \qquad (12\text{-}182)$$

$$\sqrt{t^2 - t_z^2} \approx t\left(1 - \frac{t_z^2}{2t^2}\right)$$

$$\approx t - \frac{t_z^2}{2t} \qquad t \gg t_z \qquad (12\text{-}183)$$

These results indicate a gradually diminishing oscillation at the wave-guide cutoff frequency. Application of the asymptotic expansions to H_x and H_z is less straightforward, but the final-value theorem may be applied readily:

$$\lim_{t\to\infty}[f(t)] = \lim_{s\to 0}[sF(s)] \qquad (5\text{-}86)$$

$$H_x(z,\infty) \to \left[H_{00}\sin\left(\frac{\pi x}{a}\right)\exp\left(-t_z\sqrt{s^2 + \omega_{c10}^2}\right)\right]_{s\to 0}$$

$$\to H_{00}\sin\left(\frac{\pi x}{a}\right)\exp(-t_z\omega_{c10})$$

Equations 12-162 and 12-165 may be multiplied together to simplify the exponent:

$$t_z \omega_{c10} = \frac{\pi z}{a}$$

Hence
$$H_x(z, \infty) \to H_{00} \sin\left(\frac{\pi x}{a}\right) \epsilon^{-(\pi z/a)} \tag{12-184}$$

Similarly,
$$E_y(z, \infty) \to 0 \tag{12-185}$$

$$H_z(z, \infty) \to -H_{00} \cos\left(\frac{\pi x}{a}\right) \epsilon^{-(\pi z/a)} \tag{12-186}$$

The results indicated by Eqs. 12-184 and 12-186 could be achieved physically only if the waveguide surfaces had infinite conductivity, so that the magnetic field would remain confined to the dielectric interior.

12-6 ATTENUATION OF TE_{10} PROPAGATING MODE

Physical materials used for the conducting surfaces of waveguides have finite conductivities, but much effort has been directed toward fabrication techniques which will minimize losses.[1, 12] Power will be absorbed from the traveling fields and their magnitudes will thereby be attenuated with distance traveled. As indicated earlier, the usual metals for this purpose, copper, silver, or brass, have such high conductivities that the field configurations are essentially the same as if the conductivity were infinite, and this simplifies the solution greatly.

The procedure which will be followed is to compare the power loss for a short length Δz with the power transmitted. In the discussion of Poynting's vector it was shown (Eq. 10-40) that the time-averaged power diminishes with distance traveled in proportion to $\epsilon^{-2\alpha z}$. Hence

$$\frac{P_{\text{loss}} + P_{\text{tr}}}{P_{\text{tr}}} = \epsilon^{2\alpha \Delta z}$$

If $P_{\text{loss}} \ll P_{\text{tr}}$,

$$\frac{P_{\text{loss}}}{P_{\text{tr}}} + 1 \approx 1 + 2\alpha\,\Delta z$$

$$\alpha \approx \frac{P_{\text{loss}}}{2P_{\text{tr}}\,\Delta z} \tag{12-187}$$

The expressions for $\boldsymbol{E}$ and $\boldsymbol{H}$ given in Eqs. 12-8 and 12-9 will be the bases for substitutions in the equations which follow. The time-averaged transmitted power of this wave set was found to be

$$P_{\text{tr}} = \frac{A_1^2\beta_{10}ab}{4\omega\mu} \tag{12-23}$$

Loss per unit of waveguide length may be computed by means of Poynting's vector normal to each conducting surface, as in Sec. 10-4. Equation 10-38 reduces to the following:

$$P_{\text{nor}} = E_{\tan}H_{\tan} \tag{12-188}$$

The expression for the $\boldsymbol{E}$ field in the ideally conducting waveguide, Eq. 12-8, contains no tangential component at the conducting surface; for the surface of high, yet finite, conductivity it may be approximated by the following relationship to $H_{\tan}$:

$$E_{\tan} = \text{Re}\left[H_{0\tan}\eta_M\epsilon^{j\omega t}\right]$$

$$H_{\tan} = \text{Re}\left[H_{0\tan}\epsilon^{j\omega t}\right] \tag{12-189}$$

where
$$\eta_M = \sqrt{\frac{\omega\mu}{\sigma}}\,\epsilon^{j\pi/4} \tag{10-55}$$

$$= R_S + jR_S$$

Here R_S is the surface resistance, Eq. 10-75. Poynting's vector may be written separately for the sidewalls and the horizontal surfaces.

Because of symmetry, Poynting's vector will have the same magnitude on both sidewalls; the vector for the wall for which $x = 0$ will be designated by $\boldsymbol{P}_W$. The tangential $\boldsymbol{E}$ field found from Eqs. 12-9 and 12-189, is

$$E_y = \frac{-A_1\pi}{\omega\mu a}|\eta_M|\sin(\omega t + 45° - \beta z) \tag{12-190}$$

Substitution of Eqs. 12-9 and 12-190 into Eq. 10-38 gives

$$\begin{aligned}
\boldsymbol{P}_W &= E_y H_z \boldsymbol{a}_x \\
&= -\left(\frac{A_1\pi}{\omega\mu a}\right)^2 |\eta_M|\sin(\omega t + 45° - \beta z)\sin(\omega t - \beta z)\boldsymbol{a}_z \\
&= -\left(\frac{A_1\pi}{\omega\mu a}\right)^2 |\eta_M|\{[\cos 45° \sin^2(\omega t - \beta z) \\
&\qquad + \sin 45° \sin(\omega t - \beta z)\cos(\omega t - \beta z)]\}\boldsymbol{a}_x \\
&= -\left(\frac{A_1\pi}{\omega\mu a}\right)^2 |\eta_M|\left\{\frac{\cos 45°}{2}[1 - \cos 2(\omega t - \beta z)]\right. \\
&\qquad \left. + \frac{\sin 45°}{2}\sin 2(\omega t - \beta z)\right\}\boldsymbol{a}_x
\end{aligned}$$

The time-averaged value $\boldsymbol{P}_{W\,\mathrm{av}}$ is

$$\begin{aligned}
\boldsymbol{P}_{W_{\mathrm{av}}} &= -\left(\frac{A_1\pi}{\omega\mu a}\right)^2 |\eta_M|\frac{\cos 45°}{2}\boldsymbol{a}_x \\
&= -\left(\frac{A_1\pi}{\omega\mu a}\right)^2 \frac{R_S}{2}\boldsymbol{a}_x
\end{aligned}$$

This should be integrated over a strip extending from 0 to b in the $\boldsymbol{a}_y$ direction and from z_1 to $z_1 + \Delta z$ in the $\boldsymbol{a}_z$ direction:

$$\begin{aligned}
P_{\mathrm{wall}} &= \int_0^b \int_{z_1}^{z_1+\Delta z_1} \boldsymbol{P}_{W\,\mathrm{av}} \cdot (-dy\,dz\,\boldsymbol{a}_x) \\
&= \left(\frac{A_1\pi}{\omega\mu a}\right)^2 \frac{R_S b\,\Delta z}{2}
\end{aligned} \tag{12-191}$$

For a point on the top or bottom surface, Poynting's vector ($\boldsymbol{P}_T$ or $\boldsymbol{P}_B$, respectively) is more complicated. The tangential component of $\boldsymbol{H}$ includes both H_x and H_z:

$$\boldsymbol{P}_T = (E_z H_x - E_x H_z)\boldsymbol{a}_y \qquad y = b$$

The expression $\boldsymbol{P}_B$ is the negative of that for $\boldsymbol{P}_T$ because they are oppositely directed. Let

$$H_x = \mathrm{Re}\left[H_{0x}\epsilon^{j\omega t}\right]$$

$$E_z = \mathrm{Re}\left[\eta_M H_{0x}\epsilon^{j\omega t}\right]$$

$$H_z = \mathrm{Re}\left[H_{0z}\epsilon^{j\omega t}\right]$$

$$E_x = \mathrm{Re}\left[-\eta_M H_{0z}\epsilon^{j\omega t}\right]$$

Terms from Eq. 12-9 may be substituted for H_{0x} and H_{0z}, and E_x and E_z written in forms corresponding to Eq. 12-190. The time-averaged value for $\boldsymbol{P}_T$ may be obtained by steps paralleling those for $\boldsymbol{P}_W$:

$$\boldsymbol{P}_{T\,\mathrm{av}} = \frac{1}{4}\left(\frac{A_1}{\omega\mu}\right)^2\left\{\beta^2\left[1 - \cos\left(\frac{2\pi x}{a}\right)\right] + \left(\frac{\pi}{a}\right)^2\left[1 + \cos\left(\frac{2\pi x}{a}\right)\right]\right\}R_S\boldsymbol{a}_y$$

This is to be integrated over a strip extending from 0 to a in the $\boldsymbol{a}_x$ direction and from z_1 to $z_1 + \Delta z$ in the $\boldsymbol{a}_z$ direction:

$$P_{\mathrm{top}} = \int_0^a \int_{z_1}^{z_1+\Delta z} \boldsymbol{P}_{T\,\mathrm{av}} \cdot (dx\,dz\,\boldsymbol{a}_y)$$

$$= \frac{1}{4}\left(\frac{A_1}{\omega\mu}\right)^2\left[\beta^2 a\,\Delta z + \left(\frac{\pi}{a}\right)^2 a\,\Delta z\right]R_S$$

Substitution of Eq. 12-6 reduces this to the following:

$$P_{\mathrm{top}} = \frac{1}{4}\left(\frac{A_1}{\omega\mu}\right)^2 \omega^2\mu\epsilon a\,\Delta z\,R_S$$

$$= \frac{A_1^2\epsilon R_S a\,\Delta z}{4\mu} \qquad (12\text{-}192)$$

The power loss on the bottom surface is equal to that on the top, and the losses in the two walls are equal. Hence the total loss in distance Δz along the waveguide is

$$P_{\text{loss}} = 2P_{\text{top}} + 2P_{\text{wall}}$$

$$= A_1^2 R_S \left[\frac{a\epsilon}{2\mu} + \frac{\pi^2 b}{\omega^2 \mu^2 a^2} \right] \Delta z \tag{12-193}$$

Equations 12-23 and 12-193 may be substituted in Eq. 12-187 to yield an expression for α:

$$\alpha = \left[\frac{a\epsilon\omega}{ab\beta} + \frac{\pi^2}{\omega\mu a^3 \beta} \right] R_S \tag{12-194}$$

By substitution of Eqs. 12-6, 12-10, and 10-23, this may be reduced to the following:

$$\alpha = \frac{R_S}{\eta\sqrt{1 - (f_c/f)^2}} \left[\frac{1}{b} + \frac{2}{a} \left(\frac{f_c}{f} \right)^2 \right] \tag{12-195}$$

For frequencies much greater than cutoff, Eq. 12-195 approaches the following asymptote:

$$\alpha \approx \frac{1}{b} \sqrt{\frac{\pi\epsilon f}{\sigma}} \qquad f \gg f_c \tag{12-196}$$

The value of f/f_c which yields the minimum attenuation for given values of a and b may be found by differentiation (see Prob. 12-12); the result is as follows:

$$\left(\frac{f}{f_c} \right)^2 = \frac{3}{2} \left(1 + \frac{2b}{a} \right) + \sqrt{\frac{9}{4} \left(1 + \frac{2b}{a} \right)^2 - \frac{2b}{a}} \tag{12-197}$$

Commercially standardized designs for rectangular waveguides fix dimension b at equal to or slightly less than $a/2$. For $b = a/2$, the ratio f/f_c for minimum attenuation becomes 2.41.

In the interest of preventing multiple moding, it is desirable to operate a waveguide at a frequency somewhat less than twice the cutoff value, even though this is appreciably less than the minimum-attenuation frequency just noted. Recommended frequency ranges given in current commercial literature are from about 1.25 to 1.90 times the cutoff frequency.[21]

12-7 CONCLUSIONS

Electromagnetic waves may be transmitted through hollow rectangular metallic waveguides in field configurations in which either the magnetic field or the electric field has a longitudinal component. Each distinct configuration is known as a *mode* and is characterized by a *cutoff frequency* below which it cannot exist as a traveling wave. The number of possible modes is infinite, but they may be classified as *transverse electric* (TE) or *transverse magnetic* (TM). In the former class the electric field is confined to planes perpendicular to the waveguide axis, but a longitudinal component of magnetic field is necessary. Conversely, in TM modes the magnetic field is confined to planes perpendicular to the waveguide axis, but a longitudinal component of electric field is present. For modes of both types the cutoff frequency is

$$f_{cmn} = \frac{1}{2\pi\sqrt{\mu\epsilon}}\sqrt{\left(\frac{m\pi}{a}\right)^2 + \left(\frac{n\pi}{b}\right)^2} \tag{12-82}$$

Here m and n are integers, and a and b are the longer and shorter transverse dimensions, respectively.

The *dominant* mode, the one with the lowest cutoff frequency, is the TE_{10} mode. In this mode the fields are independent of distance in the shorter transverse direction in the waveguide, although the $\boldsymbol{E}$ field is everywhere parallel to the shorter transverse dimension and is proportional to $\sin \pi x/a$. Here x is the coordinate corresponding to a. This is the usual mode for signal transmission.

Standing-wave field patterns may be formed by superposition of traveling-wave fields moving in opposite directions.

Traveling-wave or standing-wave fields in a waveguide are accompanied by surface-current-density fields in the walls, which are directed perpendicular to the tangential magnetic field. For the TE_{10} mode, no current flows across the longitudinal axis of either of the wider surfaces.

Modes of higher cutoff frequency than the excitation frequency may exist in the vicinity of waveguide discontinuities in *evanescent* or *nonprop-*

agating form. In such modes the electric and magnetic fields are in time quadrature with each other.

Absolute waveguide impedance may be defined as the ratio of the phasor magnitude of the transverse component of electric field to the phasor magnitude of the transverse component of magnetic field. *Normalized waveguide impedances* are related to the standing-wave plot of $|\boldsymbol{E}|$ of the dominant mode as a function of longitudinal distance in the same manner as normalized impedance and the voltage standing wave are related on a transmission line.

The apparent velocity of each sinusoidal component, or *phase velocity*, depends on frequency, and exceeds that for a TEM wave in the same medium as that enclosed in the waveguide. Amplitude-modulated-signal envelopes appear to travel at a lower speed, known as the *group velocity*, and also undergo progressive changes in shape.

A transverse current sheet, impressed as a step-function in time, will yield transient fields, the wave front of which propagates without attenuation at the TEM velocity for the medium. After passage of the wave front the fields oscillate with diminishing amplitude at the cutoff frequency of the respective mode.

PROBLEMS

12-1. Verify that the expression for the magnetic field of the TE_{m0} mode given in Eq. 12-9 satisfies Maxwell's equation in div B.

12-2. Solve Eq. 12-13 by separation of variables, and reduce the answer to Eq. 12-14.

12-3. Write the differential equation for the direction lines of the surface-current-density field on the top and bottom surfaces of a rectangular waveguide with a TE_{10} propagating mode in it. Separate the variables and solve. Determine what range of values of the arbitrary constant corresponds to lines which connect to the sidewalls. Sketch several direction lines for the case $\beta_{10}a/\pi = 1$.

12-4. Write Poynting's vector for the rectangular waveguide with a TE_{10} propagating mode and a short-circuiting end plate (Eqs. 12-27 and 12-28). Find the time-averaged value, if any.

12-5. Sketch the direction-line patterns for the $\boldsymbol{E}_{SC}$, $\boldsymbol{H}_{SC}$ and $\boldsymbol{J}_{DSC}$ fields (Eqs. 12-27, 12-28, and 12-29) at ωt equal to (a) zero, (b) 45°, (c) 90°, and (d) 135°.

12-6. See Fig. 12-10. The coupling unit is matched with zero reflection coefficient to the coaxial line and to the waveguide, but it has a transmission loss of 0.7 dB. Find the SWR in the coaxial line and in the waveguide. If the waveguide has a cross section of 2 cm by 1 cm, and the frequency is 10.0

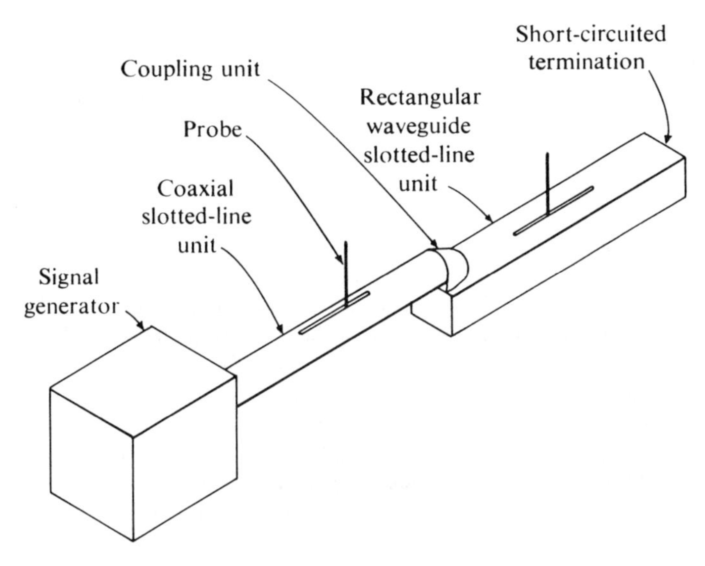

Figure 12-10. Tandem connection of rigid coaxial line and waveguide (see Prob. 12-6).

GHz, what is the longitudinal distance between ***E*** field minima in (a) the waveguide, and (b) the coaxial line?

12-7. Refer to Sec. 12-2b. Suppose that the waveguide has a cross section of 2 cm by 1 cm, that the medium for $z < 0$ is air, and that for $z > 0$ is polyethylene (relative permittivity 2.25; frequency 10.0 GHz). Find the waveguide characteristic impedances of the two sections, as indicated in Sec. 12-2c. Assuming that the polyethylene section has a reflectionless termination, find the normalized impedance in the air-filled section one eighth-wavelength from the interface. (Solution for the last part by the Smith chart is indicated in Fig. 12-3.)

12-8. Given the function $E(x, y, z, t) = E_x \boldsymbol{a}_x + E_y \boldsymbol{a}_y + E_z \boldsymbol{a}_z$. Find, by substituting in Eqs. A-25, A-26, and A-28, the Laplacian of the vector field ***E*** as given by Eq. 12-53. Verify, with the aid of Eq. A-29, the identity stated in Eq. 12-54.

12-9. A given rectangular waveguide has a cross section of 2 cm by 1.25 cm. Compute the cutoff frequencies for the following modes: TE_{10}, TE_{20}, TE_{01}, TE_{11}, and TM_{11}. If the waveguide is driven at a frequency of 10 GHz, find the corresponding βs or αs.

12-10. Examine the TE_{10} fields as given in Eqs. 12-130 through 12-137 for frequency (a) approaching the cutoff value, and (b) equal to the cutoff value.

12-11. Verify that $\nabla \cdot \boldsymbol{B} = 0$ for the magnetic field in the transient analysis of the TE_{10} mode, considering (a) the transform equations, 12-166 and 12-169, and (b) the time-domain solutions, Eqs. 12-172 and 12-175. *Hint*: Equation

5-78 for the derivative of an integral with respect to one of its limits may be helpful.

12-12. Verify Eq. 12-197 for the frequency of minimum attenuation of the TE_{10} mode.

REFERENCES

1. Allison, J. and Benson, F. A., Surface roughness and attenuation of precision drawn, chemically polished, electropolished, electroplated and electroformed waveguides, *Proc. IEE (London)*, 102, part B, 251, 1955.
2. Barrow, W. L., Transmission of electromagnetic waves in hollow tubes of metal, *Proc. IRE*, 24, 1298, October 1936.
3. Carson, J. R., Mead, S. P., and Schelkunoff, S. A., Hyperfrequency wave guides—Mathematical theory, *Bell Syst. Tech. J.*, 15, 310, April 1936.
4. Collin, R. E., *Field Theory of Guided Waves*, 2nd ed. IEEE Press, New York, 1991.
5. Collin, R. E., *Foundations for Microwave Engineering*, McGraw-Hill, New York, 1966.
6. Cotte, M., Electromagnetic waves—Propagation of a disturbance in a waveguide, *Comptes Rendues de l'Academie des Sciences (Paris)*, 221, 538, November 1945. (In French.)
7. Deschamps, G. A., Determination of the reflection coefficients and insertion loss of a waveguide junction, *J. Appl. Phys.*, 24, 1046, August 1953.
8. Elliott, R. S., Pulse waveform degradation due to dispersion in waveguides, *IRE Trans. Microwave Theory Tech.*, MTT-5, no. 4, 254, 1957.
9. Ghose, R. N., *Microwave Circuit Theory and Analysis*, McGraw-Hill, New York, 1963.
10. Grivet, P., *The Physics of Transmission Lines at High and Very High Frequencies*, vol. 2, *Microwave Circuits and Amplifiers*, Academic, New York, 1974.
11. Harrington, R. F., *Time-Harmonic Electromagnetic Fields*, McGraw-Hill, New York, 1961.
12. Harvey, A. F., Mechanical design and manufacture of microwave structures, *IRE Trans. Microwave Theory Tech.*, MTT-7, no. 4, 402, 1959.
13. Harvey, A. F., *Microwave Engineering*, Academic, New York, 1963.
14. Karbowiak, A. E., Propagation of transients in waveguides, *Proc. IEE (London)*, 104, part C, 339, 1957 (also Monograph No. 224R).
15. Knop, C. M., Pulsed electromagnetic wave propagation in dispersive media, *IEEE Trans. Antennas and Propagation*, AP-12, no. 4, 494, 1964.
16. Marcuvitz, N., *Waveguide Handbook*, McGraw-Hill, New York, 1951.
17. Obaid, A. A. S., Maclean, T. S. M., and Razaz, M., Propagation characteristics of rectangular corrugated waveguides, *Proc. IEE*, 132, part H, no. 7, 413, 1985.
18. Packard, K. S., The origins of waveguides: A case of multiple rediscovery, *IEEE Trans. Microwave Theory and Techniques*, MTT-32, no. 9, 961, 1984.

19. Ramo, S., Whinnery, J. R., and Van Duzer, T., *Fields and Waves in Communication Electronics*, Wiley, New York, 1965.
20. Rayleigh, Lord (J. W. Strutt), On the passage of electric waves through tubes, or the vibrations of dielectric cylinders, *Phil. Mag*, *Series 5*, 43, no. 261, 125, February 1897.
21. *Reference Data for Radio Engineers*, 5th ed., H. P. Westman, ed., Howard W. Sams & Co., New York, 1968, pp. 23-8, 24-1 to 24-9.
22. Roberts, G. E. and Kaufman, H., *Table for Laplace Transforms*, W. B. Saunders, Philadelphia, 1966, 250.
23. Schulz-DuBois, E. O., Sommerfeld pre- and postcursors in the context of waveguide transients, *IEEE Trans. Microwave Theory Techniques*, MTT-18, no. 8, 455, 1970.
24. Smythe, W. R., *Static and Dynamic Electricity*, 3rd ed., McGraw-Hill, New York, 1968, 506.
25. Southworth, G. C., Hyper-frequency wave guides—General considerations and experimental results, *Bell Syst. Tech. J.*, 15, 284, April 1936.
26. Southworth, G. C., Some fundamental experiments with wave guides, *Proc. IRE*, 25, 807, July 1937.

CHAPTER 13

Fields in Hollow Cylindrical Waveguides

Electromagnetic-wave propagation is possible within tubular cylindrical conductors in modes similar to those in hollow rectangular waveguides; several other types of guides of circular cross section will also propagate waves. Among the latter are closely wound helices, wires coated with dielectric material, and solid or hollow dielectric rods.[1] Detailed mathematical derivations will, for simplicity, be limited to the hollow circular metallic waveguide of high conductivity. These derivations will follow the same general approach as for the rectangular waveguide. The circular cylindrical coordinate system is the most suitable here because of the ease in fitting boundary conditions.

A coaxial line is geometrically similar to the circular waveguide, and its field components may be described with the same functions. It too will support propagating waveguide modes if the frequency is high enough. In the usual TEM applications, transmission in waveguide modes is parasitic; the cutoff frequency of the dominant waveguide mode thus limits the useful frequency range of a coaxial line of a given size.

Hollow metallic elliptical waveguides are used extensively; they have modes similar to those of circular waveguides. Because of the complexity of the wave functions in the elliptical cylindrical coordinate system, only an intuitive discussion will be presented.

13-1 SOLUTION OF MAXWELL'S EQUATIONS IN CYLINDRICAL COORDINATES

The circular waveguide will be assumed to be a perfectly conducting cylindrical tube with internal radius r_a.

a. Equation of Traveling-Wave Function

Derivation of the scalar wave equation for the propagating fields in a circular waveguide may begin with the vector wave equation, Eq. 12-52:

$$\nabla \times (\nabla \times \boldsymbol{E}) = -\mu\epsilon \frac{\partial^2 \boldsymbol{E}}{\partial t^2} \tag{12-52}$$

The following identity used in Chap. 12 is useful here:

$$-\nabla \times (\nabla \times \boldsymbol{E}) + \nabla(\nabla \cdot \boldsymbol{E}) = \nabla^2 \boldsymbol{E} \tag{12-53}$$

If one inserts the cylindrical-coordinate expressions for curl (A-37), divergence (A-36), gradient (A-38), and Laplacian of a scalar field (A-39) into Eq. 12-53 and separates the coefficients of $\boldsymbol{a}_z$, the following immediately useful result is obtained:

$$-[\nabla \times (\nabla \times \boldsymbol{E})]_z + [\nabla(\nabla \cdot \boldsymbol{E})]_z = \nabla^2 \boldsymbol{E}_z \tag{13-1}$$

The r component of Eq. 12-53 is a function of both $\boldsymbol{E}_r$ and $\boldsymbol{E}_\phi$, and the same is true of the ϕ component. In other words, the variables do not separate for those components. (Should $\boldsymbol{E}$ not be a function of ϕ, the variables will separate, but the remaining terms will not reduce to $\nabla^2 \boldsymbol{E}_r$ or $\nabla^2 \boldsymbol{E}_\phi$; see Prob. 13-1.)

The procedure here will be to find the functional form of $\boldsymbol{E}_z$ and to derive, from Maxwell's equations, supplementary equations which will enable one to find the other field components from $\boldsymbol{E}_z$.

The region within the waveguide is assumed to be free of charge; hence $\nabla \cdot \boldsymbol{E} = 0$. Substituting Eq. 13-1 in Eq. 12-52 gives

$$\nabla^2 \boldsymbol{E}_z = \mu\epsilon \frac{\partial^2 \boldsymbol{E}_z}{\partial t^2} \tag{13-2}$$

$\boldsymbol{E}_z$ by itself is a scalar field, and Eq. A-39, giving the Laplacian in terms of cylindrical coordinates, may be substituted:

$$\frac{1}{r}\frac{\partial}{\partial r}\left(r \frac{\partial \boldsymbol{E}_z}{\partial r}\right) + \frac{1}{r^2}\frac{\partial^2 \boldsymbol{E}_z}{\partial \phi^2} + \frac{\partial^2 \boldsymbol{E}_z}{\partial z^2} = \mu\epsilon \frac{\partial^2 \boldsymbol{E}_z}{\partial t^2} \tag{13-3}$$

As in the preceding chapter, it is convenient to assume that $\boldsymbol{E}_z$ is a sinusoidal traveling-wave function:

$$\boldsymbol{E}_z = \text{Re}\left[\boldsymbol{E}_{0z}(r, \phi)\epsilon^{j(\omega t - \beta z)}\right] \tag{13-4}$$

Equation 13-4 may be substituted in Eq. 13-3, the Re operator deleted from all terms, and the factor $\epsilon^{j(\omega t - \beta z)}$ cancelled throughout:

$$\frac{1}{r}\frac{\partial}{\partial r}\left(r\frac{\partial \boldsymbol{E}_{0z}(r,\phi)}{\partial r}\right) + \frac{1}{r^2}\frac{\partial^2 \boldsymbol{E}_{0z}(r,\phi)}{\partial \phi^2} - \beta^2 \boldsymbol{E}_{0z}(r,\phi)$$
$$= -\mu\epsilon\omega^2 \boldsymbol{E}_{0z}(r,\phi) \tag{13-5}$$

b. Solution by Separation of Variables

This partial differential equation in $\boldsymbol{E}_{0z}$ may be solved by separation of variables with the product-function technique. Let

$$\boldsymbol{E}_{0z}(r,\phi) = R(r)\Phi(\phi) \tag{13-6}$$

The procedure is similar to that followed in Sec. 12-3. Equation 13-6 is substituted into Eq. 13-5, and the resulting equation is divided by $R(r)$ and $\Phi(\phi)$ and multiplied by r^2 to separate the variables:

$$\frac{r}{R(r)}\frac{d}{dr}\left(r\frac{dR(r)}{dr}\right) + \frac{1}{\Phi(\phi)}\frac{d^2\Phi(\phi)}{d\phi^2} = (\beta^2 - \mu\epsilon\omega^2)r^2 \tag{13-7}$$

The term immediately to the left of the equality sign is the only one which is a function of ϕ, and it is not a function of r. For the equation to be satisfied for independently chosen combinations of r and ϕ, this term must equal a constant. Let

$$\frac{1}{\Phi(\phi)}\frac{d^2\Phi(\phi)}{d\phi^2} = -m^2 \tag{13-8}$$

This may be reduced to

$$\frac{d^2\Phi(\phi)}{d\phi^2} + m^2\Phi(\phi) = 0 \tag{13-9}$$

(1) *Solution for Φ Function.* Equation 13-9 may be recognized as a homogeneous linear differential equation of the second order with constant coefficients, the solution to which is

$$\Phi(\phi) = C_1 \sin m\phi + C_2 \cos m\phi \tag{13-10}$$

Here C_1 and C_2 are arbitrary constants.

The possible values of the assumed constant m are limited by the boundary condition that the function $\boldsymbol{E}_{0z}$, and hence $\Phi(\phi)$, must recur at intervals of 2π in ϕ

$$\Phi(\phi + 2\pi) = \Phi(\phi) \tag{13-11}$$

This will be satisfied if the constant m is any integer.

(2) *Solution for R*(r) Function. Substitution of Eq. 13-8 in Eq. 13-7 leaves the ordinary differential equation

$$\frac{r}{R(r)}\frac{d}{dr}\left(r\frac{dR(r)}{dr}\right) - m^2 = (\beta^2 - \mu\epsilon\omega^2)r^2 \tag{13-12}$$

This is a form of Bessel's differential equation. It may be reduced to standard form by multiplying it by $R(r)/r^2$ and making the change of variable

$$p = r\sqrt{\mu\epsilon\omega^2 - \beta^2} \tag{13-13}$$

Then

$$\frac{d^2R(p)}{dp^2} + \frac{1}{p}\frac{dR(p)}{dp} + R(p)\left(1 - \frac{m^2}{p^2}\right) = 0 \tag{13-14}$$

Equation 13-14 corresponds to Eq. B-1, and the general solution to 13-14 parallels Eq. B-2:

$$R(p) = A_m J_m(p) + B_m Y_m(p) \tag{13-15}$$

As elaborated in Appendix B, the J_m and Y_m are known as Bessel functions of the first and second kinds. Functions J_0 and J_1 are of particular interest in the study of the circular waveguide and are plotted in Fig. 13-1. A_m and B_m are constants which are chosen in accordance with boundary conditions.

In order to state the Bessel functions concisely in terms of the variable r, the following parameter h will be used:

$$h = \sqrt{\mu\epsilon\omega^2 - \beta^2} \tag{13-16}$$

Thus

$$R(r) = A_m J_m(hr) + B_m Y_m(hr) \tag{13-17}$$

The function $Y_m(hr)$ will approach minus infinity as the argument approaches zero. Within the hollow waveguide the fields must be finite;

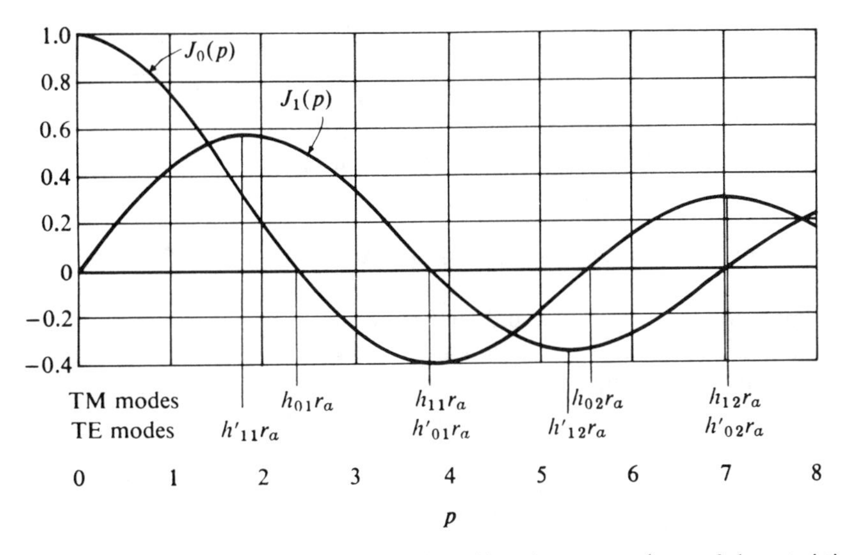

Figure 13-1. Bessel functions of the first kind with real arguments (roots of characteristic equations for waveguide modes are shown on abscissa; see Table 13-1).

hence this function is excluded from the solution:

$$B_m = 0 \tag{13-18}$$

c. Longitudinal-Component Functions

If the results obtained from Eqs. 13-10, 13-17, and 13-18 are substituted in Eq. 13-6, the following results:

$$\boldsymbol{E}_{0z}(r, \phi) = A_m J_m(hr)(C_1 \sin m\phi + C_2 \cos m\phi) \tag{13-19}$$

The solution to this point may be duplicated, step for step, in terms of the longitudinal component of a postulated $\boldsymbol{H}$ field:

$$\boldsymbol{H}_z = \text{Re}\left[\boldsymbol{H}_{0z}(r, \phi)\epsilon^{j(\omega t - \beta z)}\right] \tag{13-20}$$

$$\boldsymbol{H}_{0z}(r, \phi) = A'_m J_m(hr)(C'_1 \sin m\phi + C'_2 \cos m\phi) \tag{13-21}$$

A longitudinal $\boldsymbol{E}$ component, such as $\boldsymbol{E}_z$ in Eq. 13-4, is characteristic of transverse-magnetic modes, whereas a longitudinal $\boldsymbol{H}$ component (Eq. 13-20) is characteristic of transverse-electric modes. The values of the parameters h and β in both cases are fixed by the boundary conditions that the components of $\boldsymbol{E}$ which are parallel to the conducting guide wall and the component of $\boldsymbol{B}$, and hence $\boldsymbol{H}$, which is normal to the wall (the radial component of $\boldsymbol{H}$), shall vanish along that conducting surface. The mathematical application of these boundary conditions to the two types of modes differs, so they will be considered separately.

d. Transverse Components as Related to Longitudinal Components

Expressions for the radial and circumferential components of the fields in terms of the longitudinal components may be derived from Maxwell's equations involving the curl of $\boldsymbol{H}$, Eq. 12-51, and the curl of $\boldsymbol{E}$, Eq. 9-34. The r and ϕ components of those vector equations are

$$\frac{1}{r}\frac{\partial \boldsymbol{H}_z}{\partial \phi} - \frac{\partial \boldsymbol{H}_\phi}{\partial z} = \epsilon \frac{\partial \boldsymbol{E}_r}{\partial t} \tag{13-22}$$

$$\frac{\partial \boldsymbol{H}_r}{\partial z} - \frac{\partial \boldsymbol{H}_z}{\partial r} = \epsilon \frac{\partial \boldsymbol{E}_\phi}{\partial t} \tag{13-23}$$

$$\frac{1}{r}\frac{\partial \boldsymbol{E}_z}{\partial \phi} - \frac{\partial \boldsymbol{E}_\phi}{\partial z} = -\mu \frac{\partial \boldsymbol{H}_r}{\partial t} \tag{13-24}$$

$$\frac{\partial \boldsymbol{E}_r}{\partial z} - \frac{\partial \boldsymbol{E}_z}{\partial r} = -\mu \frac{\partial \boldsymbol{H}_\phi}{\partial t} \tag{13-25}$$

If Eq. 13-22 is differentiated with respect to z, and Eq. 13-25 is differentiated with respect to t and multiplied by ϵ, the derivative involving $\boldsymbol{E}_r$ may be eliminated from the pair of equations, and the following expression for $\boldsymbol{H}_\phi$ results:

$$\frac{\partial^2 \boldsymbol{H}_\phi}{\partial z^2} - \mu\epsilon \frac{\partial^2 \boldsymbol{H}_\phi}{\partial t^2} = \frac{1}{r}\frac{\partial}{\partial z}\left(\frac{\partial \boldsymbol{H}_z}{\partial \phi}\right) - \epsilon \frac{\partial}{\partial t}\left(\frac{\partial \boldsymbol{E}_z}{\partial r}\right) \tag{13-26}$$

As with the rectangular waveguide analysis, this equation may be greatly simplified if the fields are assumed to be sinusoidal traveling-wave functions, in accordance with Eqs. 13-4 and 13-20. The remaining components will be assumed to have the same form, and the functions $\boldsymbol{E}_{0r}(r, \phi)$, $\boldsymbol{E}_{0\phi}(r, \phi)$, $\boldsymbol{H}_{0\phi}(r, \phi)$, and $\boldsymbol{H}_{0\phi}(r, \phi)$ will be introduced for that purpose.

Equation 13-26 may then be restated as

$$(\mu\epsilon\omega^2 - \beta^2)\boldsymbol{H}_{0\phi} = -\frac{j\beta}{r}\frac{\partial \boldsymbol{H}_{0z}}{\partial\phi} - j\omega\epsilon\frac{\partial \boldsymbol{E}_{0z}}{\partial r} \tag{13-27}$$

The left-hand side of this equation may be simplified by substituting Eq. 13-16:

$$h^2\boldsymbol{H}_{0\phi} = -\frac{j\beta}{r}\frac{\partial \boldsymbol{H}_{0z}}{\partial\phi} - j\omega\epsilon\frac{\partial \boldsymbol{E}_{0z}}{\partial r} \tag{13-28}$$

Corresponding equations for the other components are

$$h^2\boldsymbol{E}_{0r} = -\frac{j\omega\mu}{r}\frac{\partial \boldsymbol{H}_{0z}}{\partial\phi} - j\beta\frac{\partial \boldsymbol{E}_{0z}}{\partial r} \tag{13-29}$$

$$h^2\boldsymbol{E}_{0\phi} = -\frac{j\beta}{r}\frac{\partial \boldsymbol{E}_{0z}}{\partial\phi} + j\omega\mu\frac{\partial \boldsymbol{H}_{0z}}{\partial r} \tag{13-30}$$

$$h^2\boldsymbol{H}_{0r} = \frac{j\omega\epsilon}{r}\frac{\partial \boldsymbol{E}_{0z}}{\partial\phi} - j\beta\frac{\partial \boldsymbol{H}_{0z}}{\partial r} \tag{13-31}$$

13-2 PROPAGATION MODES IN HOLLOW CIRCULAR WAVEGUIDE

The remaining boundary conditions for the hollow waveguide apply at the surface $r = r_a$. Here the tangential components of the electric field, $\boldsymbol{E}_z$, and $\boldsymbol{E}_\phi$, and the normal component of the magnetic field, $\boldsymbol{H}_r$, all vanish.

a. TM modes

For TM modes, $\boldsymbol{E}_z$ is described by Eqs. 13-4 and 13-19, and $\boldsymbol{H}_z$ is zero. If the latter condition is substituted in Eqs. 13-28 through 13-31, the following ratios may be obtained by dividing one equation by another:

$$\frac{\boldsymbol{E}_{0\phi}}{\boldsymbol{H}_{0r}} = \frac{-\beta}{\omega\epsilon} \tag{13-32}$$

$$\frac{\boldsymbol{E}_{0r}}{\boldsymbol{H}_{0\phi}} = \frac{\beta}{\omega\epsilon} \tag{13-33}$$

These results are consistent with the concept of waveguide characteristic impedance for a TM mode as given in Eqs. 12-41 and 12-90. The transverse components of $\boldsymbol{E}$ may be written in detail by substituting Eq. 13-19 in Eq. 13-27:

$$\boldsymbol{E}_{0\phi}(r,\phi) = -\frac{j\beta m A_m}{h^2 r} J_m(hr)(C_1 \cos m\phi - C_2 \sin m\phi) \qquad (13\text{-}34)$$

$$\boldsymbol{E}_{0r}(r,\phi) = -\frac{j\beta A_m}{h^2}\frac{dJ_m(hr)}{dr}(C_1 \sin m\phi + C_2 \cos m\phi) \qquad (13\text{-}35)^*$$

It may be noted that if $m = 0$, $\boldsymbol{E}_{0\phi}$ and $\boldsymbol{H}_{0r}$ vanish, and the remaining components are uniform circumferentially. Correspondingly, if $m \neq 0$, each of the field components has m longitudinal planes passing through the axis over which the function is zero. The two constants C_1 and C_2 serve to make the orientation of those null planes independent of the original choice of the plane for which $\phi = 0$.

(1) *Surface Boundary Condition and Cutoff Frequency.* The requirements that $\boldsymbol{E}_z$, $\boldsymbol{E}_\phi$, and $\boldsymbol{H}_r$ all vanish at the conducting surface ($r = r_a$) will be met simultaneously if the following is true:

$$J_m(hr_a) = 0 \qquad (13\text{-}37)$$

Values of the argument hr_a which will satisfy Eq. 13-37 are known as zeros or roots of the Bessel function, and have been tabulated extensively.[11] Some are presented in Table 13-1 and those within the scope of Fig. 13-1 have been marked there. They are transcendental numbers and are not related to one another by any simple arithmetic rule. For each function order m, an infinite number of zeros exist, and they are conventionally designated by counting them in sequence of location n from the origin. The designator $n = 1$ is assigned to the zero which is closest to, but not at, the origin. Field-descriptive equations may be made specific in this regard by attaching the subscripts m and n to h, β, and related quantities, and the same subscripts are used in the mode designator, TM_{mn}.

The minimum frequency which will yield a real rather than an imaginary value for β, in other words, the cutoff frequency, is determined, in

---

*The derivative of the zero-order Bessel function is as follows:

$$\frac{dJ_0(hr)}{dr} = -hJ_1(hr) \qquad (13\text{-}36)$$

TABLE 13-1. Zeros of Bessel Functions[a]

$J_m(h_{mn}r_a) = 0$ (TM Modes)			$\left[\frac{dJ_m(h'_{mn}r)}{dr}\right]_{r=r_a} = 0$ (TE Modes)		
m	n	$h_{mn}r_a$	m	n	$h'_{mn}r_a$
			1	1	1.841
0	1	2.405			
			2	1	3.054
1	1	3.832	0	1	3.832
			3	1	4.201
2	1	5.136			
			4	1	5.317
			1	2	5.331
0	2	5.520			
3	1	6.380			
			5	1	6.416
			2	2	6.706
1	2	7.016	0	2	7.016
			6	1	7.501
4	1	7.588			
			3	2	8.015
2	2	8.417			
			1	3	8.536
			7	1	8.578
0	3	8.654			

[a]Adapted from Marcuvitz.[11]

accordance with Eq. 13-16, by the parameter h. It in turn is fixed by the waveguide radius r_a and the Bessel function root of the mode selected:

$$\omega_{cmn} = \frac{h_{mn}}{\sqrt{\mu\epsilon}} \tag{13-38}$$

Equation 13-16 may be restated as

$$\beta_{mn} = \sqrt{\omega^2\mu\epsilon - h_{mn}^2} \tag{13-39}$$

$$\beta_{mn} = \sqrt{\mu\epsilon\left(\omega^2 - \omega_{cmn}^2\right)} \tag{13-40}$$

Equation 13-40 corresponds to Eq. 12-85 for the rectangular waveguide.

(2) *Equations of TM_{01} Mode.* Among the TM modes of the circular waveguide, the TM_{01} mode is the only one with important applications. It has been used in rotating joints as a means of transferring microwave signals between a fixed base and a rotating antenna. The circular symmetry of the field pattern is most helpful for that use. Equation 13-37 is satisfied for $m = 0$ and $n = 1$ by the argument value

$$J_0(2.405) = 0 \tag{13-41}$$

$$h_{01} r_a = 2.405 \tag{13-42}$$

Substitution of this value and $m = 0$ in Eqs. 13-19, 13-35, and 13-36 yields the following:

$$\boldsymbol{E}_{0z}(r) = A_0 J_0\left(2.405 \frac{r}{r_a}\right) \tag{13-43}$$

$$\boldsymbol{E}_{0r}(r) = \frac{j\beta_{01} A_0}{h_{01}} J_1\left(2.405 \frac{r}{r_a}\right) \tag{13-44}$$

$$\boldsymbol{H}_{0\phi}(r) = -\frac{j\omega\epsilon A_0}{h_{01}} J_1\left(2.405 \frac{r}{r_a}\right) \tag{13-45}$$

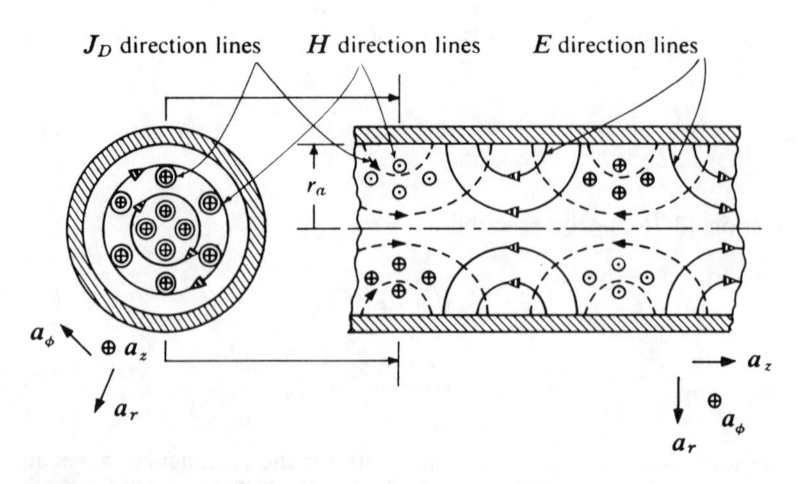

Figure 13-2. TM_{01}-mode field pattern in circular waveguide.

These fields are sketched in Fig. 13-2. The values of $\boldsymbol{E}_{0r}$ and $\boldsymbol{H}_{0\phi}$ at the waveguide surface may be found by substituting

$$J_1(2.405) = 0.5191 \tag{13-46}$$

Other aspects are dealt with in Prob. 13-3.

b. TE Modes

For TE modes, $\boldsymbol{H}_z$ is described by Eqs. 13-20 and 13-21, and $\boldsymbol{E}_z$ is zero. The following ratios may be obtained from Eq. 13-28 through 13-31:

$$\frac{\boldsymbol{E}_{0\phi}}{\boldsymbol{H}_{0r}} = -\frac{\omega\mu}{\beta} \tag{13-47}$$

$$\frac{\boldsymbol{E}_{0r}}{\boldsymbol{H}_{0\phi}} = \frac{\omega\mu}{\beta} \tag{13-48}$$

These correspond with Eq. 12-90 for the characteristic impedance of TE modes in a rectangular waveguide.

From Eqs. 13-29 and 13-21,

$$\boldsymbol{E}_{0\phi}(r,\phi) = \frac{j\omega\mu A'_m}{h^2}\frac{dJ_m(hr)}{dr}(C_1 \sin m\phi + C_2 \cos m\phi) \tag{13-49}$$

$$\boldsymbol{E}_{0r}(r,\phi) = -\frac{j\omega\mu m A'_m}{h^2 r}J_m(hr)(C_1 \cos m\phi - C_2 \sin m\phi) \tag{13-50}$$

(1) *Surface Boundary Condition and Cutoff Frequency.* The $\boldsymbol{E}_\phi$ and $\boldsymbol{H}_r$ components will vanish along the waveguide surface if the following is true:

$$\left[\frac{dJ_m(hr)}{dr}\right]_{r=r_a} = 0 \tag{13-51}$$

To distinguish the argument values which satisfy Eq. 13-51 from those for the zeros of $J_m(hr_a)$, the prime mark will be added as a superscript: h'_{mn}, β'_{mn}, etc. The corresponding mode designator is TE_{mn}. Some values of $h'r_a$ are listed in Table 13-1, and those within the scope of Fig. 13-1 have been marked there.

In accordance with Eq. 13-16, the cutoff frequency of a TE mode is

$$\omega'_{cmn} = \frac{h'_{mn}}{\sqrt{\mu\epsilon}} \tag{13-52}$$

$$\beta' = \sqrt{\mu\epsilon\left(\omega^2 - \omega_c'^2\right)} \tag{13-53}$$

(2) TE_{11} Mode — Application of Wavemeter. The TE_{11} mode has the lowest value of $h'_{mn}r_a$, and hence, from Eq. 13-52, the lowest cutoff frequency of any mode in a given circular waveguide. It is therefore the dominant mode. From Table 13-1,

$$h'_{11}r_a = 1.841 \tag{13-54}$$

The field equations of this mode include all five components and are essentially as complicated as the general forms (Eqs. 13-21, 13-49, and 13-50, supplemented by Eqs. 13-47 and 13-48); hence they will not be listed separately.

The field pattern is shown in Fig. 13-3. It is not circularly symmetrical, and there is no preferred orientation with respect to ϕ. Transmission in

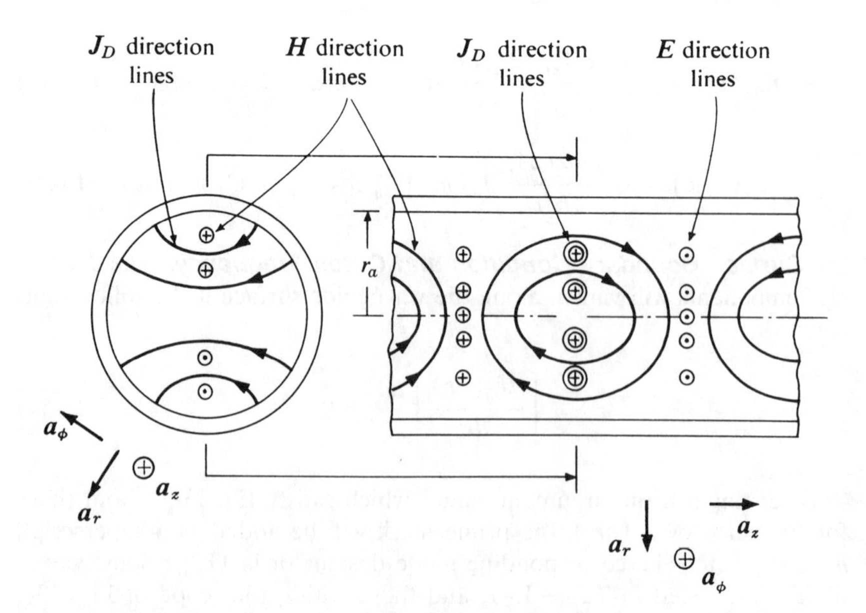

Figure 13-3. TE_{11}-mode field pattern in circular waveguide.

this mode is practical for straight, stiff sections of waveguide, such that a uniform circular cross section, with a straight longitudinal axis, prevails except at the terminals. There transition units convert between the circular TE_{11} mode and other transmission modes, such as coaxial TEM or rectangular TE_{10}. One important application is in vertical runs in towers for microwave antennas. In such assemblies it is practical to transmit, in one waveguide, two signals in TE_{11} modes which are polarized perpendicularly with respect to each other.

Standing waves in the TE_{11} mode form the basis of operation of one type of *resonant-cavity wavemeter*. This is basically a short section of circular waveguide with a short-circuiting plate on one end and an adjustable plunger, which effectively presents a second short-circuiting surface, within the waveguide, thereby forming the cavity. The system whose frequency is to be measured is coupled loosely to the cavity by means of a small aperture or some type of probe, and a detector system is also coupled loosely to the cavity by a separate aperture or probe.*

A lossless cavity, if shock energized, would oscillate naturally at those frequencies for which one or more half-wavelengths in the TE_{11} mode correspond to the longitudinal distance between the two short-circuiting surfaces. Let $\boldsymbol{H}_{SC}$ designate the magnetic field in the cavity, and let the fixed short-circuiting plate be at $z = 0$. According to boundary conditions,

$$\boldsymbol{H}_{zSC}(0) = 0 \tag{13-55}$$

Equation 13-20 for $\boldsymbol{H}_z$ in the traveling-wave mode may be modified by adding an oppositely moving traveling-wave field which will yield a resultant value of zero at the origin in z:

$$\begin{aligned} \boldsymbol{H}_{zSC} &= \mathrm{Re}\left(\boldsymbol{H}_{0z}(r,\phi)\left\{\exp\left[j(\omega t - \beta'_{11}z)\right] - \exp\left[j(\omega t + \beta'_{11}z)\right]\right\}\right) \\ &= -\mathrm{Re}\left[\boldsymbol{H}_{0z}(r,\phi)\, j2\sin(\beta'_{11}z)\epsilon^{j\omega t}\right] \end{aligned} \tag{13-56}$$

If the other short-circuiting surface is at $z = z_1$, the condition for resonance is

$$\beta'_{11}z_1 = n\pi \tag{13-57}$$

(Here n is any integer, although the usual design is on the assumption that $n = 1$.) Equations 13-52 and 13-53 may be substituted into Eq. 13-57 to

*The resonant-cavity form of wavemeter construction just described is also used for circular-waveguide designs which operate in the TE_{01} mode, and for coaxial units which operate in the TEM mode.[4, 14]

yield an expression for the resonant radian frequency ω_r:

$$\sqrt{\omega_r^2 \mu \epsilon - h_{11}'^2} = \frac{n\pi}{z_1}$$

Hence, after substituting Eq. 13-54,

$$\omega_r = \frac{1}{\sqrt{\mu \epsilon}} \sqrt{\left(\frac{n\pi}{z_1}\right)^2 - \left(\frac{1.841}{r_a}\right)^2} \tag{13-58}$$

When driven at a resonant frequency, a practical low-loss cavity will oscillate with fields of much larger amplitude, and will thereby transfer a much stronger signal to the detector, than when it is driven at some other frequency.

(3) *TE_{01} Mode.* The simplest circularly symmetrical TE mode, TE_{01}, was the subject of considerable development work as a means of transmitting microwave signals over distances of many miles. This is because of its peculiar attenuation function, one which drops continuously as frequency is raised. The resonant cavity configuration, operating in the TE_{01} mode, has been employed in wavemeters and in research equipment for measuring the properties of materials at microwave frequencies.

(a) *Field Pattern*. Equation 13-51 is satisfied for $m = 0$ and $n = 1$ by the argument value

$$J_1(3.8317) = 0$$
$$h_{01}' r_a = 3.8317 \tag{13-59}$$

Equation 13-21 may be rewritten

$$H_{0z} = A_{01}' J_0(h_{01}' r) \tag{13-60}$$

From Eqs. 13-29 through 13-31, and 13-36,

$$\boldsymbol{E}_{0\phi} = \frac{-j\omega\mu A_{01}'}{h_{01}'} J_1(h_{01}' r) \tag{13-61}$$

$$\boldsymbol{H}_{0r} = \frac{j\beta_{01}' A_{01}'}{h_{01}'} J_1(h_{01}' r) \tag{13-62}$$

These may be rewritten as explicit functions of time in anticipation of attenuation calculations:

$$\boldsymbol{H}_z = A'_{01} J_0(h'_{01} r)\cos(\omega t - \beta'_{01} z) \tag{13-63}$$

$$\boldsymbol{E}_\phi = \frac{\omega\mu A'_{01}}{h'_{01}} J_1(h'_{01} r)\sin(\omega t - \beta'_{01} z) \tag{13-64}$$

$$\boldsymbol{H}_r = \frac{-\beta'_{01} A'_{01}}{h'_{01}} J_1(h'_{01} r)\sin(\omega t - \beta'_{01} z) \tag{13-65}$$

The surface-current-density field is of particular interest in that it is purely circumferential. It is equal in magnitude to the magnetic field at radius r_a:

$$J_S = A'_{01} J_0(h'_{01} r_a)\cos(\omega t - \beta'_{01} z)\boldsymbol{a}_\phi \tag{13-66}$$

$$\begin{aligned} J_0(h'_{01} r_a) &= J_0(3.8317) \\ &= -0.4028 \end{aligned} \tag{13-67}$$

Thin transverse cuts may be made in the waveguide without interfering with the propagation of TE_{0n} modes, and this is one method of obstructing the propagation of modes outside this class.

The TE_{01} field pattern is illustrated in Fig. 13-4.

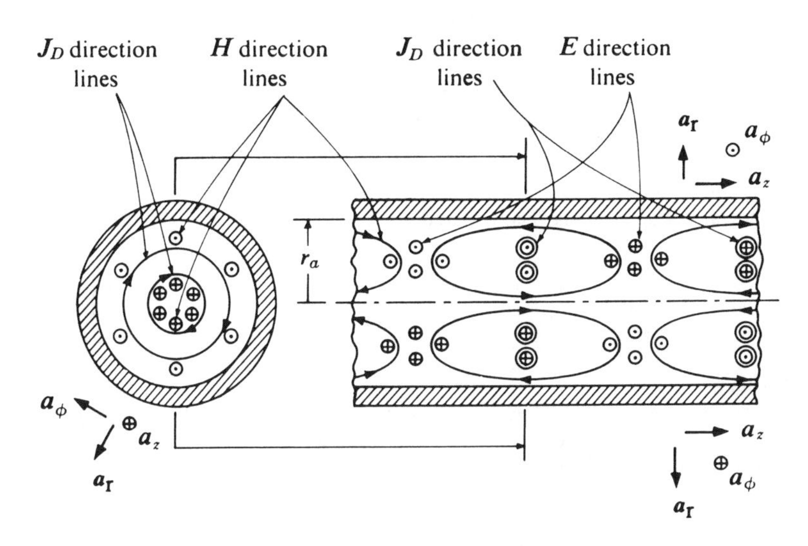

Figure 13-4. TE_{01}-mode field pattern in circular waveguide.

(b) *Attenuation Function*. The attenuation function of the TE_{01} mode may be derived on the same basis as for the rectangular waveguide in Sec. 12-6. For determination of the transmitted power, the longitudinal component of Poynting's vector may be written

$$P_{1z} = -\boldsymbol{E}_\phi \boldsymbol{H}_r$$
$$= \frac{\beta'_{01}\omega\mu A'^2_{01}}{h'^2_{01}} J_1^2(h'_{01}r)\sin^2(\omega t - \beta'_{01}z)$$
$$= \frac{\beta'_{01}\omega\mu A'^2_{01}}{2h'^2_{01}} J_1^2(h'_{01}r)[1 - \cos 2(\omega t - \beta'_{01}z)] \qquad (13\text{-}68)$$

When this function is time-averaged over an integral number of cycles it reduces to

$$P_{\text{lav}} = \frac{\beta'_{01}\omega\mu A'^2_{01}}{2h'^2_{01}} J_1^2(h'_{01}r)\,a_z \qquad (13\text{-}69)$$

The transmitted power P_{tr} may be found by integrating this over the transverse cross section of the waveguide:

$$P_{\text{tr}} = \int_0^{2\pi}\int_0^{r_a} P_{\text{lav}} \cdot (r\,dr\,d\phi\,a_z)$$
$$= \frac{\beta'_{01}\omega\mu A'^2_{01}\pi}{h'^2_{01}} \int_0^{r_a} J_1^2(h'_{01}r)\,r\,dr \qquad (13\text{-}70)^*$$
$$= \frac{\beta'_{01}\omega\mu A'^2_{01}\pi r_a^2 J_0^2(h'_{01}r_a)}{2h'^2_{01}} \qquad (13\text{-}71)$$

---

*The integral in Eq. 13-70 may be evaluated as follows:

$$\int_0^{r_a} J_1^2(h'_{01}r)\,r\,dr = \frac{r^2}{2}\left\{\left(\frac{dJ_1(h'_{01}r)}{dr}\,\frac{1}{h'_{01}}\right)^2 - \left(1 - \frac{1}{h'^2_{01}r^2}\right)J_1^2(h'_{01}r)\right\}\Bigg]_0^{r_a}$$

(McLachlan, 12, (p. 104.)

The second term vanishes at the upper limit in accordance with the boundary condition of Eq. 13-59, and the derivative of J_1 may be replaced by Eq. B-11, which also contains $J_1(h'_{01}r)$. Thus the integral reduces to the following:

$$\int_0^{r_a} J_1^2(h'_{01}r)\,r\,dr = \frac{r_a^2}{2} J_0^2(h'_{01}r_a)$$

The power dissipated in a short section Δz of the waveguide wall is found in the same general manner as for the rectangular waveguide. Thus by squaring Eq. 13-63, with $r = r_a$, and multiplying by R_S, the time-averaged value of Poynting's vector directed into the guide wall is found to be

$$\boldsymbol{P}_{W\,\mathrm{av}} = \tfrac{1}{2} A_{01}'^2 J_0^2(h_{01}' r_a) R_S \boldsymbol{a}_r \tag{13-72}$$

This should be integrated over a cylindrical strip of length Δz:

$$\begin{aligned} P_{\mathrm{loss}} &= \int_0^{2\pi} \int_{z_1}^{z_1+\Delta z} \boldsymbol{P}_{W\,\mathrm{av}} \cdot (r_a \, d\phi \, dz \, \boldsymbol{a}_r) \\ &= \pi r_a A_{01}'^2 J_0^2(h_{01}' r_a) R_S \, \Delta z \end{aligned} \tag{13-73}$$

Substitution of Eqs. 13-71 and 13-73 into Eq. 12-187 gives the following expression for the attenuation function:

$$\alpha = \frac{R_S h_{01}'^2}{\omega \mu \beta_{01}' r_a} \tag{13-74}$$

Equation 13-52 may be substituted for h_{01}' in the numerator and Eq. 13-53 for β_{01}' in the denominator:

$$\begin{aligned} \alpha &= \frac{R_S \omega_{c01}'^2}{\omega \sqrt{\omega^2 - \omega_{c01}'^2}} \sqrt{\frac{\epsilon}{\mu}} \\ &= R_S \left(\frac{\omega_{c01}'}{\omega} \right)^2 \sqrt{\frac{\epsilon}{\mu}} \frac{1}{\sqrt{1 - (\omega_{c01}'/\omega)^2}} \end{aligned} \tag{13-75}$$

$$R_S = \sqrt{\frac{\omega \mu}{2\sigma}} \tag{10-75}$$

At frequencies much greater than cutoff, the second radical in Eq. 13-75 approaches unity, and hence the attenuation function approaches the following asymptote, after substitution of Eq. 10-75:

$$\alpha \approx \sqrt{\frac{\epsilon}{2\sigma}} \frac{\omega_{c01}'^2}{\omega^{3/2}} = \sqrt{\frac{\pi \epsilon}{\sigma}} \frac{f_{c01}'^2}{f^{3/2}} \tag{13-76}$$

Thus the attenuation drops continuously as the frequency is raised above the cutoff value.

c. Low-Attenuation Transmission: Mode Conversion

Equation 13-76 suggests that if the TE_{01} mode were utilized at a carrier frequency many times the cutoff frequency, an exceeding low attenuation could be reached.

Realization of a practical communication system of this type was the object of extensive development work during past years.[6–8, 13] A major problem is that of *mode conversion*, or the transfer of part of the signal energy from the intended TE_{01} mode into other modes, followed by partial transfer back to the TE_{01} mode at near or distant locations. The TE_{11}, TE_{21}, and TM_{01} modes have lower cutoff frequencies, and the TM_{11} mode has the same cutoff frequency as the TE_{01} mode; they, and others, depending on how high the signal frequency was, would be *propagating* parasitic modes rather than evanescent ones.

Fiber optical transmission has since proved to be a much more adaptable technique for long distance applications than waveguide transmission. A useful by-product of the waveguide research work was the development of helical waveguides*; this structure has found application in resonant cavities.[17]

13-3 PARASITIC PROPAGATING MODES IN COAXIAL CABLE[3, 11, 14, 16]

A coaxial cable has the properties of a waveguide in addition to those of a TEM transmission line. The basic traveling-wave solution for the hollow waveguide (Sec. 13-1) is applicable here, but the boundary conditions differ in that the fields do not extend without interruption to the center. Rather E_ϕ and H_r must vanish at two conductor surfaces, at radii r_a and r_b in Fig. 11-1.

TEM is the intended mode of operation, and one wishes to avoid the presence of any propagating parasitic modes; hence the dominant waveguide mode is the one of primary interest in this analysis. That proves to be the TE_{11} mode; accordingly the boundary conditions will be stated explicitly only for the TE modes.

---

*A practical helical waveguide consists of a narrow strip or wire of conducting material formed into a closely spaced circular helix. To obtain mechanical support it is surrounded by a layer of insulating or poorly conducting material. Longitudinal conduction current is essential to modes outside the TE_{0n} class, and this will be thwarted by insulation between adjacent turns of the helix. The TE_{0n} mode configurations are approached very closely if the pitch, or ratio of center-to-center distance between successive turns compared to the perimeter, is very small, and if the metal has a high conductivity.[8, 15, 18, 19]

Beginning with Eq. 13-17 for the radial function $R(r)$, Eq. 13-18 is not applicable here, and the expression for $E_{0\phi}(r, \phi)$ given in Eq. 13-49 would be replaced by

$$E_{0\phi}(r, \phi) = \frac{j\omega\mu}{h^2}\left[A'_m \frac{dJ_m(hr)}{dr} + B'_m \frac{dY_m(hr)}{dr}\right] \times (C_1 \sin m\phi + C_2 \cos m\phi) \tag{13-77}$$

The constraints corresponding to Eq. 13-51 would be

$$\left[A'_m \frac{dJ_m(hr)}{dr} + B'_m \frac{dY_m(hr)}{dr}\right]_{r=r_a} = 0 \tag{13-78}$$

$$\left[A'_m \frac{dJ_m(hr)}{dr} + B'_m \frac{dY_m(hr)}{dr}\right]_{r=r_b} = 0 \tag{13-79}$$

If Eqs. 13-78 and 13-79 are each divided by their respective second terms, one equation may be subtracted from the other, the constants A'_m and B'_m cancelled, and the following then obtained:

$$\left[\frac{dJ_m(hr)}{dr}\right]_{r=r_a}\left[\frac{dY_m(hr)}{dr}\right]_{r=r_b} - \left[\frac{dJ_m(hr)}{dr}\right]_{r=r_b}\left[\frac{dY_m(hr)}{dr}\right]_{r=r_a} = 0 \tag{13-80}$$

Numerical solutions for this transcendental equation have been tabulated for various values of the ratio r_b/r_a. For the TE_{11} mode, over the range in r_b/r_a from unity to 10.0,

$$(r_a + r_b)h_{11} \approx 2 \tag{13-81}$$

Substitution in Eq. 13-16 ($\beta = 0$) yields

$$\omega_c = \frac{2}{(r_a + r_b)\sqrt{\mu\epsilon}} \qquad (TE_{11})$$

or

$$f_c = \frac{1}{\pi(r_a + r_b)\sqrt{\mu\epsilon}} \tag{13-82}$$

The magnetic and displacement-current-density fields, for a transverse plane in which $\boldsymbol{H}$ is purely longitudinal, are sketched in Fig. 13-5.

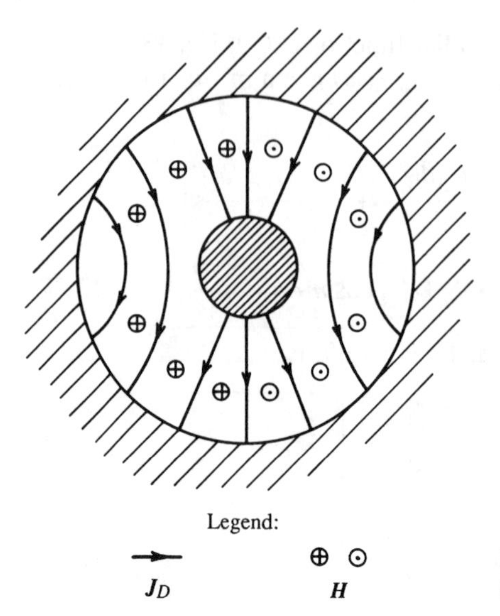

Figure 13-5. Field pattern of TE_{11} mode in coaxial line.

Cables which are not filled with solid or foam dielectric material need support structures (disc or helical rods, etc., as in Sec. 11-5b) for the inner conductor; these change the mode properties locally, and, as a result, they lower the effective cutoff frequency for satisfactory transmission. Applicable values are given in catalogs.

13-4 ELLIPTICAL CYLINDRICAL WAVEGUIDES[2, 9–12]

Waveguides of elliptical cross section are widely used as alternatives to rectangular guides. The mathematical details of modal analysis in this geometrical situation are lengthy and burdensome compared to the circular cylindrical case; accordingly only a summary of results will be presented here.

The elliptical cylindrical coordinate system is based on orthogonally intersecting sets of confocal elliptical cylinders, confocal hyperbolic cylinders, and transverse planes. The vector wave equation for $\boldsymbol{E}$ or $\boldsymbol{H}$ (Eq. 12-52) and the substitution of the vector Laplacian (Eq. 12-53) are valid in this case; the axial component reduces to the scalar wave equation as in Eqs. 13-1 and 13-2. A traveling-wave solution may be postulated, as in Eq. 13-4. When the variables are separated, forms of Mathieu's equation result.

The boundary conditions parallel those for the circular cylinder, namely, that the field functions be finite throughout the interior of the waveguide,

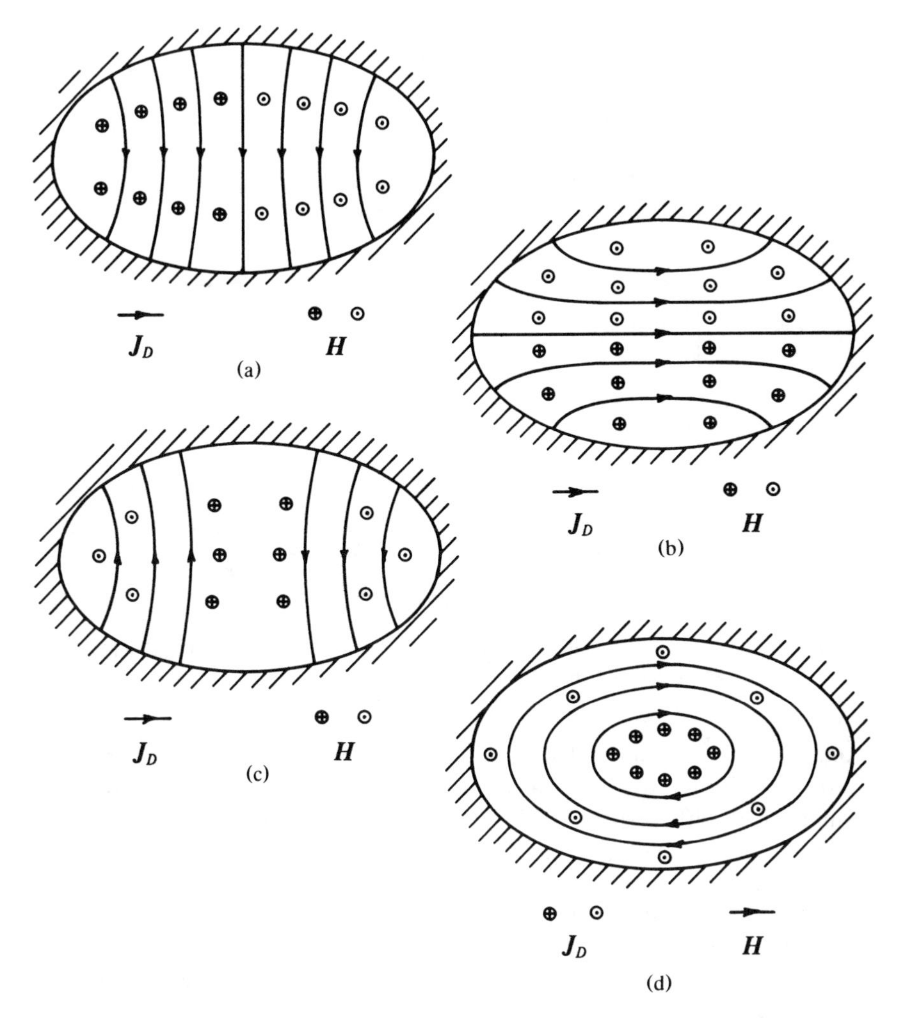

Figure 13-6. Patterns of magnetic field and displacement current density for elliptical waveguide. (a) TE_{c11} mode. (b) TE_{s11} mode. (c) TE_{c21} mode. (d) TM_{c01} mode.

and that the tangential component of $\boldsymbol{E}$ and the normal component of $\boldsymbol{B}$ vanish at the waveguide surface. Distinct modes and modal cutoff frequencies are thereby defined.

Qualitatively the corresponding field patterns (see Fig. 13-6) resemble those of the circular waveguide, but their orientations are stabilized with respect to the major and minor axes. Conventionally three subscripts are used to identify a mode. Those modes for which the longitudinal field

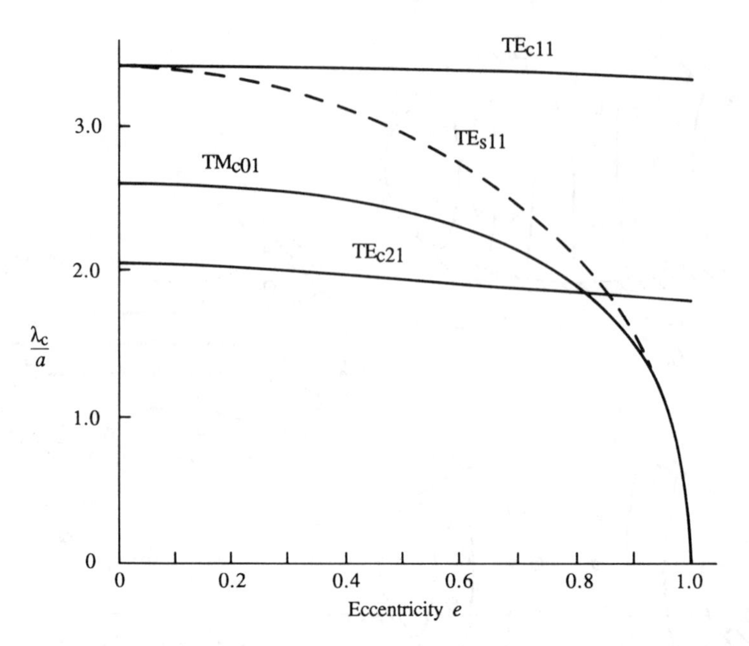

Figure 13-7. Cutoff wavelengths for four lowest-order modes in elliptical waveguide (from Kretzschmar[9], copyrighted 1970 IEEE).

component is symmetrical with respect to the major axis are assigned an initial subscript *c* (cosine); those for which the longitudinal component is symmetrical with respect to the minor axis, but not with respect to the major axis, are assigned the subscript *s* (sine). The assignment of the second and third subscripts follows the same general plan as for the modal subscripts of the circular waveguide.

Let the major axis of the elliptical cross section be designated by $2a$, and the minor axis by $2b$. A dimensionless parameter which is routinely used is e, the *eccentricity*:

$$e = \sqrt{1 - (b/a)^2} \tag{13-83}$$

Various authors have compiled numerical values for cutoff frequencies; a comprehensive set of results has been computed and assembled in graphical form by Kretzschmar.[9] Some of these are presented in Fig. 13-7 in terms of the ratio λ_c/a as a function of the eccentricity; λ_c is the equivalent TEM wavelength of the particular cutoff frequency:

$$\lambda_c = \frac{1}{f_c\sqrt{\mu\epsilon}} \tag{13-84}$$

The TE_{c11} mode is clearly dominant. A ratio of approximately 0.5 for b/a is common in standard designs of elliptical waveguides. Substitution of this value in Eq. 13-83 yields $e = 0.8660$; Kretzschmar lists a tabular value of $\lambda_c/a = 3.35$ for the TE_{c11} mode, and $\lambda_c/a = 1.84$ for TE_{c21}. The latter is the mode with the next highest cutoff wavelength in a guide of this eccentricity.

The corresponding cutoff frequencies are, assuming that the guide is air filled,

$$f_c = \frac{3 \times 10^8}{2a} \cdot \frac{2}{3.35}$$

$$= \frac{1.79 \times 10^8}{2a} \text{ Hz} \qquad (TE_{c11}) \tag{13-85}$$

$$f_c = \frac{3 \times 10^8}{2a} \cdot \frac{2}{1.84}$$

$$= \frac{3.26 \times 10^8}{2a} \text{ Hz} \qquad (TE_{c21}) \tag{13-86}$$

Thus an ample operating frequency range exists within which TE_{c11} is the only propagating mode.

It may be further noted from Fig. 13-7 that the variation with respect to eccentricity of the TE_{c11}-mode value for λ_c/a is slight; the value at $e = 0$ is 3.41 (see Prob. 13-9), which differs by less than 2% from that for $e = 0.866$ (3.35).

13-5 CONCLUSIONS

A circular or elliptical tube of highly conducting material will transmit electromagnetic waves of sufficiently high frequency much as a rectangular waveguide will. Both transverse-electric (TE) and transverse-magnetic (TM) modes are possible. Maxwell's curl equations in circular or elliptical cylindrical coordinates may be reduced to a scalar wave equation in terms of the longitudinal component of $\boldsymbol{E}$ or $\boldsymbol{H}$.

a. Circular Waveguide

The solution of the wave equation involves Bessel functions dependent on radial distance; boundary conditions restrict the solution to integral-order functions of the first kind.

The cutoff frequency ω_c for each mode is inversely proportional to the waveguide radius r_a and is fixed by the boundary condition that the tangential component of $\boldsymbol{E}$ vanish at the cylindrical conducting surface. (The radial component of $\boldsymbol{B}$ also vanishes there, but this is not an independent constraint.) Specifically, the Bessel function argument at $r = r_a$ is $\omega_c \sqrt{\mu \epsilon}\, r_a$; for TM_{mn} modes this argument must have a value which will make the mth-order Bessel function equal to zero, and for TE_{mn} modes the argument must have a value which will make the derivative of the mth-order Bessel function equal to zero.

Modes for which $m = 0$ have circularly symmetrical fields, whereas the fields of other modes vary circumferentially. The TE_{0n} modes are characterized by surface-current patterns which are purely circumferential in direction, and attenuation functions which decrease indefinitely with increasing frequency. Mode conversion at waveguide discontinuities has been an impediment to commercial exploitation of this property.

A mode which is not circularly symmetrical has no preferred orientation, and any deviation from precise circular cylindrical geometry may cause the orientation to change unpredictably.

One general type of resonant-cavity wavemeter uses a standing-wave field of the TE_{11} or TE_{01} mode in a short section of circular waveguide, which is short-circuited at one end with a fixed plate and at the other with an adjustable plunger.

A coaxial line is, inadvertently, a form of circular waveguide. It will convey, in addition to the TEM mode, one or more propagating waveguide modes when the operating frequency exceeds the respective cutoff values. Among these, the TE_{11} mode is dominant.

b. Elliptical Waveguide

Elliptical cylindrical waveguides are competitive with rectangular waveguides in many instances. In contrast to the field patterns of the circular guide, those of the elliptical guide are stabilized with respect to the major and minor axes. The cutoff frequency of the dominant mode, TE_{c11} (Fig. 13-6a), is determined primarily by major-axis width, and is influenced only slightly by the minor-axis dimension. By choosing the ratio of minor to major axes at about 0.5, all higher-order modes may be kept nonpropagating within a substantial frequency band.

PROBLEMS

13-1. Expand the function $-\nabla \times (\nabla \times)\boldsymbol{E} + \nabla(\boldsymbol{\nabla} \cdot \boldsymbol{E})$ in cylindrical coordinates to verify Eq. 13-1 and the statements following it. Partial solution:

$$[\nabla \times (\nabla \times \boldsymbol{E}) - \nabla(\nabla \cdot \boldsymbol{E})]_\phi = \frac{-2}{r^2}\frac{\partial E_r}{\partial \phi} - \nabla^2 E_\phi + \frac{E_\phi}{r^2}$$

13-2. Verify that the change of variables given in Eq. 13-13 will reduce Eq. 13-12 to the form of Bessel's equation given in Eq. 13-14.

13-3. Write expressions for (a) the electric and magnetic fields of the TM_{01} mode as explicit functions of time, (b) the surface-current density, and (c) the displacement-current density. Make a sketch supplementing Fig. 13-2 to show the surface-current field. May thin slits be cut in the waveguide without interfering with the TM_{01} mode? If so, in what direction?

13-4. At what value of $h_{01}r$ is the magnetic field in the TM_{01} mode most intense?

13-5. Equation 13-32 for E_ϕ for a TM mode contains the factor $1/r$. Investigate the behavior of E_ϕ in the vicinity of $r = 0$ for (a) $m = 0$, (b) $m = 1$, (c) $m = 2$, and (d) $m = 3$. [$J_m(p) \approx (1/m!)(p/2)^m$; $(p \ll 1)$.]

13-6. Draw sketches for the standing-wave fields in a TE_{11}-mode wavemeter at the following values of ωt: (a) zero, (b) 45°, (c) 90°, and (d) 135°.

13-7. Repeat Prob. 13-6 for a circular-waveguide wavemeter in the TE_{01} mode.

13-8. Compute the cutoff frequency for the TE_{01} mode in a circular waveguide with an inside diameter of 5.0 cm. Find the attenuation in the TE_{01} mode for (a) 35.0 GHz and (b) 75.0 GHz, if the waveguide is made of copper.

13-9. See Eq. 13-83 for the eccentricity of the cross section of an elliptical waveguide. In the limiting case of $b = a$ the cross section is circular, $e = 0$, and a is equal to the radius. Find λ_c/a from Table 13-1 for the circular TE_{11} and TE_{21} modes, and compare with the ordinate values in Fig. 13-7 at $e = 0$.

REFERENCES

1. Barlow, H. M. and Brown, J., *Radio Surface Waves*, Oxford Univ. Press, London, 1962.
2. Chu, L. J., Electromagnetic waves in elliptic hollow pipes of metal, *J. Appl. Phys.*, 9, 583, September 1938.
3. Dwight, H. B., Table of roots for natural frequencies in coaxial type cavities, *J. Math. Phys.*, 27, 84, 1948.
4. Ginzton, E. L., *Microwave Measurements*, McGraw-Hill, New York, 1957.
5. Members of the technical staff, AT & T Bell Laboratories, *A History of Engineering and Science in the Bell System. Transmission Technology (1925–1975)*, E. F. O'Neill, ed., AT & T Laboratories, Indianapolis, IN, 1985.
6. Karbowiak, A. E., System aspects of long distance communication by waveguide, Proc. IEE (London), 109, Part B, 336, 1962.
7. Karbowiak, A. E. and Solymar, L., Characteristics of waveguide for long-distance transmission, *J. Res. Natl. Bur. Std. (U.S.)*, Sec. D, Radio Propagation, 65D, No. 1, 75, 1961. Reprinted in *Elec. Commun.*, 37, no. 1, 27, 1961.
8. King, A. P., Status of low-loss waveguide and components at millimeter wavelengths, *Microwave J.*, 7, no. 3, 102, 1964.
9. Kretzschmar, J. G., Wave propagation in hollow conducting elliptical waveguides, *IEEE Trans. Microwave Theory Techniques*, MTT-18, no. 9, 547, 1970.

10. Kretzschmar, J. G., Field configuration of the TM_{c01} mode in an elliptical waveguide, *Proc. IEE*, 118, no. 9, 1187, 1971. Correction in *Proc. IEE*, 119, no. 8, 1140, 1972.
11. Marcuvitz, N., Ed., *Waveguide Handbook*, MIT Radiation Laboratory Series, vol. 10, McGraw-Hill, New York, 1951, 66.
12. McLachlan, N. W., *Theory and Applications of Mathieu Functions*, Dover, New York, 1964.
13. Miller, S. E., Waveguide as a communication medium, *Bell Syst. Tech. J.*, 33, no. 6, 1209, 1954.
14. Montgomery, C. G., *Technique of Microwave Measurements*, Radiation Laboratory Series no. 11, McGraw-Hill, New York, 1947.
15. Morgan, S. P. and Young, J. A., Helix waveguide, *Bell Syst. Tech. J.*, 35, no. 6, 1347, 1956.
16. Neubauer, H. and Huber, F. R., Higher modes in coaxial RF lines, *Microwave J.*, 12, no. 6, 57, 1969.
17. Rosenberg, C. B., Cavity resonator measurements of the complex permittivity of low-loss liquids, *Proc. IEE*, 129H, 71, April 1982.
18. Unger, H. G., Normal modes and mode conversion in helix waveguide, *Bell Syst. Tech. J.*, 40 no. 1, 255, 1961.
19. Young, D. T. and Warters, W. D., Precise 50 to 60 GHz measurements on a two-mile loop of helix waveguide, *Bell Syst. Tech. J.*, 47, no. 6, 933, 1968.

CHAPTER 14

Skin Effect in Coaxial Conductors

Frequency dependence of the distributed resistance and inductance of a transmission line was mentioned in Sec. 1-2; in Sec. 5-2a a high-frequency approximation for the distributed impedance of a coaxial cable was cited. The loss-related component of the distributed impedance depends also on the conductivity, the permeability, and the geometric cross section of the conductors. The situation may be readily analyzed by means of the conduction-current fields and the accompanying magnetic fields, in accordance with Maxwell's equations. This will serve to extend the treatment given in Sec. 11-1 for a lossless coaxial cable.

High but finite conductivity will be assumed. As in Chap. 10, this will be taken to mean that $\omega\epsilon \ll \sigma$, or in other words that displacement-current density within the conductors will be assumed negligible compared to conduction-current density. Imperfect conductivity of the conductors will alter the field pattern such that $\boldsymbol{E}$ will have a small longitudinal component similar to that for plane conductors discussed in Sec. 10-4. This component will be ignored so far as the fields between the conductors are concerned, but within the conducting regions it is significantly related to the conduction current and the magnetic field.

The inner conductor will be assumed to be a solid cylinder, and the outer one to be a tube. In other words the cable is assumed to be *circularly symmetrical*. In addition, for the basic analysis the outer conductor will be assumed to be *thick-walled*. This will mean that the wall thickness is so great that, for the frequencies of interest, the conduction-current density just within the outer surface of the outer conductor is negligible compared to that near its inner surface. (The thickness must be at least several times the skin depth δ, as defined in Eq. 10-79.) For mathematical application of boundary conditions, this is equivalent to considering the outer wall to be of infinite thickness.

If, at some frequency of interest, the skin depth is comparable to the thickness of the given outer conductor, that conductor may be termed "thin-walled." The appropriate boundary condition is given in Sec. 14-4.

Cumbersome expressions result for the fields within such a conducting region.

14.1 CONDUCTION-CURRENT-DENSITY FIELD

The conduction-current-density field will be predominantly longitudinal; for this approximation the radial component of that field will be neglected. Correspondingly, since $\boldsymbol{J} = \sigma \boldsymbol{E}$, the radial component of $\boldsymbol{E}$ within the conducting regions will be considered negligible in comparison to the longitudinal component.

a. Differential Equation for Density as a Function of Radius

A traveling-wave function in the phasor form will be postulated for $\boldsymbol{J}$. The geometry is as shown in Fig. 11-1. Because of circular symmetry one may assume that $\boldsymbol{J}$ is not a function of angle ϕ, but from the results in Sec. 10-2d variation with distance from the conductor surface (radially, in this instance) may be expected:

$$\boldsymbol{J}_1 = \mathrm{Re}\left[f_J(r) \epsilon^{j(\omega t - \beta z)} \right] \boldsymbol{a}_z \tag{14-1}$$

[The symbol $f_J(r)$ was chosen rather than $J(r)$ to avoid possible confusion with Bessel functions of the first kind, which will be used here. It is anticipated that $f_J(r)$ may be complex.]

If Eq. 14-1 is substituted successively in Maxwell's curl equations, an ordinary differential equation for $f_J(r)$ may be derived. Let Maxwell's equation in curl $\boldsymbol{E}$, Eq. 9-23, be multiplied by σ and $\boldsymbol{J}$ substituted:

$$\nabla \times \boldsymbol{J} = -\mu\sigma \frac{\partial \boldsymbol{H}_c}{\partial t} \tag{14-2}$$

The subscript c will be used with $\boldsymbol{H}$ to indicate that it is the magnetic field within one of the conductors. Subscripts a and b will be used instead when it is necessary to distinguish between the inner and outer conductors, and these subscripts will then be added to $\boldsymbol{J}$ and f_J.

When Eq. 14-1 is inserted for $\boldsymbol{J}$, Eq. 14-2 becomes

$$-\mu\sigma \frac{\partial \boldsymbol{H}_{1c}}{\partial t} = \mathrm{Re}\left[-\frac{df_J(r)}{dr} \epsilon^{j(\omega t - \beta z)} \right] \boldsymbol{a}_\phi$$

Integration of the preceding equation with respect to t yields

$$\boldsymbol{H}_{1c} = \mathrm{Re}\left[\frac{1}{j\omega\mu\sigma}\frac{df_J(r)}{dr}\epsilon^{j(\omega t - \beta z)}\right]\boldsymbol{a}_\phi \tag{14-3}$$

If Eq. 14-3 is substituted in Maxwell's equation in curl $\boldsymbol{H}$, Eq. 9-21, and the displacement current and the radial component of curl $\boldsymbol{H}$ are neglected, the following is obtained:

$$\boldsymbol{J}_1 = \mathrm{Re}\left[\frac{1}{j\omega\mu\sigma r}\frac{d}{dr}\left(r\frac{df_J(r)}{dr}\right)\epsilon^{j(\omega t - \beta z)}\right]\boldsymbol{a}_z \tag{14-4}$$

The right-hand side of Eq. 14-1 may be substituted for $\boldsymbol{J}_1$ in Eq. 14-4, and like terms cancelled:

$$f_J(r) = \frac{1}{j\omega\mu\sigma}\left[\frac{d^2 f_J(r)}{dr^2} + \frac{1}{r}\frac{df_J(r)}{dr}\right] \tag{14-5}$$

Let

$$m = \sqrt{\omega\mu\sigma} \tag{14-6}$$

$$f_J(r) = \frac{1}{jm^2}\left[\frac{d^2 f_J(r)}{dr^2} + \frac{1}{r}\frac{df_J(r)}{dr}\right] \tag{14-7}$$

An ordinary differential equation of the second order has been obtained for the current-density function as a function of radius. It is a variation of Bessel's differential equation (B-1); the following change of variables will bring Eqs. 14-7 and B-1 into agreement, if $n = 0$ in the latter equation:

$$p = \sqrt{-j}\, mr \tag{B-21}$$

As in Eq. B-25, the general solution may be written as

$$f_J(r) = A_0' J_0(j^{3/2}mr) + B_0' K_0(j^{1/2}mr) \tag{14-8}$$

The functions J_0 and K_0 are defined in Eqs. B-6 and B-19; A_0' and B_0' are constants which are determined by the boundary conditions.

b. Boundary Conditions for Current-Density Fields

Separate expressions for the current-density function for each conductor may be written by imposing the corresponding boundary conditions on Eq.

14-8. Current density for the inner conductor $\boldsymbol{J}_{1a}$ is defined for $r \leq r_a$ and must be finite throughout that range. For the outer conductor (thick-walled), current density $\boldsymbol{J}_{1b}$ is defined for $r \geq r_b$, throughout which it must be finite. In view of the fact that $K_0(j^{1/2}mr)$ will approach infinity in magnitude as mr approaches zero and that $J_0(j^{3/2}mr)$ will approach infinity in magnitude as mr approaches infinity, the current-density functions are constrained to

$$f_{Ja}(r) = A'_0 J_0(j^{3/2} m_a r) \tag{14-9}$$

$$f_{Jb}(r) = B'_0 K_0(j^{1/2} m_b r) \tag{14-10}$$

Here the constants A'_0 and B'_0 have yet to be determined, and the subscripts a and b have been attached to m because the two conductors may be composed of different materials:

$$\boldsymbol{J}_{1a} = \mathrm{Re}\left[A'_0 J_0(j^{3/2} m_a r)\epsilon^{j(\omega t - \beta z)}\right]\boldsymbol{a}_z \tag{14-11}$$

$$\boldsymbol{J}_{1b} = \mathrm{Re}\left[B'_0 K_0(j^{1/2} m_b r)\epsilon^{j(\omega t - \beta z)}\right]\boldsymbol{a}_z \tag{14-12}$$

The resultant currents within the conductors may be found by the following integrations over the respective cross sections. Let the current function be taken as $i_1(z,t)$, the same as for the derivation of the fields between the conductors, Sec. 11-1; the constants A'_0 and B'_0 of the preceding equations prove to be proportional to I_{1M}:

$$i_1(z,t) = \iint \boldsymbol{J}_{1a} \cdot d\boldsymbol{S} \tag{14-13}$$

$$-i_1(z,t) = \iint \boldsymbol{J}_{1b} \cdot d\boldsymbol{S} \tag{14-14}$$

Here $$d\boldsymbol{S} = r\,dr\,d\phi\,\boldsymbol{a}_z$$

$$i_1(z,t) = \mathrm{Re}\left[I_{1M}\epsilon^{j(\omega t - \beta z)}\right] \tag{11-9}$$

Equations 11-9, 14-11, and 14-12 may be substituted in Eqs. 14-13 and 14-14, and the operator Re and the exponential deleted from each term.

This leads to

$$I_{1M} = \int_0^{2\pi} \int_0^{r_a} A_0' J_0\left(j^{3/2} m_a r\right) r \, dr \, d\phi \tag{14-15}$$

$$-I_{1M} = \int_0^{2\pi} \int_{r_b}^{\infty} B_0' K_0\left(j^{1/2} m_b r\right) r \, dr \, d\phi \tag{14-16}$$

The following integration formulas are applicable:

$$\int p J_0(p) \, dp = p J_1(p) \tag{B-12}$$

$$\int q K_0(q) \, dq = -q K_1(q) \tag{14-17}$$

Here $J_1(p)$ and $K_1(p)$ are functions of the first order, solutions to Bessel's differential equation, B-1, if n is equal to unity. The following specific values are needed when substituting limits in the integrals:

$$J_1(0) = 0 \tag{14-18}$$

$$\lim_{q \to \infty} \left[q K_1(q) \right] = 0 \quad \text{(provided the real part of } q \text{ is positive)} \tag{14-19}$$

Hence

$$I_{1M} = \frac{2\pi A_0'}{j^{3/2} m_a} r_a J_1\left(j^{3/2} m_a r_a\right)$$

or

$$A_0' = \frac{I_{1M} j^{3/2} m_a}{2\pi r_a J_1\left(j^{3/2} m_a r_a\right)} \tag{14-20}$$

This result may be substituted into Eq. 14-11 to give a complete expression for current density in the inner conductor in terms of the maximum value of the resultant current I_{1M}:

$$\boldsymbol{J}_{1a} = \text{Re}\left[\frac{I_{1M} j^{3/2} m_a J_0\left(j^{3/2} m_a r\right)}{2\pi r_a J_1\left(j^{3/2} m_a r_a\right)} \epsilon^{j(\omega t - \beta z)} \right] \boldsymbol{a}_z \tag{14-21}$$

Similarly,

$$J_{1b} = \mathrm{Re}\left[\frac{-I_{1M}j^{1/2}m_bK_0(j^{1/2}m_br)}{2\pi r_bK_1(j^{1/2}m_br_b)}\epsilon^{j(\omega t-\beta z)}\right]\boldsymbol{a}_z \qquad (14\text{-}22)$$

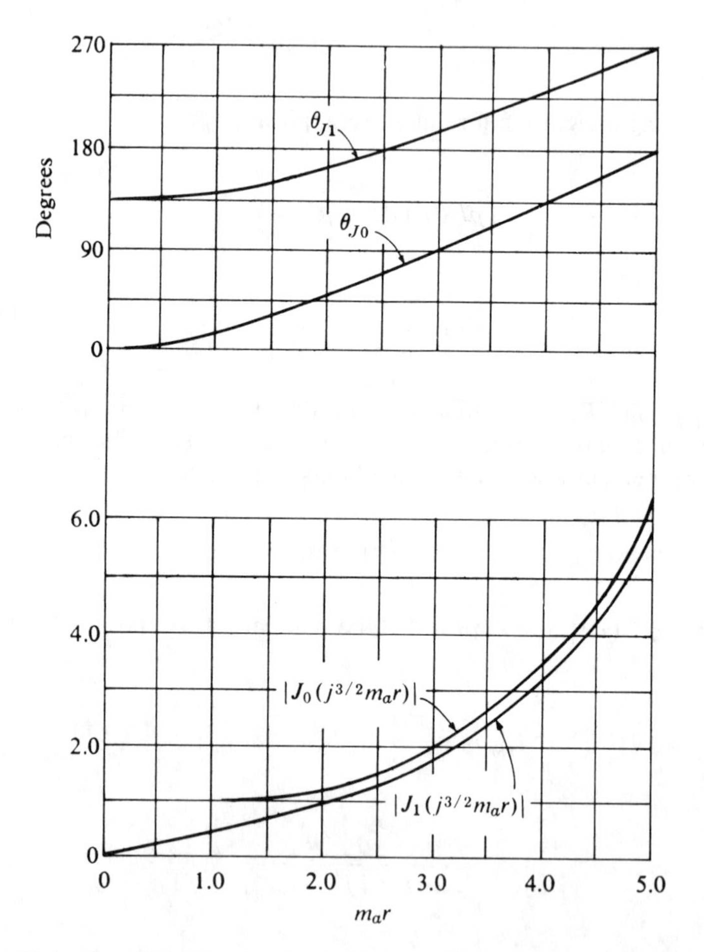

Figure 14-1. Bessel functions associated with current density and magnetic field in inner conductor of coaxial cable* (prepared from Tables 27 and 28, McLachlan,[5] pp. 227–228).

*Figures 14-1 and 14-2 were prepared from Tables 27 through 30 (pp. 227–230) in N. W. McLachlan,[5] *Bessel Functions for Engineers*, by permission of the copyright owner, Oxford University Press.

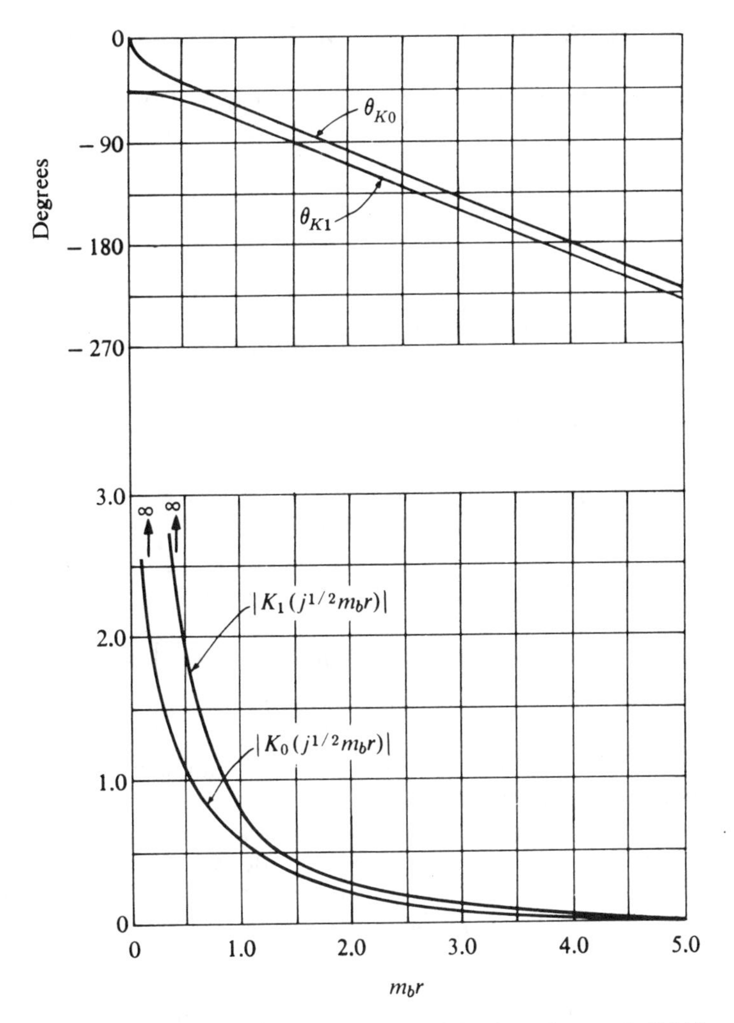

Figure 14-2. Bessel functions associated with current density and magnetic field in outer conductor (thick-walled) of coaxial cable (prepared from Tables 29 and 30, McLachlan,[5] pp. 229–230).

The Bessel functions indicated in Eqs. 14-21 and 14-22 are plotted in terms of magnitude and angle in Figs. 14-1 and 14-2. Current density as a function of radius in either conductor is directly proportional to the magnitude of the respective zero-order function. For the inner conductor, relative phase with respect to $i_1(z,t)$ is $\theta_{J0}(m_a r) - \theta_{J1}(m_a r_a) + 135°$; for the outer conductor, relative phase with respect to $-i_1(z,t)$ is $\theta_{K0}(m_b r) - \theta_{K1}(m_b r_b) + 45°$.

14-2 MAGNETIC FIELD WITHIN CONDUCTING REGIONS

The magnetic field within each conductor may be found by substituting f_J, as given by Eq. 14-21 or 14-22, in Eq. 14-3. The following derivatives are needed:

$$\frac{dJ_0(j^{3/2}mr)}{dr} = -j^{3/2}mJ_1(j^{3/2}mr) \tag{14-23}$$

$$\frac{dK_0(j^{1/2}mr)}{dr} = -j^{1/2}mK_1(j^{1/2}mr) \tag{14-24}$$

For the inner conductor,

$$\boldsymbol{H}_{1a} = \mathrm{Re}\left[\frac{I_{1M}J_1(j^{3/2}m_a r)}{2\pi r_a J_1(j^{3/2}m_a r_a)}\epsilon^{j(\omega t-\beta z)}\right]\boldsymbol{a}_\phi \tag{14-25}$$

For the outer conducting region the corresponding result is

$$\boldsymbol{H}_{1b} = \mathrm{Re}\left[\frac{I_{1M}K_1(j^{1/2}m_b r)}{2\pi r_b K_1(j^{1/2}m_b r_b)}\epsilon^{j(\omega t-\beta z)}\right]\boldsymbol{a}_\phi \tag{14-26}$$

Equations 14-25 and 14-26 may also be derived directly from Maxwell's equation, as suggested by Prob. 14-2.

The magnitude of $\boldsymbol{H}$ as a function of radius in either conductor is directly proportional to the respective first-order function in Figs. 14-1 or 14-2. Relative phase with respect to the field between the conductors is given by $\theta_{J1}(m_a r) - \theta_{J1}(m_a r_a)$ or $\theta_{K1}(m_b r) - \theta_{K1}(m_b r_b)$.

14-3 IMPEDANCE COMPONENT DUE TO FINITE CONDUCTIVITY

The derivation in Sec. 11-1 of the distributed inductance of a coaxial line was based on the assumption of infinite conductivity, a limiting condition which yields zero magnetic flux within the conductors. From the analysis of the preceding section it is apparent that unless the conductivity is infinite, some flux exists within the conductors, and this will increase the inductance beyond the value given in Eq. 11-24. Furthermore, because the conduction-current density varies both in magnitude and in relative time phase with radius, it may be expected that the resistive power loss will be greater than if the same resultant current were distributed uniformly over the transverse cross section of each conductor.

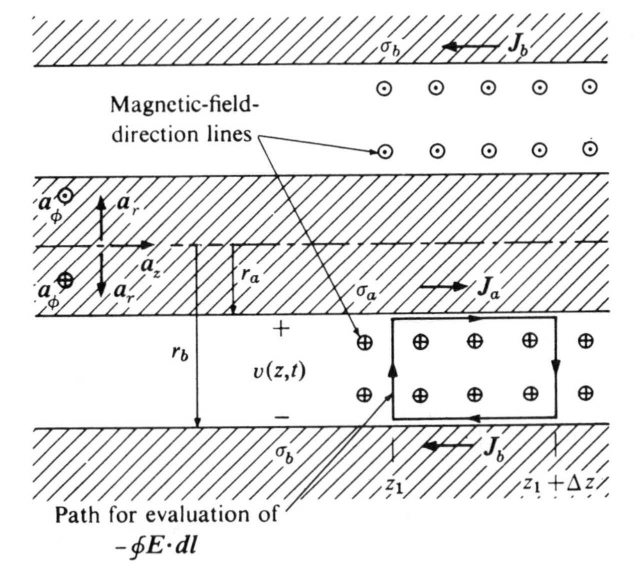

Figure 14-3. Longitudinal cross section of coaxial line.

a. Derivation of Component Impedances

Consider the voltage rises around the closed path shown in Fig. 14-3:

$$v(z_1,t) - \int_{z_1}^{z_1+\Delta z} \left(\frac{J_{1a}}{\sigma_a}\right)_z dz - v(z_1+\Delta z,t) - \int_{z_1+\Delta z}^{z_1} \left(\frac{J_{1b}}{\sigma_b}\right)_z dz$$

$$= -\oint \mathbf{E} \cdot d\mathbf{L} \qquad (14\text{-}27)$$

The difference between the voltages at z_1 and $z_1 + \Delta z$ may be stated as

$$v(z_1,t) - v(z_1+\Delta z,t) = -\int_{z_1}^{z_1+\Delta z} \frac{\partial v(z,t)}{\partial z} dz \qquad (14\text{-}28)$$

Equation 14-28 may be substituted in Eq. 14-27 and the integrals with respect to z consolidated:

$$\int_{z_1}^{z_1+\Delta z} \left[-\left(\frac{J_{1a}}{\sigma_a}\right)_z - \frac{\partial v(z,t)}{\partial z} + \left(\frac{J_{1b}}{\sigma_b}\right)_z \right] dz = -\oint \mathbf{E} \cdot d\mathbf{L} \quad (14\text{-}29)$$

Faraday's law may be substituted for the right-hand side:

$$\oint \boldsymbol{E} \cdot d\boldsymbol{L} = -\iint \frac{\partial \boldsymbol{B}}{\partial t} \cdot d\boldsymbol{S} \tag{9-11}$$

$$d\boldsymbol{S} = dr\,dz\,\boldsymbol{a}_\phi$$

$$\oint \boldsymbol{E} \cdot d\boldsymbol{L} = -\int_{z_1}^{z_1+\Delta z} \int_{r_a}^{r_b} \frac{\partial B_\phi}{\partial t}\, dr\, dz$$

The magnetic-field function between the conductors is given by Eq. 11-12; it may be substituted into the preceding equation and integration performed with respect to r:

$$\oint \boldsymbol{E} \cdot d\boldsymbol{L} = -\mathrm{Re}\left[\int_{z_1}^{z_1+\Delta z} \int_{r_a}^{r_b} \frac{j\omega\mu \boldsymbol{I}_{1M}}{2\pi r} \epsilon^{j(\omega t - \beta z)}\, dr\, dz\right]$$

$$= -\mathrm{Re}\left[\int_{z_1}^{z_1+\Delta z} \frac{j\omega\mu \boldsymbol{I}_{1M}}{2\pi} \ln\left(\frac{r_b}{r_a}\right) \epsilon^{j(\omega t - \beta z)}\, dz\right] \tag{14-30}$$

The derivative of v with respect to z, which is used in Eq. 14-29, may be put in terms of the current function $i_1(z, t)$ by substituting Eq. 11-9 in Eq. 4-3:

$$\frac{\partial v(z,t)}{\partial z} = -\mathrm{Re}\left[(r + j\omega l)\boldsymbol{I}_{1M}\epsilon^{j(\omega t - \beta z)}\right] \tag{14-31}$$

Equations 14-30 and 14-31 may be substituted in Eq. 14-29. Since every term is integrated with respect to z from z_1 to $z_1 + \Delta z$, and the location z_1 may be chosen arbitrarily on the line, the integrands must satisfy the corresponding equation. In addition, the Re operator, the term $\boldsymbol{I}_{1M}$, and the exponential are common to each term, and they may be deleted. The following results:

$$(r + j\omega l) - \frac{j^{3/2} m_a J_0(j^{3/2} m_a r_a)}{2\pi r_a \sigma_a J_1(j^{3/2} m_a r_a)} - \frac{j^{1/2} m_b K_0(j^{1/2} m_b r_b)}{2\pi r_b \sigma_b K_1(j^{1/2} m_b r_b)}$$

$$= \frac{j\omega\mu}{2\pi} \ln\left(\frac{r_b}{r_a}\right) \tag{14-32}$$

For mathematical convenience, the distributed inductance l may be split into three parts; thus

$$l = l' + l_a + l_b \tag{14-33}$$

Let

$$l' = \frac{\mu}{2\pi} \ln\left(\frac{r_b}{r_a}\right) \tag{14-34}$$

This definition corresponds to Eq. 11-24, the expression for the distributed inductance of a lossless coaxial line.

Also let

$$r = r_a + r_b \tag{14-35}$$

If one subtracts Eq. 14-34 from Eq. 14-32, the remaining terms may be apportioned as

$$r_a + j\omega l_a = \frac{j^{3/2} m_a J_0(j^{3/2} m_a r_a)}{2\pi r_a \sigma_a J_1(j^{3/2} m_a r_a)} \tag{14-36}$$

$$r_b + j\omega l_b = \frac{j^{1/2} m_b K_0(j^{1/2} m_b r_b)}{2\pi r_b \sigma_b K_1(j^{1/2} m_b r_b)} \tag{14-37}$$

TABLE 14-1. Normalized Resistance and Internal-Flux Reactance of Solid Circular Conductor[a] (See Eqs. 14-6 and 14-39)

$m_a r_a$	r_a / r_{adc}	l_a / l_{adc}	$\omega l_a / r_{adc}$
0.0	1.0000	1.0000	0.0
0.5	1.0003	0.9998	0.0313
1.0	1.0052	0.9974	0.1246
1.5	1.0258	0.9871	0.2776
2.0	1.0782	0.9611	0.4806
2.5	1.1754	0.9135	0.7136
3.0	1.3181	0.8452	0.9508
3.5	1.4920	0.7655	1.1722
4.0	1.6779	0.6863	1.3726
4.5	1.8627	0.6156	1.5583
5.0	2.0427	0.5560	1.7374
6.0	2.3936	0.4652	2.0934
7.0	2.7432	0.4002	2.4513
8.0	3.0944	0.3511	2.8086
9.0	3.4464	0.3126	3.1647
10.0	3.7986	0.2816	3.5203
15.0	5.5621	0.1882	5.2931
20.0	7.3277	0.1413	7.065

[a]Adapted from Table XXII, Rosa and Grover[6].

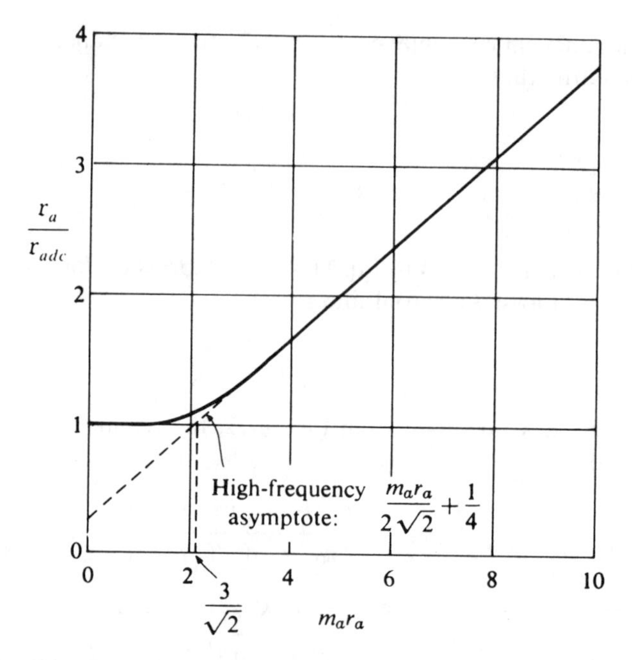

Figure 14-4. Skin-effect-resistance ratio for circular conductor (prepared from Rosa and Grover,[6] pp. 226–228) $m_a = \sqrt{\omega \mu_a \sigma_a}$. Skin-effect demarcation at $m_a r_a = 3/\sqrt{2}$. (See Eq. 14-55.)

The component of impedance due to finite conductivity of the inner conductor, Eq. 14-36, may be divided by the resistance of that conductor to direct current to yield a dimensionless ratio:

$$r_{\text{adc}} = \frac{1}{\pi r_a^2 \sigma_a} \tag{14-38}$$

$$\frac{r_a + j\omega l_a}{r_{\text{adc}}} = \frac{j^{3/2} m_a r_a J_0(j^{3/2} m_a r_a)}{2J_1(j^{3/2} m_a r_a)} \tag{14-39}$$

Numerical values for the real and imaginary parts of Eq. 14-39 are listed in Table 14-1; the real part, which is the *skin-effect-resistance ratio*, is plotted in Fig. 14-4.

b. High-Frequency Asymptotes

High-frequency approximations for Eqs. 14-37 and 14-39, derivable from asymptotic expansions of Bessel functions, are valid if $mr \ggg 1$.

(1) *Inner Conductor.* The first-order asymptotic approximation ($\tilde{J}_0$) for $J_0(j^{3/2}mr)$ was given in Eq. B-23; second-order asymptotic approximations for J_0 and J_1 are related to it as shown below (see McLachlan,[5] p. 152). The following change of variable will simplify the subsequent algebraic steps:

$$u = \frac{1}{8\sqrt{2}\,m_a r_a} \tag{14-40}$$

$$J_0\left(j^{3/2}m_a r_a\right) \approx \tilde{J}_0\left(j^{3/2}m_a r_a\right)(1+u)\epsilon^{-ju} \tag{14-41}$$

$$J_1\left(j^{3/2}m_a r_a\right) \approx j\tilde{J}_0\left(j^{3/2}m_a r_a\right)(1-3u)\epsilon^{j3u} \tag{14-42}$$

Substitution of these into Eq. 14-39 gives

$$\frac{r_a + j\omega l_a}{r_{\text{adc}}} \approx \frac{j^{3/2}m_a r_a(1+u)\epsilon^{-j4u}}{2j(1-3u)} \tag{14-43}$$

The approximations in Eqs. 14-41 and 14-42 were valid to the first power in u, and the same is true of Eq. 14-43. Power-series expansions in u may be used to simplify Eq. 14-43 (u approaches zero as $m_a r_a$ approaches infinity), but any resulting terms in u^2 or higher powers must be discarded:

$$\frac{1+u}{1-3u} \approx (1+u)(1+3u)$$

$$\approx 1 + 4u$$

$$\epsilon^{-j4u} \approx 1 - j4u$$

$$j^{1/2} = \frac{1}{\sqrt{2}}(1+j)$$

Substitution of these will reduce Eq. 14-43 to

$$\frac{r_a + j\omega l_a}{r_{adc}} \approx \frac{m_a r_a}{2\sqrt{2}}(1 + j)(1 + 4u)(1 - j4u)$$

$$\approx \frac{m_a r_a}{2\sqrt{2}}(1 + j + 8u)$$

Replacement of the variable u in terms of $m_a r_a$ gives the

$$\frac{r_a + j\omega l_a}{r_{adc}} \approx \frac{m_a r_a}{2\sqrt{2}} + j\frac{m_a r_a}{2\sqrt{2}} + \frac{1}{4} \tag{14-44}$$

The real part of this is, of course, the skin-effect-resistance ratio

$$\frac{r_a}{r_{adc}} \approx \frac{m_a r_a}{2\sqrt{2}} + \frac{1}{4} \tag{14-45}$$

This formula gives results within 1% of the true value if $m_a r_a > 6$,* and is especially useful for radio-frequency calculations.

The high-frequency asymptote for r_a/r_{adc} described by Eq. 14-45 is indicated in Fig. 14-4 with a dashed line.

(2) *Outer Conductor.* The derivation of the high-frequency approximation for $r_b + j\omega l_b$ parallels that for the inner conductor. The first-order approximation $\tilde{K}_0$ was given in Eq. B-24. Here let

$$u = \frac{1}{8\sqrt{2}\, m_b r_b} \tag{14-46}$$

$$K_0(j^{1/2} m_b r_b) \approx \tilde{K}_0(j^{1/2} m_b r_b)(1 - u)\epsilon^{ju} \tag{14-47}$$

$$K_1(j^{1/2} m_b r_b) \approx \tilde{K}_0(j^{1/2} m_b r_b)(1 + 3u)\epsilon^{-j3u} \tag{14-48}$$

---

*The following higher-order asymptotic approximation is given by Dwight[2] (p. 160):

$$\frac{r_a}{r_{adc}} \approx \frac{mr_a}{2\sqrt{2}} + \frac{1}{4} + \frac{3}{16\sqrt{2}\, mr_a} - \frac{63}{256\sqrt{2}\, m^3 r_a^3} - \frac{27}{64 m^4 r_a^4}$$

The usefulness of the last three terms (and of any following terms which might be derived) is dubious; they improve the accuracy of the approximation at large values of the argument, where Eq. 14-45 already gives results to engineering accuracy, but these terms render the series divergent at smaller values of mr_a.

Substitution of Eqs. 14-47 and 14-48 into Eq. 14-37 yields

$$r_b + j\omega l_b \approx \frac{j^{1/2} m_b (1-u)\epsilon^{j4u}}{2\pi r_b \sigma_b (1+3u)}$$

With the same approximations as before, and then replacing u by Eq. 14-46, the following results:

$$r_b + j\omega l_b \approx \frac{m_b(1+j-8u)}{2\sqrt{2}\,\pi r_b \sigma_b}$$

$$r_b + j\omega l_b \approx \frac{m_b}{2\sqrt{2}\,\pi r_b \sigma_b} - \frac{1}{4\pi r_b^2 \sigma_b} + j\frac{m_b}{2\sqrt{2}\,\pi r_b \sigma_b} \qquad (14\text{-}49)$$

Asymptotic approximations may be applied to determine the conduction-current density and magnetic field within the conductors at high frequencies (see Problem 14-5).

(3) *Resultant Asymptotic Approximation for Coaxial Cable.* The results just derived for the inner and outer conductors may be combined into a single expression for high-frequency distributed impedance. From Eqs. 14-33 and 14-35,

$$r + j\omega l = j\omega l' + r_a + j\omega l_a + r_b + j\omega l_b \qquad (14\text{-}50)$$

Substitution of Eqs. 14-34, 14-38, 14-44, 14-49, and 14-6 will yield

$$r + j\omega l \approx \frac{j\omega\mu}{2\pi}\ln\left(\frac{r_b}{r_a}\right) + \sqrt{\omega}\,(1+j)\left(\frac{1}{2\pi r_a}\sqrt{\frac{\mu_a}{2\sigma_a}} + \frac{1}{2\pi r_b}\sqrt{\frac{\mu_b}{2\sigma_b}}\right) + \frac{1}{4\pi}\left(\frac{1}{r_a^2\sigma_a} - \frac{1}{r_b^2\sigma_b}\right)$$

$$\sqrt{\mu_a\sigma_a\omega}\, r_a > 6 \quad \text{and} \quad \sqrt{\mu_b\sigma_b\omega}\, r_b > 6 \qquad (14\text{-}51)$$

Example—An air-insulated coaxial cable has a nominal (high-frequency asymptote) characteristic impedance of 50 Ω, a solid copper inner conductor with a radius of 2.5×10^{-3} m, and an annular copper outer conductor with a thickness of 6.0×10^{-4} m. (a) Find the inner and outer radii of the outer conductor, (b) evaluate the coefficients in the high-frequency-impedance approximation, (c) evaluate, in terms of frequency, the inequalities for its validity, and (d) find the distributed impedance at 1.0 MHz.

From Eqs. 11-18 and 10-24,

$$\ln\left(\frac{r_b}{r_a}\right) = 2\pi Z_0\sqrt{\frac{\epsilon}{\mu}}$$

$$= (2\pi)(50)\sqrt{\frac{10^{-9}/36\pi}{4\pi \times 10^{-7}}}$$

$$= 0.833$$

$$r_b = 2.30 r_a$$

$$= (2.30)(2.5 \times 10^{-3})$$

$$= 5.75 \times 10^{-3} \text{ m}$$

$$r_c = 5.75 \times 10^{-3} + 6.0 \times 10^{-4}$$

$$= 6.35 \times 10^{-3} \text{ m}$$

$$l' = \frac{4\pi \times 10^{-7}}{2\pi} \ln\left(\frac{5.75}{2.50}\right)$$

$$= 1.67 \times 10^{-7} \text{ H/m}$$

The conductivity of copper is 5.8×10^7 S/m; substitution in Eq. 14-51 yields

$$r + j\omega l \approx j\omega(1.67 \times 10^{-7}) + \frac{\sqrt{\omega}\,(1+j)}{2\pi\sqrt{2}}\sqrt{\frac{4\pi \times 10^{-7}}{5.8 \times 10^7}}$$

$$\times\left(\frac{1}{2.5 \times 10^{-3}} + \frac{1}{5.75 \times 10^{-3}}\right)$$

$$+ \frac{1}{4\pi(5.8 \times 10^7)}\left[\frac{1}{(2.5 \times 10^{-3})^2} - \frac{1}{(5.75 \times 10^{-3})^2}\right]$$

$$\approx j\omega(1.67 \times 10^{-7}) + \sqrt{\omega}\,(1+j)(9.50 \times 10^{-6})$$

$$+ 1.78 \times 10^{-4}\ \Omega/\text{m}$$

Substitution in the first inequality after Eq. 14-51 (in terms or r_a) yields

$$\omega(4\pi \times 10^{-7})(5.8 \times 10^{7})(2.5 \times 10^{-3})^2 > 6^2$$

$$\omega > 7.9 \times 10^4 \text{ rad/s}$$

or

$$f > 1.26 \times 10^4 \text{ Hz}$$

The second inequality (in terms of r_b) is less severe because $r_b > r_a$.
The indicated inequality is satisfied ($1.0 \times 10^6 > 1.26 \times 10^4$). Hence

$$(r + j\omega l)_{1.0 \text{ MHz}} \approx j(2\pi \times 10^6)(1.67 \times 10^{-7})$$

$$+ (1 + j)\sqrt{2\pi \times 10^6}\,(9.50 \times 10^{-6}) + 1.78 \times 10^{-4}$$

$$\approx j(1.05) + (1 + j)(2.38 \times 10^{-2}) + 1.78 \times 10^{-4}$$

$$\approx 2.40 \times 10^{-2} + j1.07\ \Omega/\text{m}$$

c. Low-Frequency Approximation for Inner Conductor

A power series approximation for $r_a + j\omega l_a$ may be found from Eq. 14-36 by substituting the power series for J_0 and J_1 (Eqs. B-6 and B-7) and performing the indicated division. In terms of the general variable p, the following is obtained if the first five terms of the series are used:

$$\frac{J_0(p)}{J_1(p)} \approx \frac{2}{p}\left[1 - \frac{1}{2}\left(\frac{p}{2}\right)^2 - \frac{1}{12}\left(\frac{p}{2}\right)^4 - \frac{1}{48}\left(\frac{p}{2}\right)^6 - \frac{1}{180}\left(\frac{p}{2}\right)^8\right] \quad (14\text{-}52)$$

Substitution of Eq. B-21 for the independent variable, and substituting the resulting expression in Eq. 14-36 yields the result

$$r_a + j\omega l_a \approx \frac{1}{\pi r_a^2 \sigma_a}\left[1 + j\frac{m_a^2 r_a^2}{8} + \frac{1}{12}\left(\frac{m_a r_a}{2}\right)^4 - j\frac{1}{48}\left(\frac{m_a r_a}{2}\right)^6 - \frac{1}{180}\left(\frac{m_a r_a}{2}\right)^8\right] \quad (14\text{-}53)$$

This expression may be rearranged into the following by substitution of Eq. 14-38:

$$r_a + j\omega l_a \approx r_{adc}\left[1 + \frac{1}{12}\left(\frac{m_a r_a}{2}\right)^4 - \frac{1}{180}\left(\frac{m_a r_a}{2}\right)^8\right]$$

$$+ j\frac{\omega\mu_a}{8\pi}\left[1 - \frac{1}{24}\left(\frac{m_a r_a}{2}\right)^4\right] \tag{14-54}$$

From the real part of this equation it appears that the skin-effect-resistance ratio (r_a/r_{adc}) has as its low-frequency asymptote the constant value of unity and departs from that as the square of frequency (m is proportional to $\sqrt{f}$, in accordance with Eq. 14-6). Similarly, from the imaginary component, the incremental inductance due to internal flux linkages has a low-frequency asymptote of $\mu_a/8\pi$ henrys per meter and departs from this asymptote as the square of frequency.

d. Skin-Effect-Demaraction Frequency

From the plot in Fig. 14-4 it appears that, in the study of conductor resistance as a function of frequency, the spectrum may logically be divided into two ranges of distinctly different behavior. These might be called, loosely, the range of negligible skin effect and the range of pronounced skin effect (resistance proportional to the square root of frequency). A similar plot may be anticipated for the sum of the resistances of the conductors in a coaxial cable or an open-wire line. *Skin-effect-demarcation frequency*, f_{skd}, is suggested as a name for the frequency corresponding to the intersection of the high-frequency asymptote with the direct-current-resistance ordinate.

(1) *Normalized Form for Solid Circular Conductor.* From Fig. 14-4 it appears that f_{skd} corresponds to $m_a r_a = 3/\sqrt{2}$. Substitution of $\sqrt{\omega\mu_a\sigma_a}$ for m_a yields

$$\omega_{skd} = \frac{9}{2\mu\sigma r_a^2}$$

or

$$f_{skd} = \frac{9}{4\pi\mu\sigma r_a^2} \tag{14-55}$$

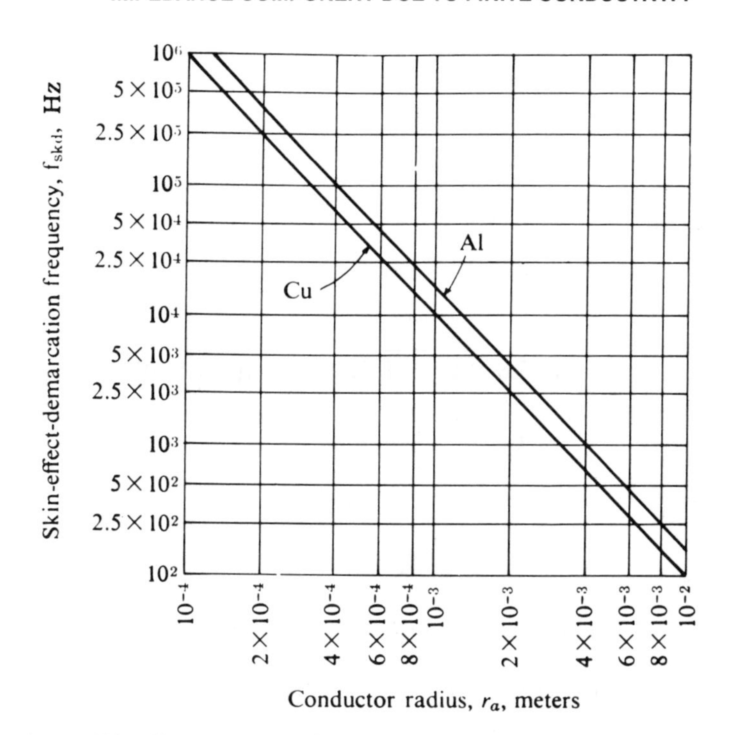

Figure 14-5. Skin-effect-demarcation frequency of solid circular conductor as a function of radius (see Eq. 14-55).

The value of the skin-effect-resistance ratio at the intersection of the asymptotes is

$$\left[\frac{r_a}{r_{adc}}\right]_{m_a r_a = 3/\sqrt{2}} = 1.097 \tag{14-56}$$

Figure 14-5 shows f_{skd} as a function of radius for solid conductors of copper and aluminum.

(2) *Example for Complete Cable.* Consider the cable which was studied in the example in Sec. 14-3b(3). Substitution of Eq. 14-38 yields

$$r_{adc} = \frac{1}{\pi(2.5 \times 10^{-3})^2(5.8 \times 10^7)}$$

$$= 8.78 \times 10^{-4}\ \Omega/\text{m}$$

Similarly

$$
\begin{aligned}
r_{bdc} &= \frac{1}{\pi\sigma_b\left(r_c^2 - r_b^2\right)} \\
&= \frac{1}{\pi\sigma_b(r_c + r_b)(r_c - r_b)} \\
&= \frac{1}{\pi(5.8 \times 10^7)(1.21 \times 10^{-2})(6.0 \times 10^{-4})} \qquad (14.57) \\
&= 7.56 \times 10^4\ \Omega/\text{m} \\
r_{dc} &= (8.78 + 7.56) \times 10^{-4} \\
&= 1.63 \times 10^{-3}\ \Omega/\text{m}
\end{aligned}
$$

The real part of the high-frequency asymptote is

$$
r \approx 9.50 \times 10^{-6}\sqrt{\omega} + 1.78 \times 10^{-4}\ \Omega/\text{m}
$$

If the last two expressions are equated and solved for ω, the result is

$$
\begin{aligned}
\sqrt{\omega_{\text{skd}}} &= \frac{(1.63 - 0.18) \times 10^{-3}}{9.50 \times 10^{-6}} \\
&= 1.53 \times 10^2 \\
\omega_{\text{skd}} &= 2.33 \times 10^4\ \text{rad/s} \\
f_{\text{skd}} &= 3.71 \times 10^3\ \text{Hz}
\end{aligned}
$$

14.4 THIN-WALLED OUTER CONDUCTOR

The outer conductor of a physical coaxial line is necessarily of finite thickness; the case in which this is of the same order of magnitude as the skin depth for a particular operating frequency has more complicated boundary-value requirements than the thick-walled conductor just discussed. Figure 14-6 shows the geometry.

The current-density function will be designated by $\boldsymbol{J}_{1bc}$, and it is defined only for the region $r_b \le r \le r_c$. The Bessel functions J_0 and K_0 are both

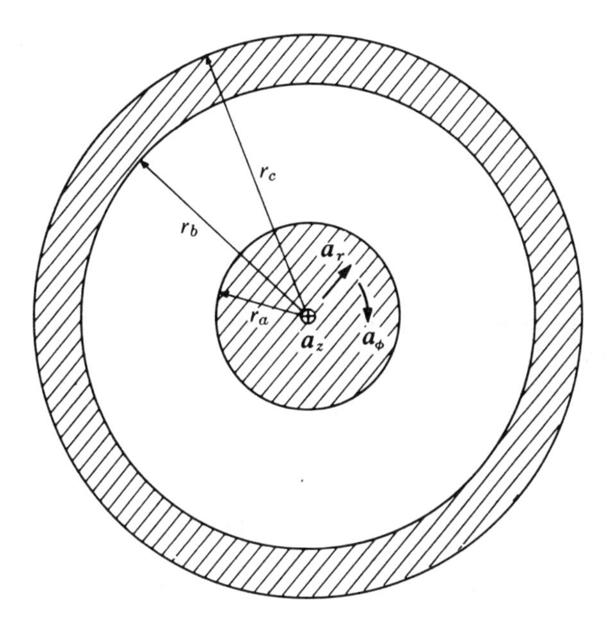

Figure 14-6. Cross section of thin-walled coaxial line.

finite within the range, and hence are valid solutions. The following expression parallels Eq. 14-12:

$$\boldsymbol{J}_{1bc} = \mathrm{Re}\left[\left\{A_0'' J_0\left(j^{3/2} m_b r\right) + B_0'' K_0\left(j^{1/2} m_b r\right)\right\} \epsilon^{j(\omega t - \beta z)}\right] \boldsymbol{a}_z \quad (14\text{-}58)$$

Two boundary conditions are needed to fix the constants A_0'' and B_0'': (1) the surface integral of $\boldsymbol{J}_{1bc}$ must equal the current in the outer conductor, $-i_1(z, t)$, as in Eq. 14-14, and (2) the magnetic field must be continuous at $r = r_b$.

The surface-integral requirement reduces to the following, which should be compared with Eq. 14-16 for the thick-walled conductor:

$$-\boldsymbol{I}_{1M} = \int_0^{2\pi} \int_{r_b}^{r_c} \left[A_0'' J_0\left(j^{3/2} m_b r\right) + B_0'' K_0\left(j^{1/2} m_b r\right)\right] r\, dr\, d\phi \quad (14\text{-}59)$$

The integrals given in Eqs. B-12 and 14-17 are applicable (see Prob. 14-7).

The magnetic field in the outer conductor may be found by substituting Eq. 14-58 in Eq. 14-3:

$$\boldsymbol{H}_{1bc} = \mathrm{Re}\left[\frac{-A_0'' j^{3/2} m_b J_1(j^{3/2} m_b r) - B_0'' j^{1/2} m_b K_1(j^{1/2} m_b r)}{j\omega\mu\sigma_b}\epsilon^{j(\omega t - \beta z)}\right]\boldsymbol{a}_\phi \tag{14-60}$$

The magnetic field between the conductors is the following whether the outer conductor is thick-walled or thin-walled:

$$\boldsymbol{H}_1 = \mathrm{Re}\left[\frac{I_{1M}}{2\pi r}\epsilon^{j(\omega t - \beta z)}\right]\boldsymbol{a}_\phi \tag{11-12}$$

The right-hand sides of Eqs. 14-60 and 11-12 may be set equal to each other with r set equal to r_b. The resulting equation and Eq. 14-59 may be solved simultaneously for the constants A_0'' and B_0'', as called for in Prob. 14-7.

14.5 CONCLUSIONS

Finite conductivity in a coaxial line limits the current density to finite values and thereby causes it to be distributed continuously over the cross section of each conductor rather than concentrated on the surface closest to the other conductor. This in turn creates a magnetic field within each conductor. The current density is not uniform, however, unless the frequency approaches zero. This makes the effective resistance to alternating current higher than that to direct current.

For a solid, homogeneous inner conductor (or any solid, homogeneous circular conductor with a circularly symmetrical current distribution) the ratio of effective resistance to DC resistance, or skin-effect-resistance ratio r_a/r_{adc} may be described by the parameter mr_a, where r_a is the conductor radius and $m = \sqrt{\omega\mu\sigma}$. The skin-effect-demarcation frequency f_{skd} will be defined as the frequency at which $mr_a = 3/\sqrt{2}$:

$$f_{skd} = \frac{9}{4\pi\mu\sigma r_a^2} \tag{14-55}$$

The skin-effect-resistance ratio at $f = f_{skd}$ is about 1.097; it is essentially unity at lower frequencies, and for higher frequencies it quickly approaches the asymptote

$$\frac{r_a}{r_{adc}} \approx \frac{mr_a}{2\sqrt{2}} + \frac{1}{4} \tag{14-45}$$

The function and its asymptotes are shown in Fig. 14-4.

For usual telephone lines, voice-frequency transmission is largely in the range below the skin-effect-demarcation frequency, whereas carrier-frequency transmission is primarily in the range above that frequency.

PROBLEMS

14-1. Verify that the change of variables specified by Eq. B-21 will reduce Eq. 14-5 to the form of Bessel's differential equation given in Eq. B-1.

14-2. Let $\boldsymbol{H} = \text{Re}[H(r)\epsilon^{j(\omega t - \beta z)}]\boldsymbol{a}_\phi$. By substitution in Maxwell's equations, derive a differential equation for $H(r)$ within the conducting regions of a coaxial cable. Change variables in accordance with Eq. B-21 and reduce to the form of Eq. B-1. *Note*: The resulting equation will have n equal to unity rather than zero, as was the case with $f_J(r)$. Write the general solution for $\boldsymbol{H}$.

(a) Substitute the following boundary conditions for the inner conductor: (i) $\boldsymbol{H}$ vanishes at the center of the conductor, and (ii) the $\boldsymbol{H}$ field is continuous at the surface of the conductor. (Use Eq. 11-12 for $\boldsymbol{H}$ outside the conductor.)

(b) Write the corresponding boundary conditions for $\boldsymbol{H}$ in the outer conductor (assuming it to be "thick-walled") and substitute in the solution for the differential equation. (Answers should agree with Eqs. 14-25 and 14-26.)

14-3. Verify the approximation for $J_0(p)/J_1(p)$ shown in Eq. 14-52.

14-4. Compare the first-order asymptotic approximation for r_a as given in Eq. 14-45 with the DC resistance of a strip of thickness δ and width $2\pi r_a$. Also find r_a in terms of the surface resistance R_S as given by Eq. 10-75.

14-5. Find asymptotic, high-frequency approximations for the magnetic field and the conduction-current density inside the conductors of a coaxial line. Let $r = r_a - \Delta r_a$ for the inner conductor, and $r = r_b + \Delta r_b$ for the outer conductor. Compare these with the corresponding fields in a good conductor with a plane surface, caused by a normally incident field, Eqs. 10-44 and 10-62, respectively.

14-6. Prepare a chart with the same ordinate variable as in Fig. 14-5, but with the resistance to direct current, in ohms per meter, as the abscissa.

14-7. Evaluate the constants A_0'' and B_0'' for the current-density function in a thin-walled outer conductor of a coaxial line (Eq. 14-58). *Partial answer*:

$$A_0'' = \frac{I_{1M} m_b K_1(j^{1/2} m_b r_c)}{2\pi r_b \sqrt{j}\left[J_1(j^{3/2} m_b r_c) K_1(j^{1/2} m_b r_b) - J_1(j^{3/2} m_b r_b) K_1(j^{1/2} m_b r_c)\right]}$$

REFERENCES

1. Dwight, H. B., Reactance and skin effect of concentric tubular conductors, *AIEE Trans.*, 61, 513, July, 1942.
2. Dwight, H. B., *Electrical Coils and Conductors*, McGraw-Hill, New York, 1945.
3. Higgins, T. J., The origins and developments of the concepts of inductance, skin effect, and proximity effect, *Amer. J. Phys.*, 9, no. 6, 337, 1941.
4. Kennelly, A. E., Laws, F. A., and Pierce, P. H., Experimental researches on skin effect in conductors, *AIEE Trans.*, 34, part II, 1953, 1915.
5. McLachlan, N. W., *Bessel Functions for Engineers*, 2nd ed., Oxford University Press, London, 1961.
6. Rosa, E. B. and Grover, F. W., Formulas and tables for the calculation of mutual and self inductance, Sci. Paper 169, *Bull. Natl. Bur. Std. (U.S.)*, 8 no. 1, 1, 1912 (see pp. 173–77, 180).

APPENDIX A

Vector Analysis — Definitions and Formulas

The following presentation is meant primarily to refresh the memory of the reader in conjunction with the derivations in Chaps. 9 through 13. Rigorous treatments may be found in the references listed at the close of this appendix.

The properties of vectors and their operations, and the physical concepts to which they are applied, exist independently of coordinate systems, and will be so examined initially. Coordinate systems are useful auxiliaries for detailed study of particular geometric assemblies. The forms to which the various functions of vector analysis reduce will be listed for the rectangular and circular cylindrical systems.

A-1 VECTOR OPERATIONS AND FUNCTIONS

Vectors are space-directed quantities; in the present text their application will be limited to functions in three-space which are descriptive of physical behavior.

a. Vector Algebra[2, 3]

Vector-algebra operations of immediate interest are (1) the equating of one vector to another; (2) addition and subtraction of vectors; (3) scalar, or dot, multiplication of two vectors; and (4) vector, or cross, multiplication of two vectors.

(1) *Equality among Vectors.* Two vectors are said to be equal to each other if their directions are parallel and their magnitudes equal. The latter property can be true in physical applications only if the two quantities represented have the same dimensions. When the vectors are resolved into components along coordinate directions, the criterion of equality requires that equality exist between corresponding components.

(2) Addition and Subtraction of Vectors. The addition of vectors may be visualized geometrically as follows: (1) represent each vector with a line segment parallel to the vector, making the length proportional to the magnitude of the vector; (2) designate the ends of each as "base" and "tip" such that the direction from base to tip is the direction of the vector represented; (3) translate one line segment in space (moving it such that it remains parallel to its original position) until the tip of one touches the base of the other; then (4) a line segment drawn from the base of the former to the tip of the latter represents the vector sum or resultant. This is shown graphically in Fig. A-1(a). If one vector is to be subtracted from another, the first vector is reversed in direction and then added to the second. The operation of addition is physically meaningful only among quantities which have the same dimensions.

(3) Scalar Multiplication. The scalar product of two vectors may be indicated as follows:

$$W = \boldsymbol{F} \cdot \boldsymbol{L} \tag{A-1}$$

It is evaluated by multiplying together the magnitudes of vectors $\boldsymbol{F}$ and $\boldsymbol{L}$ and the cosine of the angle between them. The result W is a scalar, devoid of any space-directed properties, but it has an algebraic sign which depends on the cosine of the angle between the vectors; thus it is positive if that angle is less than a right angle and negative if the angle is greater. If the angle between the two vectors is a right angle, the scalar product vanishes.

(4) Vector Multiplication. The vector product of two vectors may be indicated as follows:

$$\boldsymbol{A} \times \boldsymbol{B} = \boldsymbol{C} \tag{A-2}$$

The magnitude of $\boldsymbol{C}$ is equal to the product of the magnitudes of $\boldsymbol{A}$ and $\boldsymbol{B}$ and the sine of the angle between them, if the angle is measured so that it is less than π radians. In terms of Fig. A-1(b),

$$|\boldsymbol{C}| = |\boldsymbol{A}||\boldsymbol{B}| \sin \theta_{AB} \tag{A-3}$$

The magnitude, and hence the vector product, vanishes if vectors $\boldsymbol{A}$ and $\boldsymbol{B}$ are collinear.

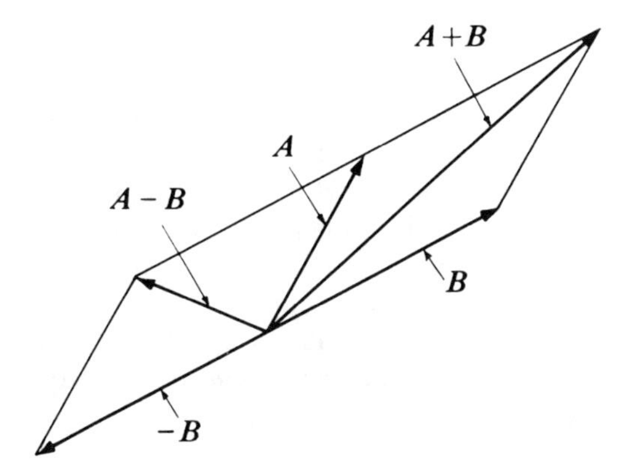

(a) Vector addition and subtraction

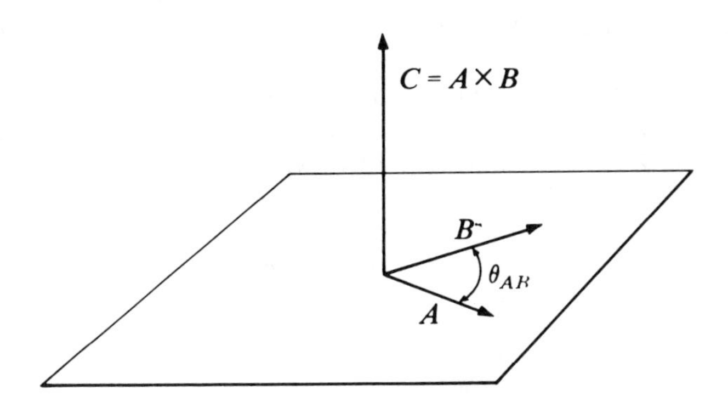

(b) Vector product of two vectors

Figure A-1. Operations in vector algebra.

The direction of $\boldsymbol{C}$ is perpendicular to the plane of vectors $\boldsymbol{A}$ and $\boldsymbol{B}$, and ambiguity is avoided by including in the definition a right-hand rule. The product vector is assigned the direction in which a right-handed screw would advance when rotated in the direction from the first-named vector ($\boldsymbol{A}$) to the second-named vector ($\boldsymbol{B}$), going through the smaller angle.

Interchanging $\boldsymbol{A}$ and $\boldsymbol{B}$ in the foregoing discussion leads to the following conclusion:

$$\boldsymbol{A} \times \boldsymbol{B} = -\boldsymbol{B} \times \boldsymbol{A} \tag{A-4}$$

b. Fields and Field Integrals

A *field* is a function which has a value at every point in space or within some designated region. If the function has both a direction and a magnitude, it is called a *vector field*; if it has magnitude only, it is called a *scalar field*. Dependence on location may be indicated explicitly by labeling them as functions of a *general position vector* $\tilde{r}$ in this manner: $\boldsymbol{A}(\tilde{r})$ for a vector field, and $V(\tilde{r})$ for a scalar field.

Three operations in field theory involving integration are (1) the *volume integral* (of a scalar field), (2) the *line integral* (of a vector field), and (3) the *surface integral* (of a vector field).

(1) *Volume Integral.* The volume integral is evaluated simply by integrating the given scalar function with respect to the coordinate directions throughout some specified volume. For example, the volume integral of the air-density function will yield the mass of air in a given volume.

(2) *Line Integral.* An example which will introduce the line integral is the calculation of the work done in moving along some prescribed path from location $\tilde{r}_A$ to $\tilde{r}_B$ in a force field $\boldsymbol{F}$, which depends solely on position in that field:

$$W_{AB} = -\int_{\tilde{r}_B}^{\tilde{r}_A} \boldsymbol{F} \cdot d\boldsymbol{L} \tag{A-5}$$

Figure A-2 shows the force vectors and differential-displacement vectors at several points along such a path.

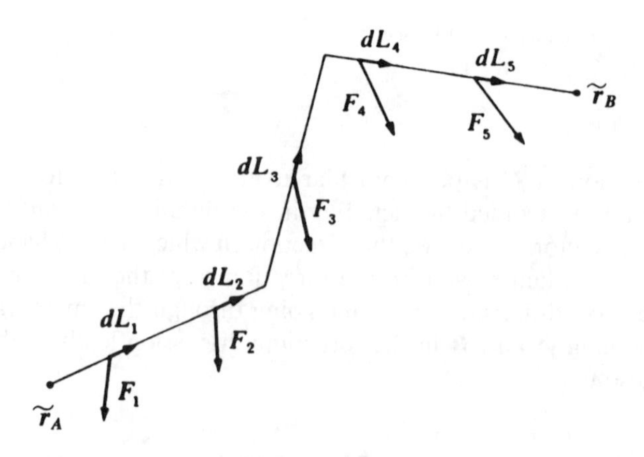

Figure A-2. Vector relationships for line integral.

The operation called for is (1) taking the scalar product of each differential segment of path and the vector field at the corresponding location, and (2) summing, by means of ordinary integration, those differential scalar products.

An important class of vector fields has the property that the line integral between any two end points is independent of the particular path traversed. Such a field is said to be *conservative*. (The gravitational field, the static $\boldsymbol{E}$ field, and those portions of the static $\boldsymbol{H}$ field that do not enclose any current, are members of this class, but time-varying $\boldsymbol{E}$, $\boldsymbol{H}$, $\boldsymbol{D}$, and $\boldsymbol{B}$ fields are not.)

A special type of line integral is that for which the path of integration is a closed curve, and this will be indicated with a small circle superposed on the integral sign. If the field is conservative, the line integral around any closed path is zero. In terms of the static $\boldsymbol{E}$ field, this result is stated as

$$\oint \boldsymbol{E} \cdot d\boldsymbol{L} = 0 \quad (\boldsymbol{E} \text{ not a function of time}) \tag{A-6}$$

(3) *Surface Integral.* An example which will introduce the surface integral is the relationship between the current density within a conducting mass and the current i entering or leaving part of it through a specified cross section. This is illustrated in Fig. A-3. The current density constitutes a vector field $\boldsymbol{J}$, the magnitude and direction of which may differ from one

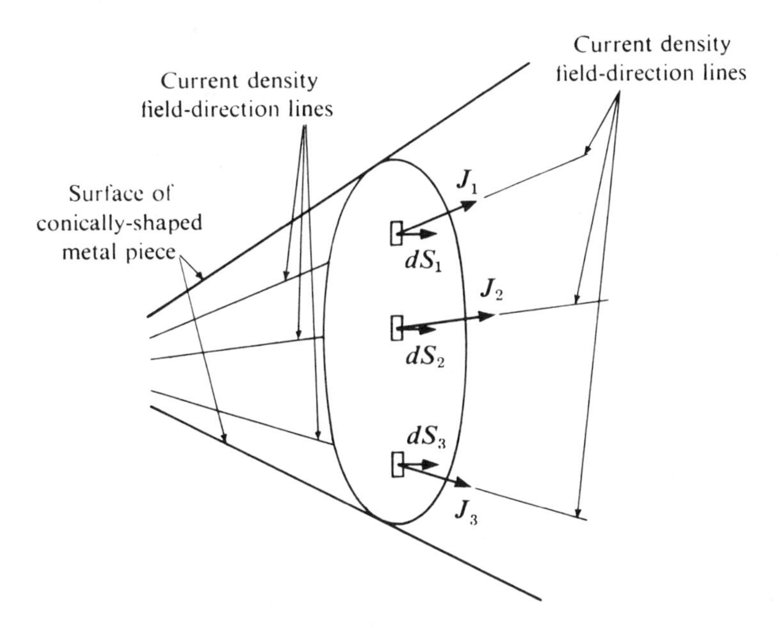

Figure A-3. Geometry for evaluation of surface integral (surface bounded by closed curve).

location to another. A given cross section may be divided into differential-area surfaces, each represented by a vector $\boldsymbol{dS}$ which is normal to its orientation plane and is assigned a magnitude equal to the differential area. The current through each differential area is equal to the current density at the point multiplied by the differential area and by the cosine of the angle between $\boldsymbol{J}$ and $\boldsymbol{dS}$. The scalar product expedites the statement of this relationship:

$$di = \boldsymbol{J} \cdot \boldsymbol{dS}$$

The sum of the differential currents for all the differential areas in the cross section is the resultant current:

$$i = \iint \boldsymbol{J} \cdot \boldsymbol{dS} \tag{A-7}$$

This is the surface integral, a scalar quantity. In this instance the limits are understood to be fixed by the closed curve formed by the intersection of the given cross-sectional surface with the walls of the conducting mass. Bases for choosing limits depend on the particular physical problem.

The same mathematical operation is applied to the $\boldsymbol{B}$ and $\boldsymbol{D}$ fields in electromagnetic theory, and the resulting scalar quantities are known as *magnetic flux* (webers) and *electric flux* (coulombs), respectively.

If the angle between the current-density (or flux-density) vector and the vector representing area is less than a right angle, the differential current or flux is positive, whereas if that angle is greater than a right angle, the differential current or flux is negative. One must decide at the outset which direction through the surface is to be considered positive.

Integrals over *enclosing* surfaces are used in the derivation of Maxwell's equations. An enclosed-surface integral will be designated by a small ellipse or oval linking the two integral signs ($\oiint$). Conventionally the direction *outward* from an enclosing surface is considered *positive*. It may be noted that flux which enters an enclosing surface offsets that which leaves it, and, if the two amounts are equal, the surface integral is zero.

c. Field-Function Derivatives

These first-order derivatives of field functions are useful: (1) the *divergence* of a vector field, (2) the *curl* of a vector field, and (3) the *gradient* of a scalar field. From these, two second-order derivatives, the *Laplacian of a scalar field* and the *Laplacian of a vector field*, have been defined.

(1) *The Divergence Function.* As an example, consider a small enclosing surface in the electric-flux density $\boldsymbol{D}$ field, in a region of finite charge density. If the volume enclosed is small enough, the charge density will be essentially uniform throughout that volume, and the volume integral becomes simply the product of the local charge density and the volume, which may be indicated by Δv. Volume may then be divided into the surface integral to yield an expression for average charge density in terms of the surface integral. Let the enclosing surface be shrunk around some discrete point. The given ratio then approaches, as a limit, the charge density at that point. This mathematical result is defined as the *divergence* of $\boldsymbol{D}$, and is written div $\boldsymbol{D}$ or $\nabla \cdot \boldsymbol{D}$. It is a scalar function of location in the $\boldsymbol{D}$ field:

$$\nabla \cdot \boldsymbol{D} = \left[\frac{\oiint \boldsymbol{D} \cdot d\boldsymbol{S}}{\Delta v} \right]_{\lim \Delta v \to 0} \tag{A-8}$$

It may be noted that the inverted delta symbol introduced above is customarily called "del" when it is designated separately. In applications *in rectangular coordinates* the symbol del resembles a vector, but it does not transform as a vector into other coordinate systems.[1] The paragraph preceding Eq. A-8 gives the *definition* of divergence, and accordingly it is preferable that the expression "$\nabla \cdot \boldsymbol{D}$" be read "divergence of D" rather than "del dot D." The same principle applies to other derivative functions with symbols which incorporate del.

(2) *The Curl Function.* Ampère's law for static fields, Eq. 9-2, provides a good example from which to develop this concept. Substitution of Eq. A-7 for i yields the following in terms of the current-density vector field $\boldsymbol{J}$:

$$\oint \boldsymbol{H} \cdot d\boldsymbol{L} = \iint \boldsymbol{J} \cdot d\boldsymbol{S}$$

If an exceedingly small planar loop within the current-carrying region is visualized as the bounding edge of the surface and as the path of line integration, the current density will be essentially uniform across that surface, and the surface integral may be replaced by the product of the area (which will be denoted by ΔS) and the component of current density normal to the plane of the loop. Next the loop may be revolved until that product is a maximum, in other words, so that the plane of the loop is normal to the local current-density vector field. If the line integral around the loop is divided by the area bounded by the loop and the limit taken by shrinking the loop onto a point, the result is the current density at that point, and the mathematical operation is defined as the *curl* of $\boldsymbol{H}$. This is written as curl $\boldsymbol{H}$ or $\nabla \times \boldsymbol{H}$. It is a vector field which is a function of location in the $\boldsymbol{H}$ field; its orientation at every point is normal to the corresponding differential surface, and it is directed such that the path taken for the line integral is clockwise when one looks along the curl vector from base to tip:

$$\nabla \times \boldsymbol{H} = \left[\frac{\oint \boldsymbol{H} \cdot d\boldsymbol{L}}{\Delta S} \right] \boldsymbol{a}_{\max_{\lim \Delta S \to 0}} \tag{A-9}$$

Here $\boldsymbol{a}_{\max}$ is a unit vector which is so directed as to maximize the coefficient by which it is multiplied.

(3) *Scalar Potential Function and Gradient.* In Sec. A-1b a conservative field was defined as one in which line integrals are independent of path. With respect to some reference "ground" point, one may locate within such a field *equipotential surfaces*, for which the line integral to every point on any such surface is the same. The static $\boldsymbol{E}$ field is of this type, and the equipotential surfaces would represent voltages. The assembly of such surfaces may be generalized into a *scalar potential field* V.

Between any two closely-spaced equipotential surfaces, the $\boldsymbol{E}$ field would be normal to those surfaces, and equal in magnitude to the potential difference between them, divided by the distance. This is expressed on a differential basis by defining a vector function known as the *gradient*, symbolized by ∇V for the scalar field V. The direction of the gradient vector at any point is that of the maximum rate of increase in the scalar field:

$$\nabla V = \frac{\partial V}{\partial u} \boldsymbol{a}_{\max} \tag{A-10}$$

$$\boldsymbol{E} = -\nabla V \quad \text{(static field)} \tag{A-11}$$

From the definition of the curl function (Eq. A-9) in terms of a closed line integral, it follows that if the line integral is independent of path in some vector field such as $\boldsymbol{E}$, then any closed line integral would equal zero, and hence curl $\boldsymbol{E}$ would also be zero. The following is a general proposition, which the reader may test for the various coordinate systems which are introduced in Sec. A-2:

$$\nabla \times (\nabla V) = 0 \tag{A-12}$$

(4) ***Laplacians of Scalar and Vector Fields.*** The following is defined as the Laplacian of a scalar field V:

$$\nabla^2 V = \nabla \cdot (\nabla V) \tag{A-13}$$

This second-order differential function is common to many physical problems, both electromagnetic and in other domains.

The Laplacian of a vector field $\boldsymbol{A}$ is defined as

$$\nabla^2 \boldsymbol{A} = -\nabla \times (\nabla \times \boldsymbol{A}) + \nabla(\nabla \cdot \boldsymbol{A}) \tag{A-14}$$

In the analysis of electromagnetism, the Laplacian of a vector field arises when Maxwell's two equations involving $\nabla \times \boldsymbol{E}$ and $\nabla \times \boldsymbol{H}$ (Eqs. 9-23 and 9-21) are solved simultaneously in general form so as to eliminate $\boldsymbol{E}$ or $\boldsymbol{H}$, as in Sec. 12-3. When the resulting vector wave equation (Eq. 12-52) is to be applied to a specific geometric situation, it may be substituted in a particular coordinate system. Scalar wave equations in terms of components of $\boldsymbol{E}$ or $\boldsymbol{H}$ may be partitioned from it, and these equations include combinations corresponding to the Laplacian of a scalar field.

d. Integral Theorems

The following are general theorems in vector analysis.[2, 3]

1. The divergence theorem:

$$\oint\!\!\!\oint \boldsymbol{A} \cdot d\boldsymbol{S} = \iiint (\nabla \cdot \boldsymbol{A})\, dv \tag{A-15}$$

2. Stokes' theorem:

$$\oint \boldsymbol{A} \cdot d\boldsymbol{L} = \iint (\nabla \times \boldsymbol{A}) \cdot d\boldsymbol{S} \tag{A-16}$$

These identities are useful in the reduction of equations which contain both a surface integral and a volume integral, as in Eq. 9-18, or both a line integral and a surface integral, as in Eqs. 9-10 and 9-11.

A-2 COORDINATE SYSTEMS AND COMPONENT RESOLUTION

Coordinate systems in three-space have been devised in order that geometrical relationships and geometrically related functions may be stated in algebraic form (or may be represented graphically by projection on defined surfaces). Many systems have been studied and applied in physical problems. An important class is composed of systems in which the three reference directions are everywhere mutually perpendicular; such systems are said to be *orthogonal.*

Two orthogonal systems, the Cartesian, or rectangular, and the circular cylindrical, adapt directly to the characteristic symmetry of the field problems which are examined in this text. A book emphasizing the mathematical aspect of vector analysis would focus attention on the derivation of the principal vector operators and operations (differential length and area, divergence, etc.) in different coordinate systems, and to their transformation from one system to another. Here the resulting forms will be given directly.

a. Rectangular Coordinates

The rectangular coordinate system is based on three mutually perpendicular coordinate axes, each axis straight and of infinite extent. The set of axes may be initially oriented, however, to suit the user. The directions of the coordinate axes may be designated with three unit vectors (having direction and unit magnitude, but no dimensions) $\boldsymbol{a}_x$, $\boldsymbol{a}_y$, and $\boldsymbol{a}_z$. These are arranged in right-handed order, which may be readily visualized if $\boldsymbol{a}_x$ is directed upwardly and $\boldsymbol{a}_y$ to the right; then $\boldsymbol{a}_z$ is directed away from the observer.

Differential length and differential volume have the following forms in rectangular coordinates:

$$\boldsymbol{dL} = dx\,\boldsymbol{a}_x + dy\,\boldsymbol{a}_y + dz\,\boldsymbol{a}_z \tag{A-17}$$

$$dv = dx\,dy\,dz \tag{A-18}$$

The specific forms of the scalar product and the vector product in terms of rectangular-coordinate components may be set forth. The various products among the three unit vectors will be noted first. In the first three of the following equations, m and n may represent x, y, or z, but it is understood that m and n may not represent the same direction:

$$\begin{aligned} \boldsymbol{a}_m \cdot \boldsymbol{a}_m &= 1 \\ \boldsymbol{a}_m \cdot \boldsymbol{a}_n &= 0 \\ \boldsymbol{a}_m \times \boldsymbol{a}_m &= 0 \end{aligned} \tag{A-19}$$

$$\begin{aligned} \boldsymbol{a}_x \times \boldsymbol{a}_y = \boldsymbol{a}_z &= -\boldsymbol{a}_y \times \boldsymbol{a}_x \\ \boldsymbol{a}_y \times \boldsymbol{a}_z = \boldsymbol{a}_x &= -\boldsymbol{a}_z \times \boldsymbol{a}_y \\ \boldsymbol{a}_z \times \boldsymbol{a}_x = \boldsymbol{a}_y &= -\boldsymbol{a}_x \times \boldsymbol{a}_z \end{aligned} \tag{A-20}$$

The scalar and vector products of two general vectors $\boldsymbol{K}$ and $\boldsymbol{L}$ are

$$\boldsymbol{K} \cdot \boldsymbol{L} = K_x L_x + K_y L_y + K_z L_z \tag{A-21}$$

$$\boldsymbol{K} \times \boldsymbol{L} = (K_y L_z - K_z L_y)\boldsymbol{a}_x + (K_z L_x - K_x L_z)\boldsymbol{a}_y + (K_x L_y - K_y L_x)\boldsymbol{a}_z \tag{A-22}$$

The vector product may also be stated in determinant form as a memory aid:

$$\boldsymbol{K} \times \boldsymbol{L} = \begin{vmatrix} \boldsymbol{a}_x & \boldsymbol{a}_y & \boldsymbol{a}_z \\ K_x & K_y & K_z \\ L_x & L_y & L_z \end{vmatrix} \tag{A-23}$$

Let a vector field $\boldsymbol{A}$ be defined as

$$\boldsymbol{A} = A_x \boldsymbol{a}_x + A_y \boldsymbol{a}_y + A_z \boldsymbol{a}_z \tag{A-24}$$

The divergence and curl of $\boldsymbol{A}$, stated in rectangular coordinates, are

$$\nabla \cdot \boldsymbol{A} = \frac{\partial A_x}{\partial x} + \frac{\partial A_y}{\partial y} + \frac{\partial A_z}{\partial z} \tag{A-25}$$

$$\nabla \times \boldsymbol{A} = \left(\frac{\partial A_z}{\partial y} - \frac{\partial A_y}{\partial z} \right) \boldsymbol{a}_x + \left(\frac{\partial A_x}{\partial z} - \frac{\partial A_z}{\partial x} \right) \boldsymbol{a}_y + \left(\frac{\partial A_y}{\partial x} - \frac{\partial A_x}{\partial y} \right) \boldsymbol{a}_z \tag{A-26}$$

The curl may also be stated in determinant form:

$$\nabla \times \boldsymbol{A} = \begin{vmatrix} \boldsymbol{a}_x & \boldsymbol{a}_y & \boldsymbol{a}_z \\ \dfrac{\partial}{\partial x} & \dfrac{\partial}{\partial y} & \dfrac{\partial}{\partial z} \\ A_x & A_y & A_z \end{vmatrix} \tag{A-27}$$

The gradient and Laplacian of a scalar function V, in rectangular coordinates, are

$$\nabla V = \frac{\partial V}{\partial x}\boldsymbol{a}_x + \frac{\partial V}{\partial y}\boldsymbol{a}_y + \frac{\partial V}{\partial z}\boldsymbol{a}_z \tag{A-28}$$

$$\nabla^2 V = \frac{\partial^2 V}{\partial x^2} + \frac{\partial^2 V}{\partial y^2} + \frac{\partial^2 V}{\partial z^2} \tag{A-29}$$

b. Cylindrical Coordinates

A set of circularly cylindrical coordinates is built around a longitudinal axis, whose orientation may be chosen at will and along which the unit vector $\boldsymbol{a}_z$ may be pointed in either direction. The radial unit vector $\boldsymbol{a}_r$ is perpendicular to the axis and points outward. The azimuthal unit vector $\boldsymbol{a}_\phi$ is perpendicular to both $\boldsymbol{a}_r$ and $\boldsymbol{a}_z$ at each given location in space. The three unit vectors form a right-handed set, when taken in the order $\boldsymbol{a}_r$, $\boldsymbol{a}_\phi$, and $\boldsymbol{a}_z$.

Differential length and differential volume have the following forms:

$$\boldsymbol{dL} = dr\,\boldsymbol{a}_r + r\,d\phi\,\boldsymbol{a}_\phi + dz\,\boldsymbol{a}_z \tag{A-30}$$

$$dv = r\,d\phi\,dr\,dz \tag{A-31}$$

It should be noted that whereas $\boldsymbol{a}_r$ and $\boldsymbol{a}_\phi$ have the same dimensions (both are dimensionless), dr and $d\phi$ do not (dr has the dimensions of length; $d\phi$ is dimensionless).

The three equations in set A-19 for scalar and vector products among unit vectors are applicable to cylindrical coordinates if m and n represent r, ϕ, or z. The other vector products among the unit vectors are

$$\begin{aligned} \boldsymbol{a}_r \times \boldsymbol{a}_\phi &= \boldsymbol{a}_z = -\boldsymbol{a}_\phi \times \boldsymbol{a}_r \\ \boldsymbol{a}_\phi \times \boldsymbol{a}_z &= \boldsymbol{a}_r = -\boldsymbol{a}_z \times \boldsymbol{a}_\phi \\ \boldsymbol{a}_z \times \boldsymbol{a}_r &= \boldsymbol{a}_\phi = -\boldsymbol{a}_r \times \boldsymbol{a}_z \end{aligned} \tag{A-32}$$

The scalar and vector products of two general vectors $\boldsymbol{K}$ and $\boldsymbol{L}$ are

$$\boldsymbol{K}\cdot\boldsymbol{L} = K_r L_r + K_\phi L_\phi + K_z L_z \tag{A-33}$$

$$\boldsymbol{K}\times\boldsymbol{L} = (K_\phi L_z - K_z L_\phi)\boldsymbol{a}_r + (K_z L_r - K_r L_z)\boldsymbol{a}_\phi + (K_r L_\phi - K_\phi L_r)\boldsymbol{a}_z \tag{A-34}$$

Let a vector field $\boldsymbol{A}$ be defined as

$$\boldsymbol{A} = A_r\boldsymbol{a}_r + A_\phi\boldsymbol{a}_\phi + A_z\boldsymbol{a}_z \tag{A-35}$$

The divergence and curl of $\boldsymbol{A}$, stated in cylindrical coordinates, are

$$\nabla\cdot\boldsymbol{A} = \frac{1}{r}\frac{\partial(rA_r)}{\partial r} + \frac{1}{r}\frac{\partial A_\phi}{\partial\phi} + \frac{\partial A_z}{\partial z} \tag{A-36}$$

$$\nabla\times\boldsymbol{A} = \left(\frac{1}{r}\frac{\partial A_z}{\partial\phi} - \frac{\partial A_\phi}{\partial z}\right)\boldsymbol{a}_r + \left(\frac{\partial A_r}{\partial z} - \frac{\partial A_z}{\partial r}\right)\boldsymbol{a}_\phi + \left(\frac{1}{r}\frac{\partial(rA_\phi)}{\partial r} - \frac{1}{r}\frac{\partial A_r}{\partial\phi}\right)\boldsymbol{a}_z \tag{A-37}$$

The gradient and Laplacian of a scalar function V are

$$\nabla V = \frac{\partial V}{\partial r}\boldsymbol{a}_r + \frac{1}{r}\frac{\partial V}{\partial\phi}\boldsymbol{a}_\phi + \frac{\partial V}{\partial z}\boldsymbol{a}_z \tag{A-38}$$

$$\nabla^2 V = \frac{1}{r}\frac{\partial(r\,\partial V/\partial r)}{\partial r} + \frac{1}{r^2}\frac{\partial^2 V}{\partial\phi^2} + \frac{\partial^2 V}{\partial z^2} \tag{A-39}$$

REFERENCES

1. Morse, P. M. and Feshback, H., *Methods of Theoretical Physics*, McGraw-Hill, New York, 1953, 44.
2. Spiegel, M. R., *Theory and Problems of Vector Analysis*, Schaum, New York, 1959.
3. Wills, A. P., *Vector Analysis, with an Introduction to Tensor Analysis*, Prentice-Hall, New York, 1931. Also Dover Publications, Inc., 1958.

REFERENCES

APPENDIX B
Bessel Functions

Bessel's differential equation is characteristic of three-space problems which involve circular cylindrical symmetry, such as wave propagation in a circular waveguide (Sec. 13-1), and the current density field in the conductors of a coaxial cable (Chap. 14). The solutions to this equation, Bessel functions, also arise in problems which have no apparent relationship to geometric form, such as the step-function response of (a) a lossy transmission line when the parameters are assumed to be frequency-independent [Sec. 5-3d(1) footnote], (b) a lumped-parameter delay line or artificial line (Sec. 8-3), or (c) a waveguide (Sec. 12-5c), and also in the frequency-modulation and velocity-modulation spectra, and elsewhere in mathematical physics.

The basic form is

$$\frac{d^2w}{dp^2} + \frac{1}{p}\frac{dw}{dp} + w\left(1 - \frac{n^2}{p^2}\right) = 0 \tag{B-1}$$

The independent variable p may be real or complex.

Power series and the leading terms of asymptotic expansions for solutions will be given in the following text, but it should be noted that *empirical* approximations, which converge more rapidly, have been derived.[1]

B-1 INDEPENDENT VARIABLE REAL

If the variable p is real, the general solution is conventionally written as

$$w(p) = A_n J_n(p) + B_n Y_n(p) \tag{B-2}$$

Here $J_n(p)$ designates the Bessel function of the first kind, nth order; $Y_n(p)$ designates the Bessel function of the second kind, nth order; and A_n and B_n are constants which are determined for each application by the boundary conditions. Some comments on Bessel functions and their properties are in order.[3]

a. Functions of the First Kind

The function of the first kind is obtainable by assuming, for the function $J_n(p)$, an infinite power series with undetermined coefficients and substituting it and its derivatives into Eq. B-1 in place of w, dw/dp, and d^2w/dp^2. As an illustration, the zero-order case $n = 0$ will be considered. Let

$$J_0(p) = a_0 + a_1 p + a_2 p^2 + a_3 p^3 + \cdots \tag{B-3}$$

Equation B-3 is differentiated twice and substituted into B-1 after n has been set equal to zero:

$$2a_2 + 6a_3 p + 12a_4 p^2 + \cdots + \frac{a_1}{p} + 2a_2 + 3a_3 p + 4a_4 p^2 + \cdots$$

$$+ a_0 + a_1 p + a_2 p^2 + \cdots = 0$$

Since the latter equation must be satisfied for all values of p, the coefficients of each power of p in the equation, when grouped together, must separately sum to zero:

$$\frac{a_1}{p} + (4a_2 + a_0) + (9a_3 + a_1)p + (16a_4 + a_2)p^2 + \cdots = 0$$

$$\begin{aligned} a_1 &= 0 \\ 4a_2 + a_0 &= 0 \\ 9a_3 + a_1 &= 0 \\ 16a_4 + a_2 &= 0 \end{aligned} \tag{B-4}$$

The inclusion of higher-order terms in Eq. B-3 will yield additional equations for set B-4. In general,

$$k^2 a_k + a_{k-2} = 0 \tag{B-5}$$

and so

$$J_0(p) = a_0\left(1 - \frac{p^2}{2^2} + \frac{p^4}{2^2 \cdot 4^2} - \frac{p^6}{2^2 \cdot 4^2 \cdot 6^2} + \cdots\right)$$

The constant a_0 might have any nonzero value; for the compilation of tables it has regularly been set equal to unity. It should be noted that the coefficients of all odd-power terms in p are zero. The preceding expression may be restated more conveniently as follows, where the k of the

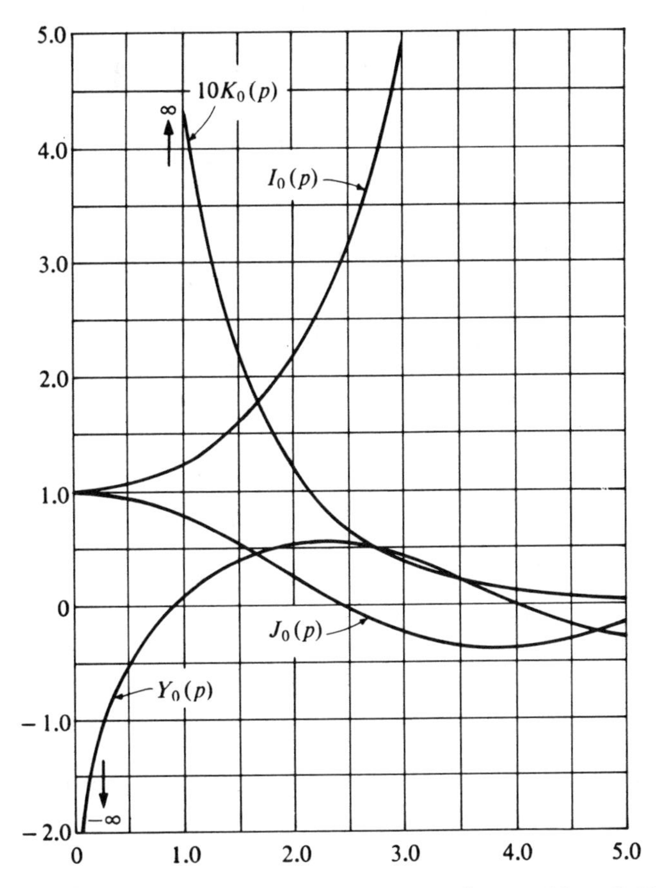

Figure B-1. Zero-order Bessel functions, J, Y, I, and K. (Prepared from G. N. Watson, *Theory of Bessel Functions*, Cambridge University Press, London, 1944, 666.)

recurrence formula is replaced by $2i$:

$$J_0(p) = 1 - \left(\frac{p}{2}\right)^2 + \left(\frac{p}{2}\right)^4 \frac{1}{(2!)^2} - \left(\frac{p}{2}\right)^6 \frac{1}{(3!)^2} + \cdots + (-1)^i \left(\frac{p}{2}\right)^{2i} \frac{1}{(i!)^2} \tag{B-6}$$

This series proves to be convergent for all values of p real or complex. The function $J_0(p)$ is plotted in Fig. B-1 along with three other Bessel functions, and $J_0(p)$ and $J_1(p)$ have been plotted in Fig. 13-1.

The general series for order n is

$$J_n(p) = \frac{(p/2)^n}{n!}\left[1 - \frac{(p/2)^2}{n+1} + \frac{(p/2)^4}{2!(n+1)(n+2)} - \frac{(p/2)^6}{3!(n+1)(n+2)(n+3)} + \cdots\right] \tag{B-7}$$

The following are useful identities:

$$\frac{dJ_0(hr)}{dr} = hJ_1(hr) \tag{B-8}$$

$$\frac{dJ_m(hr)}{dr} = [J_{m-1}(hr) - J_{m+1}(hr)]\frac{h}{2} \tag{B-9}$$

$$\frac{dJ_m(hr)}{dr} = \frac{m}{r}J_m(hr) - hJ_{m+1}(hr) \tag{B-10}$$

$$\frac{dJ_m(hr)}{dr} = -\frac{m}{r}J_m(hr) + hJ_{m-1}(hr) \tag{B-11}$$

$$\int pJ_0(p)\,dp = pJ_1(p) \tag{B-12}$$

The leading term of the asymptotic expansion may be designated by $\tilde{J}_n(p)$:

$$\tilde{J}_n(p) = \sqrt{\frac{2}{\pi p}}\cos\left(p - \frac{\pi}{4} - \frac{n\pi}{2}\right) \qquad p \gg 1 \tag{B-13}$$

b. Functions of the Second Kind

A second-order linear differential equation must have two linearly independent solutions. The derivation of the Bessel function of the second kind[4] is beyond the scope of this book, but some discussion of the function itself is appropriate. The form of solution credited to Weber [designated $Y_n(p)$ in Eq. B-2] has been generally adopted[3]:

$$Y_0(p) = \frac{2}{\pi}\left[\gamma + \ln\left(\frac{p}{2}\right)\right]J_0(p) - \frac{2}{\pi}\sum_{r=1}^{\infty}\left[(-1)^r\left(\frac{p}{2}\right)^{2r}\frac{1}{(r!)^2}\right]\left[1 + \frac{1}{2} + \frac{1}{3} + \cdots + \frac{1}{r}\right] \quad \text{(B-14)}$$

Here γ is Euler's constant, approximately 0.577216:

$$Y_n(p) = \frac{2}{\pi}\left[\gamma + \ln\left(\frac{p}{2}\right)\right]J_n(p) - \frac{1}{\pi}\sum_{r=0}^{n-1}\frac{(n-r-1)!}{r!}\left(\frac{2}{p}\right)^{n-2r} - \frac{1}{\pi}\sum_{r=0}^{\infty}\frac{(-1)^r(p/2)^{n+2r}}{r!(n+r)!}\left[1 + \frac{1}{2} + \frac{1}{3} + \cdots + \frac{1}{r} + 1 + \frac{1}{2} + \cdots + \frac{1}{n+r}\right] \qquad n \geq 1$$

The term in the last series is

$$\left[1 + \frac{1}{2} + \frac{1}{3} + \cdots + \frac{1}{n}\right] \quad \text{when } r = 0 \quad \text{(B-15)}$$

Any linear combination of independent solutions is also a solution, and this principle permits convenient latitude in the definition of the "function of the second kind." In particular it justifies a modification of the definition in Eq. B-14 when a complex argument is used.

The leading term of the asymptotic expansion, $\tilde{Y}_n(p)$, is

$$\tilde{Y}_n(p) = \sqrt{\frac{2}{\pi p}}\sin\left(p - \frac{\pi}{4} - \frac{n\pi}{2}\right) \qquad p \gg 1 \quad \text{(B-16)}$$

The function $Y_0(p)$ "teams up" well with the function of the first kind when real arguments are used, a quality which is manifested by the fact that the leading terms of the respective asymptotic expansions are diminishing sinusoids which are bounded by the same envelope and which are in phase quadrature with each other.

B-2 INDEPENDENT VARIABLE PURELY IMAGINARY[3]

If the independent variable in Eq. B-1 is purely imaginary (jq), the function of the first kind is designated as $I_n(q)$, where

$$I_n(q) = (-j)^n J_n(jq) \tag{B-17}$$

With an imaginary argument or one proportional to $j^{3/2}$, the functions J_n and Y_n all approach infinity in magnitude as the magnitude of the argument approaches infinity. For many applications the boundary conditions require a function which vanishes as the argument magnitude approaches infinity. The following linear combination of J_n and Y_n has this property for arguments proportional either to j or to $j^{3/2}$; it is customarily designated as a new function, K_n:

$$K_n(q) = \frac{\pi}{2} j^{(n+1)} [J_n(jq) + jY_n(jq)] \tag{B-18}$$

All Y_n and K_n all approach infinity in magnitude as their respective arguments approach zero.

The first few terms of the series expansions for $K_0(q)$ and $K_1(q)$ are:

$$K_0(q) = -\left[\gamma + \ln\left(\frac{q}{2}\right)\right] J_0(jq) + \left(\frac{q}{2}\right)^2 + \left(\frac{q}{2}\right)^4 \cdot \frac{1}{(2!)^2}\left(1 + \frac{1}{2}\right) + \cdots \tag{B-19}$$

$$K_1(q) = \frac{1}{q} + \left[\gamma + \ln\left(\frac{q}{2}\right)\right](-j) J_1(jq)$$

$$-\frac{1}{2}\left[\frac{q}{2} + \left(\frac{q}{2}\right)^3 \cdot \frac{1}{2!} \cdot \left(1 + 1 + \frac{1}{2}\right) + \cdots\right] \tag{B-20}$$

B-3 INDEPENDENT VARIABLE COMPLEX

In the analysis of skin effect in a coaxial cable, Bessel functions of the first and second kinds with arguments of the following types are used:

$$p = \sqrt{-j}\,mr$$

$$= j^{3/2}mr \quad (\text{argument for } J_n) \tag{B-21}$$

$$q = j^{1/2}mr \quad (\text{argument for } K_n) \tag{B-22}$$

Those functions have been tabulated in polar and rectangular forms,[1-3] and $J_0(p)$, $J_1(p)$, $K_0(q)$, and $K_1(q)$ have been plotted in their polar forms in Figs. 14-1 and 14-2.

The leading terms of the asymptotic expansions for the zero-order functions are

$$\tilde{J}_0(j^{3/2}mr) = \frac{1}{\sqrt{2\pi mr}}\exp\left[\frac{mr}{\sqrt{2}} + j\left(\frac{mr}{\sqrt{2}} - \frac{\pi}{8}\right)\right] \tag{B-23}$$

$$\tilde{K}_0(j^{1/2}mr) = \sqrt{\frac{\pi}{2mr}}\exp\left[-\frac{mr}{\sqrt{2}} - j\left(\frac{mr}{\sqrt{2}} + \frac{\pi}{8}\right)\right] \tag{B-24}$$

Additional terms of the asymptotic expansions are introduced in Eqs. 14-41, 14-42, 14-47, and 14-48.

The general solution for Eq. B-1, given in Eq. B-2 in the form appropriate for real arguments, may be rewritten for the complex arguments just indicated in terms of J_n and K_n and new constants A'_n and B'_n:

$$w(j^{3/2}mr) = A'_n J_n(j^{3/2}mr) + B'_n K_n(j^{1/2}mr) \tag{B-25}$$

REFERENCES

1. Abramowitz, M. and Stegun, I. A., *Handbook of Mathematical Functions*. New York: Dover Publications, Inc., 1965.
2. Dwight, H. B., *Tables of Integrals and Other Mathematical Data*, 4th ed., MacMillan, New York, 1961.
3. McLachlan, N. W., *Bessel Functions for Engineers*, 2nd ed., Oxford University Press, London, 1961.
4. Watson, G. N., *Theory of Bessel Functions*, Cambridge University Press, London, 1944.

APPENDIX C

Parallel-Slab Equivalent of Slotted Coaxial Line

A transmission-line cross section of particular interest from the standpoints of mathematical analysis and experimental measurement technique is the *parallel-plane-equivalent* coaxial line,[3] which is sketched in Fig. C-1. This cross section, in which the side planes extend to infinity and the center conductor is approximately elliptical, corresponds in electrical properties to the circularly symmetrical cross section shown in Fig. C-2. (The latter drawing corresponds to Fig. 11-1, except that distances have been normalized with respect to r_b, the inner radius of the outer conductor.) The mathematical relationship between the two geometrical shapes is known as a *conformal transformation*.

The cross section shown in Fig. C-1 is impractical, but for a considerable range of parameters it may be closely approximated by the one shown in Fig. C-3. Here the side conductors, which together represent the outer conductor of Fig. C-2, are of finite cross-sectional extent, and the middle conductor is circular and hence much easier to fabricate than the corresponding conductor in Fig. C-1. This modified cross section has proved useful in the design of slotted-line units for microwave measurements.

C-1 SLOTTED LINES: GENERAL

A movable antenna-like probe, with suitable detecting and indicating equipment, is a practical means of obtaining standing-wave patterns and other field measurements at microwave frequencies.

The most obvious method of placing the probe in the electromagnetic field of a coaxial line is to cut a narrow longitudinal slot through the outer conductor and mount the probe on a longitudinally sliding carriage such that the probe projects through the slot some adjustable distance. This

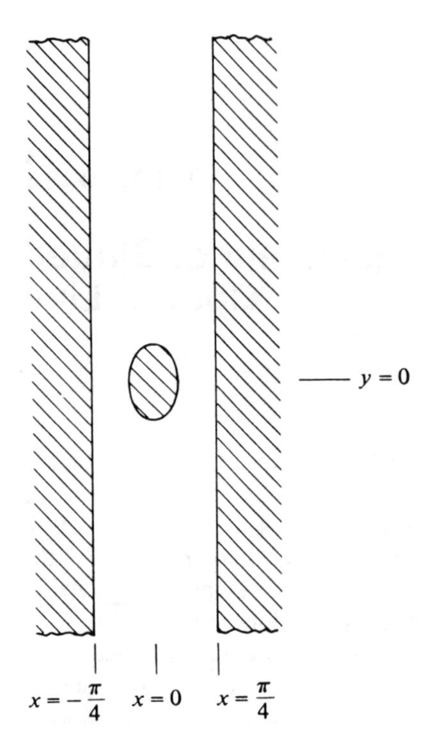

Figure C-1. Parallel-plane-equivalent coaxial line—ideal. (Coordinates shown are for complex z-plane.)

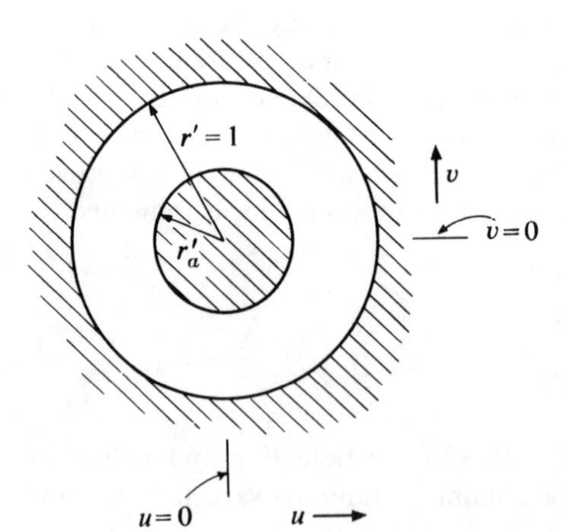

Figure C-2. Coaxial line with normalized w-plane coordinates.

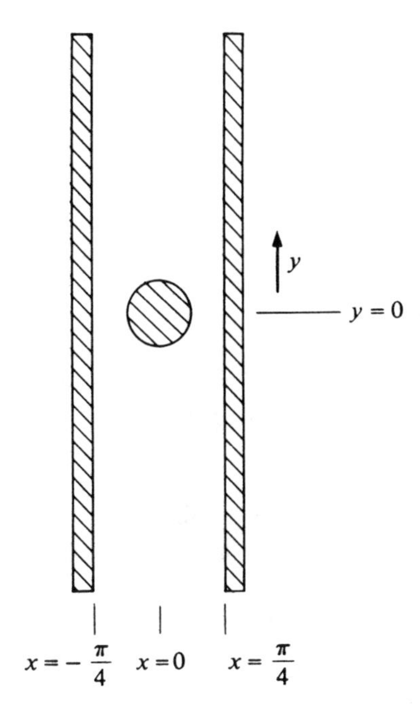

Figure C-3. Parallel-slab equivalent slotted coaxial line.

arrangement has the following disadvantages: (1) the characteristic impedance and the probe-unit output are sensitive to deviations in mechanical proportions, such as eccentric location of the inner conductor, or a slight bend in its axis, or nonparallelism between the probe movement and the inner conductor; and (2) a slot of practical width permits appreciable radiation of energy. This has the effect of adding a reactive component to the characteristic impedance and of altering the field patterns through attenuation, such that they are not precisely repetitive at half-wavelength intervals.

The slab-type-equivalent slotted line has less severe requirements as to mechanical alignment and dimensional tolerances than its circular prototype, and it has the equivalent of an extremely narrow slot from the standpoint of radiation losses. The latter feature may be readily examined analytically.

C-2 CONFORMAL TRANSFORMATIONS: BASIC PROPERTIES[1]

In this method the transverse cross section of the circular coaxial line and its parallel-plane equivalent are represented in two two-dimensional fields, in which the coordinates of points are stated in complex-number form. The field representing the cross section of the circular coaxial line is known as the w plane, and that of its parallel-plane equivalent, the z plane. The respective Cartesian coordinates are u and v, and x and y:

$$w = u + iv \tag{C-1}$$

$$z = x + iy \tag{C-2}$$

(Here the symbol i will be used for the complex operator rather than j, to minimize the possibility of confusing vectors and distances in two-dimensional space with phasors representing sinusoidal functions of time.)

If w is defined as some continuous function of z, each point in the z plane has a corresponding point in the w plane. For multiple-valued functions, a given point in one plane may correspond to more than one point (possibly an infinite number of points) in the other plane, but in those instances it is possible to divide one or both planes into regions such that a one-for-one correspondence exists between any point in a given region on one plane and a functionally related point in a specified region on the other plane.

By restricting the functional relationship between w and z such that the *Cauchy–Riemann criteria* are met, an important property of differential geometric correspondence is obtained between figures drawn on one plane and the respective "mapped" figures on the other. The transformation is then said to be *conformal*. The Cauchy–Riemann equations are

$$\frac{\partial u}{\partial x} = \frac{\partial v}{\partial y} \tag{C-3}$$

$$\frac{\partial u}{\partial y} = -\frac{\partial v}{\partial x} \tag{C-4}$$

The key concept in conformal mapping is that angles between intersecting lines in one plane are faithfully reproduced in the other plane on a "microscopic" basis. More specifically, if one (1) draws lines C_1 and C_2 (which may be curved) in the z plane such that they intersect at a point z_0, (2) locates the corresponding point w_0 and a succession of points in the w plane which will define curves C_1' and C_2' corresponding to curves C_1 and C_2, (3) draws straight lines through the point z_0 tangent to C_1 and C_2, and through point w_0 tangent to C_1' and C_2', then (4) the angles between the pairs of tangents are equal (see Fig. C-4). This property is applicable throughout the field except at points where dw/dz is zero or where w approaches infinity; no such points exist in the region of interest of the particular function to be investigated here.

Thus a geometric figure of differential extent in the neighborhood of point z_0 and the functionally corresponding figure in the neighborhood of

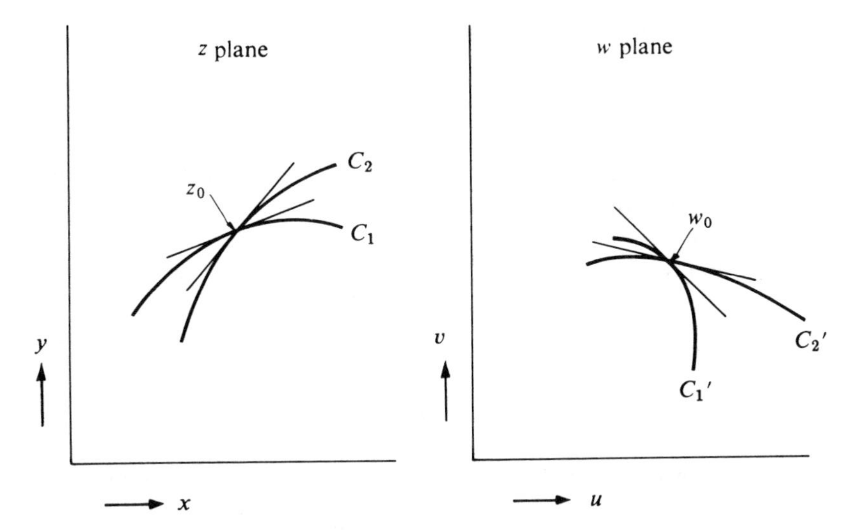

Figure C-4. Conformal transformation.

point w_0 will be similar (corresponding angles equal, corresponding sides proportional), but they would not necessarily be of the same size nor would their corresponding sides bear the same orientation to the respective coordinate axes. Furthermore, if the first geometric figure were moved without rotation to the neighborhood of some other point z_1, the functionally corresponding figure in the neighborhood of point w_1 would in general differ in size from the previously mapped figure near w_0 and would be differently oriented.

A complementary feature pertains to two-dimensional vector fields. If the boundary requirements of such a field may be realized by a set of equipotential curves in the w plane, the corresponding lines in the z plane may likewise be regarded as equipotentials, and the complete pattern of equipotential lines and field-direction lines is transformable from the w plane to the z plane. This is applicable to the electric field of the TEM mode on a transmission line as viewed in any transverse plane; for the coaxial line studied in Sec. 11-1 the equipotential boundary curves are circles of radii r_a and r_b (Fig. 11-1).

C-3 TRANSFORMATION FUNCTION FOR COAXIAL LINE

Useful transformation functions have probably been found primarily by judicious trial-and-error prospecting. The following proved to yield results useful in the present physical problem[3]:

$$w = \tan z \tag{C-5}$$

This is a multiple-valued function in that an infinite number of values of z, separated by multiples of π, all have the same tangent. Accordingly attention will be limited to a single region within the z plane, namely, a strip defined by $-\pi/4 \le x \le \pi/4$ and $-\infty < y < \infty$.

Substitution of Eqs. C-1 and C-2 into Eq. C-5 yields

$$u + iv = \tan(x + iy)$$

$$= \frac{\tan x + \tan(iy)}{1 - \tan x \tan(iy)}$$

As may be shown from Eq. 6-51, $\tan iy$ may be replaced by $i \tanh y$:

$$u + iv = \frac{\tan x + i \tanh y}{1 - i \tan x \tanh y} \tag{C-6}$$

If Eq. C-6 is multiplied by its conjugate, an expression is obtained for loci in the z plane corresponding to circles in the w plane, concentric with the origin:

$$u^2 + v^2 = \frac{\tan^2 x + \tanh^2 y}{1 + \tan^2 x \tanh^2 y} \tag{C-7}$$

It may be noted that if $\tan x$ is unity in magnitude, the right-hand side of Eq. C-7 becomes independent of y and, further, it reduces to unity. Thus a circle of unit radius in the w plane transforms into two lines parallel to the y axis but displaced from it by $\pm\tan^{-1}(1)$ or $\pm\pi/4$. Circles of radii less than unity will be found to transform into closed curves which are symmetric about both coordinate axes and which have x intercepts of less than $\pi/4$.

In order to make the parallel planes in the z plane ($x = \pm\pi/4$) correspond to the outer conductor of the circular coaxial line, the unit radius in the w plane should correspond to the inner radius of the outer conductor r_b. In other words distance in the w plane corresponds to distance in the circular coaxial cross section normalized with respect to r_b. For a radius $r < r_b$,

$$r' = \frac{r}{r_b} \tag{C-8}$$

$$r' = \sqrt{u^2 + v^2} \tag{C-9}$$

The normalized radius of the inner conductor may be designated by r'_a:

$$r'_a = \frac{r_a}{r_b} \tag{C-10}$$

The semiminor axis of the inner conductor in the z plane, x_a, may be found by substituting Eq. C-9, in Eq. C-7 and setting y equal to zero.

$$x_a = \tan^{-1} r'_a \tag{C-11}$$

For the semimajor axis y_a, the same equations are used, but x is set equal to zero:

$$y_a = \tanh^{-1} r'_a \tag{C-12}$$

Example—To illustrate the foregoing, the principal proportions of a 50-Ω, air dielectric line will be calculated:

$$Z_0 = \frac{1}{2\pi}\sqrt{\frac{\mu}{\epsilon}}\ \ln\left(\frac{r_b}{r_a}\right) \tag{11-18}$$

The intrinsic impedance ($\sqrt{\mu/\epsilon}$) of air is 120π Ω. If Eq. C-10 is substituted in Eq. 11-18, the following value is obtained:

$$r'_a = 0.435$$

Substitution of this value in Eqs. C-11 and C-12 gives

$$x_a = 0.410 \text{ rad} \qquad y_a = 0.466$$

The distance in the z plane between the planes representing the outer conductor is $\pi/2$, from whence the following proportions may be written:

$$\frac{\text{Minor axis}}{\text{Slab separation}} = \frac{(0.410)(2)(2)}{\pi}$$

$$= 0.522$$

$$\frac{\text{Major axis}}{\text{Slab separation}} = \frac{(0.466)(2)(2)}{\pi}$$

$$= 0.594$$

A convenient identity* enables one to express x and y as explicit functions of u and v. From Eq. C-5,

$$z = \tan^{-1} w \tag{C-13}$$

$$\tan^{-1}(u + iv) = \frac{1}{2}\tan^{-1}\left(\frac{2u}{1 - u^2 - v^2}\right) + \pi k + \frac{i}{4}\ln\left[\frac{u^2 + (1 + v)^2}{u^2 + (1 - v)^2}\right] \tag{C-14}$$

Substitution of Eq. C-9 may be advantageous:

$$x + iy = \frac{1}{2}\tan^{-1}\left(\frac{2u}{1 - r'^2}\right) + \frac{i}{4}\ln\left(\frac{1 + r'^2 + 2v}{1 + r'^2 - 2v}\right) \tag{C-15}$$

Problems C-1, C-2, and C-3 extend the analysis of the conformal transformation.

A circular middle conductor in the equivalent line transforms into an approximately elliptical figure in the w plane, and the characteristic impedance for that configuration may be computed by approximation methods.[3]

Necessary auxiliaries for a slab slotted-line section are transition sections[2] to change the field patterns from those of the circular coaxial line to those of the slab section and back again, without producing reflections.

PROBLEMS

C-1. Suppose that a parallel-slab equivalent line has a ratio of slab width to slab separation of 5.5. What are the u and v coordinates corresponding to the upper edge of the right-hand slab? What is the angle between the edges of a slot in the circular line which is equivalent to the opening between the edges of the slabs in the z plane?

C-2. Find the parametric equation for the ***E***-field-direction lines in a parallel-slab line. *Hint*: These correspond to radial lines in the w-plane; for such a line the ratio v/u is a constant. Rationalize the right-hand side of Eq. C-6 to obtain expressions for u and v. *Answer*:

$$\frac{v}{u} = \frac{\sinh 2y}{\sin 2x}$$

*Dwight, H. B., *Tables of Integrals and Other Mathematical Data*, 4th ed., Macmillan, New York, 1961, 120.

C-3. Make a drawing similar to Figs. C-1 and C-2, but show only one quadrant of each transmission line, and omit the inner conductor.

(a) Sketch ***E***-field-direction lines spaced 22.5° apart in the coaxial-line drawing, and plot the corresponding lines in the drawing of the parallel-slab line.

(b) Plot in each drawing the equipotential line corresponding to the surface of the inner conductor of a 50-Ω line.

(c) Assuming that the inner conductor found in part (b) is at potential V_0 with respect to the outer conductor, plot an equipotential line in each drawing corresponding to $0.5V_0$.

REFERENCES

1. Churchill, R. V., An Introduction to Complex Variables and Applications, McGraw-Hill, New York, 1948.
2. Ginzton, E. L., Microwave Measurements. McGraw-Hill, New York, 1957, 253.
3. Wholey, W. B. and Eldred, W. W., A new type of slotted line section, *Proc. IRE*, 38, 244, 1950.

APPENDIX D

Earth Currents and Distributed Impedances

The physical earth has a conductivity which is low compared with those of metals, but it provides current paths of enormous cross-sectional areas compared with those of cables, wires, and stranded conductors. All transmission systems are, at least inadvertently, in proximity to the earth. In some instances, significant coupling may exist.

D-1 EARTH-CURRENT CHARACTERISTICS

Use of the earth as a return path for telegraph lines, and, to a lesser extent, for telephone lines, was initially a common practice as a measure of economy. Two undesirable features were inherent because of the diffuse pattern of the resulting current density field: (1) attenuation, phase velocity, and characteristic impedance varied widely with frequency, and (2) impedance coupling was pronounced between circuits which were in proximity to each other, causing *crosstalk*, or interfering signals. These effects could be mitigated with appropriate terminal networks, but all-metallic lines proved more satisfactory and have been generally adopted. Likewise power systems are designed on an all-metallic-transmission basis.*

Shields and unshielded wires are inductively and capacitively coupled to the earth. For safety reasons, many components of electrical systems are

---

*Electrified railroads use the track rails for a return path; currents may migrate to, and then at other locations from, nearby utility pipes. Electrolytic corrosion sometimes results.[17] Direct-current transmission lines, under emergency conditions (disabled converter units, for example) may be operated with one metallic conductor (or conductor bundle) and ground return.

connected conductively to the earth.* Ground-path currents are of particular interest in power system planning; such currents may be expected whenever a conductor is short-circuited to the earth. [Some common causes of short circuits are (a) lightning strokes, (b) deterioration of insulation by contaminants, chemical reactions, or living organisms, and (c) human carelessness.]

D-2 DISTRIBUTED IMPEDANCES OF GROUND-RETURN CIRCUITS

Two basic impedances for transmission-line circuit analyses involving ground currents are (a) the distributed self-impedance Z_{1g} of an overhead-wire-and-ground-return circuit, and (b) the distributed mutual impedance Z_{12} between two such circuits with parallel courses. The classic reference in the English language for the derivation of those impedances is the 1926 paper by John Carson.[2] He assumed the ground conductivity to be uniform.† The distributed impedances may then be stated in terms of standard transcendental functions; formulas commonly given for applications[6] are series approximations of the latter.

a. General Formulas Based on Carson's Analysis

Carson assumed a quasi-TEM mode for the fields above the earth's surface; that is, one in which the rms values of the transverse components of the $\boldsymbol{E}$ and $\boldsymbol{H}$ fields are everywhere much larger than the rms values of the longitudinal components of those fields. Within the frequency range for which the quasi-TEM assumption is valid, $\sigma \gg \omega\varepsilon$ below the surface of the earth; there the $\boldsymbol{E}$ and $\boldsymbol{H}$ fields are assumed to diffuse downwardly. The transverse components of $\boldsymbol{E}$ in the ground are assumed to be very small compared to the longitudinal component, but the reverse is assumed to be true of the $\boldsymbol{H}$ field there.

The original results were given in cgs units; the equations below have been restated in terms of SI units and present-day definitions of functions.

*A phenomenon which was identified many years ago in telegraph lines, and which has recently become an acute problem in power systems, is that of *geomagnetically induced currents*.[10] These currents, associated with solar storms, commonly consist of surges of several seconds duration. Power transmission lines with neutral-grounded wye-connected transformers provide low-resistance alternate paths for currents induced in the earth, and DC saturation of the transformer cores may result.

†The ground-impedance problem has been studied under assumptions of stratification and other modes of variation of conductivity with depth.[14, 15, 17]

The heights of conductors 1 and 2 above an assumed plane earth surface of infinite extent will be designated by x_1 and x_2, the horizontal separation between them by y_1, and the radius of conductor 1 by r_a (all in meters). The following dimensionless variables will be used:

$$x'_1 = x_1\sqrt{\omega\mu_0\sigma} \tag{D-1}$$

$$x'_2 = x_2\sqrt{\omega\mu_0\sigma} \tag{D-2}$$

$$y'_1 = y_1\sqrt{\omega\mu_0\sigma} \tag{D-3}$$

$$d' = x'_1 + x'_2 - jy'_1 \tag{D-4}$$

$$d'^* = x'_1 + x'_2 + jy'_1 \tag{D-5}$$

The resistance of conductor 1 will be designated by r_1, and l_{1i} will represent the partial inductance which corresponds to the flux linkages within conductor 1:

$$\mu_0 = 4\pi \times 10^{-7}\ \mathrm{H/m}$$

The following is an average value for ground conductivity, adequate for order-of-magnitude estimating:

$$\sigma \approx 10^{-2}\ \mathrm{S/m}$$

The basic derivation is intricate and lengthy, and will not be repeated here. The following equation is one form for Z_{1g}:

$$Z_{1g} = r_1 + j\omega l_{1i} + \frac{j\omega\mu_0}{2\pi}\ln\left(\frac{2x_1}{r_a}\right) + \frac{\omega\mu_0}{\pi}\int_0^\infty \exp(-2a'x'_1)\left(\sqrt{a'^2 + j} - a'\right)da' \tag{D-6}$$

Computed values[9] of the infinite integral as a function of $2x'_1$ (Eq. D-1) have been assembled in Table D-1; numerical values of the ground-return

TABLE D-1. Computed Data for Evaluation of Z_{1g}: Real and Imaginary Parts of the Infinite Integral in Eq. D-6 as Functions of $2x_1'$[a]

$2x_1'$	Re[ ]	Im[ ]	$2x_1'$	Re[ ]	Im[ ]	$2x_1'$	Re[ ]	Im[ ]
0.01	0.3904	2.6129	0.51	0.3076	0.7537	1.1	0.2480	0.4740
0.02	0.3881	2.2687	0.52	0.3064	0.7460	1.2	0.2401	0.4465
0.03	0.3859	2.0683	0.53	0.3051	0.7384	1.3	0.2327	0.4220
0.04	0.3837	1.9268	0.54	0.3039	0.7310	1.4	0.2257	0.4000
0.05	0.3816	1.8175	0.55	0.3027	0.7237	1.5	0.2192	0.3802
0.06	0.3795	1.7286	0.56	0.3014	0.7166	1.6	0.2130	0.3623
0.07	0.3774	1.6539	0.57	0.3002	0.7097	1.7	0.2071	0.3459
0.08	0.3754	1.5894	0.58	0.2990	0.7029	1.8	0.2015	0.3309
0.09	0.3734	1.5328	0.59	0.2978	0.6963	1.9	0.1963	0.3172
0.10	0.3714	1.4824	0.60	0.2967	0.6897	2.0	0.1913	0.3045
0.11	0.3695	1.4370	0.61	0.2995	0.6833	2.1	0.1865	0.2928
0.12	0.3676	1.3957	0.62	0.2944	0.6771	2.2	0.1819	0.2819
0.13	0.3657	1.3579	0.63	0.2932	0.6710	2.3	0.1776	0.2718
0.14	0.3638	1.3231	0.64	0.2921	0.6650	2.4	0.1735	0.2624
0.15	0.3620	1.2908	0.65	0.2909	0.6591	2.5	0.1695	0.2536
0.16	0.3602	1.2608	0.66	0.2898	0.6533	2.6	0.1657	0.2453
0.17	0.3584	1.2327	0.67	0.2887	0.6476	2.7	0.1621	0.2375
0.18	0.3566	1.2063	0.68	0.2876	0.6421	2.8	0.1586	0.2302
0.19	0.3549	1.1814	0.69	0.2865	0.6366	2.9	0.1553	0.2234
0.20	0.3531	1.1580	0.70	0.2854	0.6312	3.0	0.1521	0.2169
0.21	0.3514	1.1358	0.71	0.2843	0.6260	3.2	0.1461	0.2050
0.22	0.3497	1.1147	0.72	0.2833	0.6208	3.4	0.1405	0.1943
0.23	0.3481	1.0946	0.73	0.2822	0.6157	3.6	0.1353	0.1846
0.24	0.3464	1.0755	0.74	0.2812	0.6107	3.8	0.1305	0.1758
0.25	0.3448	1.0572	0.75	0.2801	0.6058	4.0	0.1260	0.1678
0.26	0.3432	1.0397	0.76	0.2791	0.6010	4.2	0.1218	0.1604
0.27	0.3416	1.0230	0.77	0.2781	0.5962	4.4	0.1179	0.1537
0.28	0.3400	1.0069	0.78	0.2770	0.5916	4.6	0.1142	0.1474
0.29	0.3384	0.9915	0.79	0.2760	0.5870	4.8	0.1108	0.1417
0.30	0.3368	0.9766	0.80	0.2750	0.5825	5	0.1075	0.1362
0.31	0.3353	0.9623	0.81	0.2740	0.5780	6	0.0936	0.1148
0.32	0.3338	0.9485	0.82	0.2731	0.5736	7	0.0828	0.0991
0.33	0.3323	0.9352	0.83	0.2721	0.5694	8	0.0742	0.0871
0.34	0.3308	0.9224	0.84	0.2711	0.5651	9	0.0672	0.0776
0.35	0.3293	0.9099	0.85	0.2701	0.5609	10	0.0614	0.0700
0.36	0.3279	0.8979	0.86	0.2692	0.5568	20	0.0330	0.0353
0.37	0.3265	0.8862	0.87	0.2682	0.5528	50	0.0137	0.0141
0.38	0.3250	0.8749	0.88	0.2672	0.5488	100	0.0070	0.0071
0.39	0.3236	0.8640	0.89	0.2663	0.5449			
0.40	0.3222	0.8534	0.90	0.2654	0.5410			
0.41	0.3208	0.8430	0.91	0.2645	0.5372			
0.42	0.3194	0.8330	0.92	0.2636	0.5334			
0.43	0.3181	0.8232	0.93	0.2626	0.5297			
0.44	0.3167	0.8137	0.94	0.2617	0.5261			
0.45	0.3154	0.8045	0.95	0.2608	0.5225			
0.46	0.3141	0.7955	0.96	0.2599	0.5189			
0.47	0.3127	0.7867	0.97	0.2590	0.5154			
0.48	0.3115	0.7782	0.98	0.2591	0.5120			
0.49	0.3102	0.7698	0.99	0.2573	0.5086			
0.50	0.3089	0.7617	1.00	0.2564	0.5053			

[a]From Hylten-Cavallius and Gjerløv.[9]

impedance of a given conductor as a function of frequency, for example, may be readily compiled.

Alternatively, the infinite integral may be evaluated* in terms of the Bessel function of the second kind and first order (Y_1), and the Struve function of the first order ($\mathbf{H}_1$):

$$Z_{1g} = r_1 + j\omega l_{1i} + \frac{j\omega\mu_0}{2\pi}\ln\left(\frac{2x_1}{r_a}\right)$$

$$+ \frac{\omega\mu_0\sqrt{j}}{4x_1'}\left[\mathbf{H}_1\left(\sqrt{j}\,2x_1'\right) - Y_1\left(\sqrt{j}\,2x_1'\right) - \frac{1}{\pi x_1'\sqrt{j}}\right]\ \Omega/\mathrm{m} \quad (\text{D-7})^{\dagger}$$

$$\mathbf{H}_1(u) = \frac{2}{\pi}\left(\frac{u^2}{1^2\cdot 3} - \frac{u^4}{1^2\cdot 3^2\cdot 5} + \frac{u^6}{1^2\cdot 3^2\cdot 5^2\cdot 7} - \cdots\right) \quad (\text{D-8})$$

The corresponding expression for the mutual impedance is

$$Z_{12} = \frac{j\omega\mu_0}{4\pi}\ln\left[\frac{y_1^2 + (x_1 + x_2)^2}{y_1^2 + (x_1 - x_2)^2}\right]$$

$$+ \frac{\omega\mu_0\sqrt{j}}{4d'}\left[\mathbf{H}_1\left(\sqrt{j}\,d'\right) - Y_1\left(\sqrt{j}\,d'\right) - \frac{2}{\pi\sqrt{j}\,d'}\right]$$

$$+ \frac{\omega\mu_0\sqrt{j}}{4d'^*}\left[\mathbf{H}_1\left(\sqrt{j}\,d'^*\right) - Y_1\left(\sqrt{j}\,d'^*\right) - \frac{2}{\pi\sqrt{j}\,d'^*}\right]\ \Omega/\mathrm{m} \quad (\text{D-9})$$

For the special case in which both wires are on the surface of the earth ($x_1 = 0$; $x_2 = 0$), the latter equation may be reduced through identities (ref. 13, p. 81, Eq. 8; p. 198, Eq. 128; and p. 204, Eq. 206) to

$$Z_{12} = \frac{1}{\pi\sigma y_1^2}\left[1 - \sqrt{j}\,y_1' K_1\left(\sqrt{j}\,y_1'\right)\right]\ \Omega/\mathrm{m} \quad (\text{D-10})$$

---

*A table of Laplace transforms $F(s) = \int_0^\infty \exp(-st)f(t)\,dt$ may be used for an integral of the type at hand, if one considers a' the equivalent of t, and $2x_1'$ the equivalent of s in the Laplace transformation. (Roberts and Kaufman,[16] p. 342, #25.10.1-1.)

---

†It should be noted that the function $K_1(u)$ as used by Carson is the equivalent of $-(\pi/2)Y_1(u)$ in the notation of this text, which corresponds to that of McLachlan.[13]

An alternative form in terms of the tabulated functions ker′ and kei′ may be found from the following identity from McLachlan,[13] (p. 211, Eq. 320):

$$-\sqrt{j}\,K_1\!\left(\sqrt{j}\,y_1'\right) = \frac{d}{dy_1'}K_0\!\left(\sqrt{j}\,y_1'\right)$$

$$= \mathrm{ker}'(y_1') + j\,\mathrm{kei}'(y_1') \qquad \text{(D-11)}$$

$$Z_{12} = \frac{1}{\pi\sigma y_1^2}\left\{1 + y_1'\left[\mathrm{ker}'(y_1') + j\,\mathrm{kei}'(y_1')\right]\right\}\ \Omega/\mathrm{m} \qquad \text{(D-12)}$$

b. Power Series (Low-Frequency) Approximations

The power series for $\mathbf{H}_1$, Y_1, and K_1 converge rapidly if the corresponding arguments have magnitudes appreciably less than unity. (Unity magnitude for $2x_1'$, Eq. D-1, would correspond, for $\sigma = 10^{-2}$ S/m and $f =$ 60 Hz, to $x_1 = 205$ m.) Substitution of Eqs. D-8, B-15, and B-20 into Eqs. D-7, D-9, and D-10 yields the following, in which γ is Euler's number, 0.577216 . . . :

$$Z_{1g} = r_1 + j\omega l_{1i} + \frac{j\omega\mu_0}{2\pi}\left[\ln\!\left(\frac{2}{r_a\sqrt{j\omega\mu_0\sigma}}\right) - \gamma + \frac{1}{2}\right]$$

$$-(1-j)\frac{\omega\mu_0\sqrt{2}\,x_1'}{3\pi} + \cdots$$

$$\ln\!\left(\frac{2}{r_a\sqrt{j\omega\mu_0\sigma}}\right) = \ln\!\left(\frac{2}{r_a\sqrt{\omega\mu_0\sigma}}\right) - \frac{j\pi}{4}$$

$$Z_{1g} = r_1 + j\omega l_{1i} + \frac{j\omega\mu_0}{2\pi}\left[\ln\!\left(\frac{2}{r_a\sqrt{\omega\mu_0\sigma}}\right) - \gamma + \frac{1}{2}\right] + \frac{\omega\mu_0}{8}$$

$$-(1-j)\frac{\omega\mu_0\sqrt{2}\,x_1'}{3\pi} + \cdots\ \Omega/\mathrm{m} \qquad \text{(D-13)}$$

$$Z_{12} = \frac{j\omega\mu_0}{4\pi}\left\{\ln\!\left[\frac{4}{(x_1' - x_2')^2 + y_1'^2}\right] + 1 - 2\gamma\right\} + \frac{\omega\mu_0}{8}$$

$$-(1-j)\frac{\omega\mu_0(x_1' + x_2')}{3\sqrt{2}\,\pi} + \cdots\ \Omega/\mathrm{m} \qquad \text{(D-14)}$$

For the special case in which $x_1 = 0$ and $x_2 = 0$,

$$Z_{12} = \frac{j\omega\mu_0}{4\pi}\left[\ln\left(\frac{2}{y_1'}\right) + \frac{1}{2} - \gamma\right] + \frac{\omega\mu_0}{8} + \frac{\omega\mu_0 y_1'^2}{16\pi}\left[\ln\left(\frac{y_1'}{2}\right) + \gamma - \frac{5}{4}\right]$$

$$+ \frac{j\omega\mu_0 y_1'^2}{64} + \cdots \quad \Omega/\text{m} \qquad \text{(D-15)}$$

c. Asymptotic (High-Frequency) Approximations

The following asymptotic approximation is given by Abramowitz and Stegun[1] (p. 497, #12.1.31):

$$\mathbf{H}_1(u) - Y_1(u) \approx \frac{2}{\pi}\left(1 + \frac{1}{u^2} - \frac{1^2 \cdot 3}{u^4} + \frac{1^2 \cdot 3^2 \cdot 5}{u^6} - \cdots\right) \qquad \text{(D-16)}$$

Substitution in Eqs. D-7 and D-9 yields the leading terms

$$Z_{1g} \approx r_1 + j\omega l_{1i} + \frac{j\omega\mu_0}{2\pi}\ln\left(\frac{2x_1}{r_a}\right) + (1+j)\frac{\omega\mu_0}{\pi 2\sqrt{2}\,x_1'} - \frac{\omega\mu_0}{4\pi x_1'^2} + \cdots$$

$$x_1' \gg 1 \qquad \text{(D-17)}$$

$$Z_{12} \approx \frac{j\omega\mu_0}{4\pi}\ln\left[\frac{y_1^2 + (x_1 + x_2)^2}{y_1^2 + (x_1 - x_2)^2}\right] + (1+j)\frac{\omega\mu_0(x_1' + x_2')}{\pi\sqrt{2}\left[y_1'^2 + (x_1' + x_2')^2\right]}$$

$$- \frac{\omega\mu_0\left[(x_1 + x_2)^2 - y_1^2\right]}{\pi\left[y_1'^2 + (x_1' + x_2')^2\right]^2} + \cdots \qquad |x_1' + x_2' + jy_1'| \gg 1 \qquad \text{(D-18)}$$

The component of inductance represented by the logarithmic term in Eq. D-17 may be compared with the distributed inductance l_{cg} in Sec. 11-3a for a conductor above a lossless ground plane. From Eqs. 11-77 and 11-83, Fig. 11-4, and the definition of x_1,

$$l_{cg} = \frac{\mu_0}{2\pi}\ln\left(\frac{k'}{r_a}\right) \qquad \text{(D-19)}$$

$$= \frac{\mu_0}{2\pi}\ln\left(\frac{2x_1}{r_a}\right) \qquad \text{(D-20)}$$

This is in agreement with Eq. D-17 if σ, and hence x_1', approach infinity.

d. An All-Frequency Approximation

The following formula is a restatement of an empirical approximation proposed by Sunde[17]:

$$Z_{1g} \approx r_1 + j\omega l_{1i} + \frac{j\omega\mu_0}{2\pi} \ln\left(2\frac{x_1\sqrt{j\omega\mu_0\sigma} + 1}{r_a\sqrt{j\omega\mu_0\sigma}}\right) \tag{D-21}$$

$$\approx r_1 + j\omega l_{1i} + \frac{j\omega\mu_0}{2\pi} \ln\left(2\frac{x_1 + \left(1/\sqrt{j\omega\mu_0\sigma}\right)}{r_a}\right) \tag{D-22}$$

If Eq. 10-79 for skin depth δ is substituted, and the argument of the logarithm is converted to polar form, the following may be obtained:

$$x_1 + \frac{1}{\sqrt{j\omega\mu_0\sigma}} = x_1 + \frac{\delta}{\sqrt{2j}}$$

$$= x_1 + \frac{\delta}{2} - j\frac{\delta}{2}$$

$$= \sqrt{\left(x_1 + \frac{\delta}{2}\right)^2 + \left(\frac{\delta}{2}\right)^2} \exp\left[j\tan^{-1}\left(\frac{-\delta/2}{x_1 + \delta/2}\right)\right] \tag{D-23}$$

The magnitude of this term corresponds to the distance above a fictitious ground plane located between $\delta/2$ (at high frequencies, when $\delta \ll x_1$) and $\delta/\sqrt{2}$ (at low frequencies, when $\delta \gg x_1$) below the surface of the earth.

Agreement between Eq. D-22 and the asymptotic-approximation terms shown in Eq. D-17 may be demonstrated by rearranging the argument of the logarithm and substituting Eq. D-1:

$$Z_{1g} \approx r_1 + j\omega l_{1i} + \frac{j\omega\mu_0}{2\pi} \ln\left[\left(\frac{2x_1}{r_a}\right)\left(1 + \frac{1}{x_1'\sqrt{j}}\right)\right]$$

$$\approx r_1 + j\omega l_{1i} + \frac{j\omega\mu_0}{2\pi} \left[\ln\left(\frac{2x_1}{r_a}\right) + \ln\left(1 + \frac{1}{x_1'\sqrt{j}}\right)\right] \tag{D-24}$$

The following power series is applicable:

$$\ln(1+u) \approx u - 0.5u^2 + \cdots \qquad |u| < 1 \tag{D-25}$$

Hence

$$Z_{1g} \approx r_1 + j\omega l_{1i} + \frac{j\omega\mu_0}{2\pi}\left[\ln\left(\frac{2x_1}{r_a}\right) + \frac{1}{x_1'\sqrt{j}} - \frac{1}{2jx_1'^2}\right] \tag{D-26}$$

Equation D-26 reduces to Eq. D-17.

The behavior at low frequencies may be examined as follows:

$$Z_{1g} \approx r_1 + j\omega l_{1i} + \frac{j\omega\mu_0}{2\pi}\ln\left[\left(\frac{2}{r_a\sqrt{j\omega\mu_0\sigma}}\right)\left(1 + x_1\sqrt{j\omega\mu_0\sigma}\right)\right]$$

$$\approx r_1 + j\omega l_{1i} + \frac{j\omega\mu_0}{2\pi}\left[\ln\left(\frac{2}{r_a\sqrt{j\omega\mu_0\sigma}}\right) + \ln\left(1 + x_1'\sqrt{j}\right)\right]$$

$$\approx r_1 + j\omega l_{1i} + \frac{j\omega\mu_0}{2\pi}\left[\ln\left(\frac{2}{r_a\sqrt{\omega\mu_0\sigma}}\right) - j0.25\pi + x_1'\sqrt{j}\right]$$

$$\approx r_1 + j\omega l_{1i} + \frac{j\omega\mu_0}{2\pi}\ln\left(\frac{2}{r_a\sqrt{\omega\mu_0\sigma}}\right) + \frac{\omega\mu_0}{8} + (-1+j)\frac{x_1'\omega\mu_0}{2\sqrt{2}\,\pi} \tag{D-27}$$

This result may be compared with Eq. D-13.

Chen and Damrau[3] have presented calculated data which indicate close agreement between impedances predicted by the approximation in Eq. D-22 and those from Eq. D-6 and Table D-1.

D-3 CONCLUSIONS

The earth is used only occasionally as a transmission-line component in the conveyance of signals or power, but accidental short circuits from line conductors to the earth occur "randomly yet routinely." Also, geophysically induced earth currents may interfere with transmission-system performance. Analysis of potential ground currents is an essential part of the design of protective systems.

Results derived by Carson for (a) the distributed self-impedance of an overhead wire with ground return, and (b) the distributed mutual impedance between two such circuits, have been assembled in the preceding pages. Those derivations were based on the assumption of a uniform earth, of such conductivity that displacement current in the earth would be negligible compared to conduction current density, and within a frequency range such that the electromagnetic fields above the earth could be assumed to be quasi-TEM.

REFERENCES

1. Abramowitz, M. and Stegun, I. A., Eds., *Handbook of Mathematical Functions*, Dover, New York, 1965.
2. Carson, J. R., Wave propagation in overhead wires with ground return, *Bell Syst. Tech. J.*, 5, 539, October 1926.
3. Chen, K. C. and Damrau, K. M., Accuracy of approximate transmission line formulas for overhead wires, *IEEE Trans. Electromagnetic Compatibility*, 31, 4, 396, 1989.
4. Coupling factors for ground-return circuits—General considerations and methods of calculation, Engineering Report No. 14, Joint Subcommittee on Development and Research, National Electric Light Association and Bell Telephone System, May 14, 1931, *Engineering Reports*, 2 (Reports 9–15), 121, April 1932.
5. Dwight, H. B., *Tables of Integrals and Other Mathematical Data*, 4th ed., MacMillan, New York, 1961. Extensive tabulation of ker′ and kei′.
6. *Electrical Transmission and Distribution Reference Book*, Central Station Engineers of the Westinghouse Electric Corporation, Pittsburgh, 1950, 41.
7. Forsman, M. E., Zero-sequence current density in the earth, *AIEE Trans. Part III-B, Power Systems and Apparatus*, 78, 1525, 1959.
8. Hedman, D. E., Propagation on overhead transmission lines II—Earth-conduction effects and practical results, *IEEE Trans. Power Apparatus and Systems*, PAS-84, no. 3, 205, 1965.
9. Hyltén-Cavallius, N. and Gjerløv, P., Distortion of traveling waves in high-voltage power lines, *ASEA Research*, no. 2, 147, 1959.
10. Kappenman, J. G. and Albertson, V. D., Bracing for the geomagnetic storms, *IEEE Spectrum*, 27, no. 3, 27, 1990.
11. Kimbark, E. W., *Electrical Transmission of Power and Signals*, Wiley, New York, 1949, 85.
12. Klewe, H. R. J., *Interference Between Power Systems and Telecommunication Lines*, Edward Arnold, London, 31, 1958. Extensive bibliography.
13. McLachlan, N. W., *Bessel Functions for Engineers*, 2nd ed., Oxford University Press, London, 1955, 76, 81, 84, 191, 197, 198, 203, 204.

14. Nakagawa, M. and Iwamoto, K., Earth-return impedance for the multi-layer case, *IEEE Trans. on Power Apparatus and Systems*, PAS-95, no. 2, 671, 1976.
15. Riordan, J. and Sunde, E. D., Mutual impedance of grounded wires for horizontally stratified two-layer earth, *Bell Syst. Tech. J.*, 12, 162, 1933.
16. Roberts, G. E. and Kaufman, H., *Table of Laplace Transforms*, Saunders, Philadelphia, 1966, 342.
17. Romanoff, M., *Underground Corrosion*, *National Bureau of Standards Circular* 579, United States Department of Commerce, Washington, DC, 1957.
18. Sunde, E. D., *Earth Conduction Effects in Transmission Systems*, Van Nostrand, New York, 1949.
19. Wait, J. R., On the impedance of long wire suspended over the ground, *Proc. IRE*, 49, 1576, 1961.

APPENDIX E

Low-Temperature Impedance Effects

The resistivities of most metals vary approximately linearly with temperature over the ordinary climatic range, and become very low at temperatures approaching absolute zero. But the low-temperature impedance of a metal cannot be predicted by mere extrapolation in accordance with its mode of variation in the ambient range. As was mentioned in a footnote at the beginning of Chap. 9, two principal aspects of deviation from such estimated performance are (1) superconductivity and (2) anomalous skin effect.

E-1 SUPERCONDUCTIVITY[5, 8–10, 16, 17, 19, 25]

The phenomenon of superconductivity was discovered by Onnes in 1911 while he was studying the electrical properties of mercury. Since World War II it has been a major area of research.

Superconductivity is characterized by a transition temperature T_c, which differs among various materials, below which the resistivity is many orders of magnitude lower than it is at higher temperatures. This is evidenced by the fact that currents which have been induced in superconducting loops have been found to persist for months. Lead and niobium have two of the higher known values of T_c for a pure metal, about 7.2 K and 9.2 K, respectively. (These values are only approximate, as the purity of the material, presence of mechanical strain, and other factors influence the T_c and related properties of a specimen.) Aluminum has a much lower T_c, about 1.18 K, whereas copper has not been proven to have the property of superconductivity even though it has been sought for at temperatures as low as 0.05 K.

Various alloys and intermetallic compounds developed in the 1950s have higher values of T_c; two which may be noted are niobium–tin, Ni_3Sn ($\approx$ 18 K), and niobium-germanium, Ni_3Ge ($\approx$ 23.3 K). Mixed metallic oxides (*perovskites*) developed in the latter 1980s have yet higher values of

T_c; combinations of yttrium, barium, and copper in the proportion of YBa_2Cu_3 (hence the acronym "1-2-3"), with varying proportions of oxygen, are superconducting up to about 93 K.

The range of temperature within which a given material is superconducting will be reduced below T_c if a magnetic field is present, and if the material is conducting current. The limits of magnetic field strength and current magnitude at which superconductivity is extinguished vary widely among materials; the intermetallic compounds and mixed metallic oxides are markedly superior to pure niobium in these regards.[11]

Potential engineering utility is also dependent on mechanical properties; Ni_3Sn and the 1-2-3 compounds, in their natural states, resemble ceramics and are brittle. On the other hand, niobium–titanium, NiTi ($T_c \approx 10$ K), is ductile and is used in the windings of high-field magnets.

"Electromagnetically long line" applications of superconductivity include (1) cables used in instrumentation,[1, 2, 13, 18, 20, 22] and (2) resonant cavities.[6, 7, 14, 23, 24] As indicated by the analysis in Chap. 5 on lossy lines, distortion over a given distance of travel is reduced if the line resistance is lowered. Resonant cavities were mentioned in Sec. 13-2b; in such units the sharpness of tuning is a function of the losses. Sharply tuned resonators are useful in frequency-stabilizing circuits, and resonant cavities lined with various test metals have provided experimental means of investigating the microwave-frequency properties of those metals.

Feasibility studies have been made regarding the possibility of power transmission in superconducting cables.[5] The entire area of superconductivity is one of great technological potential.

Physical theories as to the mechanism of superconductivity have been advanced, but they are outside the scope of this review.[3, 15, 23] It may be noted that the depth of penetration of the current sheet at microwave frequencies is estimated to be in the order of tens of nanometers.

E-2 ANOMALOUS SKIN EFFECT[4, 13, 14, 23, 26]

From inspection of Eq. 10-75 it would appear that the surface resistance at any given frequency should decrease markedly as temperature is lowered from the ambient level. Experimental measurements have shown that such a decrease does take place initially but, if the temperature is decreased sufficiently, the subsequent decrease in surface resistance is not so pronounced as Eq. 10-75 predicts (unless the superconducting temperature is reached). Eventually the surface resistance reaches a lower limit, and does not decrease with further decrease in temperature. The higher the frequency, the higher the temperature at which deviation from Eq. 10-75 occurs, and the higher the limiting value just referred to.

In solid-state theory the electrical conduction behavior of a metal may be simulated by a vibrating lattice structure with free electrons moving within it. As temperature is decreased, the amplitude of the vibrations decreases, and the lattice then obstructs the movement of the free electrons to a lesser extent, thereby lengthening their mean free path. In electric circuit terminology, the conductivity has been increased.

The electron mean free path in copper at 300 K has been estimated at 5×10^{-8} m. This is small compared to the skin depth at 100 MHz, which, for a conductivity of 5.8×10^{7} mhos per meter, is about 6.7×10^{-5} m. (Skin depth is a convenient measuring stick in this analysis, since it is the distance within which the amplitude of a given macroscopic field decreases by a factor of $1/\epsilon$.) Hence the concept of a continuous conduction current density which is, at every point, directly proportional to the $\boldsymbol{E}$ field at the same point, is reasonable. On the other hand, the electron mean free path in copper at 4 K is approximately 10^{-3} m, and the conductivity about 10^{12} mhos/m. A frequency as low as 1 Hz will give a skin depth that is less than the mean free path. When the electron mean free path is comparable to or greater than the skin depth, the $\boldsymbol{E}$ field changes sufficiently over the distances between successive collisions of an electron to make the point-field relationship $\boldsymbol{J} = \sigma \boldsymbol{E}$ inadequate.

Reuter and Sondheimer[21] and Mathis and Bardeen[12] have examined this problem.

REFERENCES

1. Allen, R. J., Long superconducting delay lines, *Proc. IEEE*, 52, no. 5, 615, 1964.
2. Allen, R. J. and Nahman, N. S., Analysis and performance of superconductive coaxial transmission lines, *Proc. IEEE*, 52, no. 10, 1147, 1964.
3. Bardeen, J., Critical fields and currents in superconductors, *Rev. Mod. Phys.*, 34, 667, 1962.
4. Casimir, H. B. G. and Ubbink, J., The skin effect, *Philips Tech. Rev.*, 28, nos. 9, 10, and 12, 271, 300, 366, 1967.
5. Fitzgerald, K., Superconductivity: Fact vs. fancy, *IEEE Spectrum*, 25, no. 5, 30, 1988.
6. Gittleman, J. I. and Rosenblum, B., Microwave properties of superconductors, *Proc. IEEE*, 52, no. 10, 1138, 1964.
7. Hahn, H., Halama, H. J., and Foster, E. H., Measurement of the surface resistance of superconducting lead at 2.868 GHz, *J. Appl. Phys.*, 39, 2606, May 1968.
8. Hinken, J. H., *Superconductor Electronics*, Springer, Berlin, 1989.

9. Hornak, L. A., Tewksbury, S. K., and Hatamian, M., The impact of high-T_c superconductivity on system communications, *IEEE Trans. Components, Hybrids, Manufacturing Tech.*, 11, no. 4, 412, 1988.
10. Hunt, D. V., *Superconductivity Sourcebook*, Wiley, New York, 1989.
11. Larbalestier, D., Fisk, G., Montgomery, B., and Hawksworth, D., High-field superconductivity, *Physics Today*, 39, no. 3, 24, 1986.
12. Mathis, D. C. and Bardeen, J., Theory of anomalous skin effect in normal and superconducting metals, *Phys. Rev.*, 111, no. 2, 412, 1958.
13. Matick, R. E., Transmission Lines for Digital Communication Networks, McGraw-Hill, New York, 1969, 130, 211.
14. Maxwell, E., Superconducting resonant cavities, in Progress in Cryogenics, vol. 4, K. Mendelssohn, ed., Heywood, London, 1964, 125.
15. McCaa, W. D. and Nahman, N. S., Frequency and time domain analysis of a superconductive coaxial cable using the two-fluid model, *J. Appl. Phys.*, 39, 2592, May 1968.
16. Articles on Superconducting devices and Superconductivity, in *McGraw-Hill Encyclopaedia of Science and Technology*, 6th ed., vol. 17, McGraw-Hill, New York, 599, 609 pp. 609-17.
17. *McGraw-Hill Yearbook of Science and Technology, 1989*, McGraw-Hill, New York, 1988, 387.
18. Nahman, N. S. and Gooch, G. M., Nanosecond response and attenuation characteristics of a superconductive coaxial line, *Proc. IRE*, 48, no. 11, 1852, 1960.
19. Proceedings of the 1988 Applied Superconductivity Conference, in *IEEE Trans. Magn.*, 25, no. 2, 740, 1989.
20. Rathbun, D. K. and Jensen, H. J., Nuclear test instrumentation with miniature superconductive cables, *IEEE Spectrum*, 5 no. 9, 1968.
21. Reuter, G. E. H. and Sondheimer, E. H., The theory of the anomalous skin effect in metals, *Proc. Roy. Soc.* (*London*), A195, 336, December 1948.
22. Rose, C. and Gans. M. J., A dielectric-free superconducting coaxial cable, *IEEE Trans. Microwave Theory Techniques*, 38, no. 2, 166, 1990.
23. Rosenberg, H. M., Low Temperature Solid State Physics, Oxford University Press, London, 1963, 102.
24. Simon, I., Surface impedance of superconducting tin, mercury and lead at 9200 Mc/Sec, *Phys. Rev.*, 77, no. 3, 384, 1950.
25. Tinkham, M., Superconductivity, 75th anniversary, *Physics Today*, 39, no. 3, 22, 1986.
26. Tischer, F. J. and Choung, Y. H., Anomalous temperature dependence of the surface resistance of copper at 10 GHz, *IEE Proc.*, 129, part H, no. 2, 56, 1982.

APPENDIX F

Table of Physical Constants

Absolute permeability of free space (μ_0)	$4\pi \times 10^{-7}$ H/m
Relative permeabilities	
Air, solid dielectrics, nonferrous metals	1.00
Absolute permittivity of free space (ϵ_0)	8.854×10^{-12} F/m
	$\approx \dfrac{10^{-9}}{36\pi}$ F/m
Relative permittivities	
Air	1.00
Polyethylene	≈ 2.26
Velocity of propagation of TEM waves in free space	2.998×10^8 m/s
Conductivities at 20°C (S/m)	
Aluminum[a]:	3.5×10^7
Copper[b]:	5.8×10^7 (International Annealed Copper Standard)

The resistivity of copper increases (conductivity decreases) about 0.00393 per unit degree Celsius; the corresponding factor for aluminum is 0.00403. Conductivity is sensitive to the presence of alloying elements or impurities, and is affected by mechanical working such as hard drawing.

1 statute mile = 1609 meters (m)

1 neper = 8.686 decibels (dB)

[a] *Aluminum Wire Tables*, National Bureau of Standards Handbook 109, C. Peterson, J. L. Thomas, and H. Cook, Eds., U.S. Government Printing Office, Washington, DC, February 1972.

[b] *Copper Wire Tables*, National Bureau of Standards Handbook 100, U.S. Government Printing Office, Washington, DC, February 1966.

AUTHOR INDEX

SUBJECT INDEX